The Chinese City

China's cities are home to 10 percent of the world's population today. They display unprecedented dynamism under the country's surging economic power. Their remarkable transformation builds on immense traditions, having lived through feudal dynasties, semicolonialism, and socialist commands. Studying them offers a lens into both the complex character of the changing city and the Chinese economy, society, and environment.

This text is anchored in the spatial sciences to offer a comprehensive survey of the evolving urban landscape in China. It is divided into four parts, with 13 chapters that can be read together or as stand-alone material. Part I sets the context, describing the geographical setting, China's historical urban system, and traditional urban forms. Part II covers the urban system since 1949, the rural–urban divide and migration, and interactions with the global economy. Part III outlines the specific sectors of urban development, including economic restructuring, social–spatial transformation, urban infrastructure, and urban land and housing. Finally, Part IV showcases urbanism through the lens of the urban environment, lifestyle and social change, and urban governance.

The Chinese City offers a critical understanding of China's urbanization, exploring how the complexity of the Chinese city both conforms to and defies conventional urban theories and experience of cities elsewhere around the world. This comprehensive book contains a wealth of up-to-date statistical information, case studies, and suggested further reading to demonstrate the diversity of urban life in China.

Weiping Wu is a Professor of Urban and Environmental Policy and Planning at Tufts University. She conducts research on migration and urbanization, university–industry linkages, and China's urban development.

Piper Gaubatz is a Professor of Urban Geography at the University of Massachusetts Amherst. She specializes in the study of urban change, development, and environmental issues in East Asia.

The Chinese City

Weiping Wu and Piper Gaubatz

Routledge
Taylor & Francis Group

LONDON AND NEW YORK

First published 2013
by Routledge
2 Park Square, Milton Park, Abingdon, Oxon OX14 4RN

Simultaneously published in the USA and Canada
by Routledge
711 Third Avenue, New York, NY 10017

Routledge is an imprint of the Taylor & Francis Group, an informa business

British Library Cataloguing in Publication Data
A catalogue record for this book is available from the British Library

Library of Congress Cataloging in Publication Data
A catalog record for this book has been requested

ISBN: 978–0–415–57574–4 (hbk)
ISBN: 978–0–415–57575–1 (pbk)
ISBN: 978–0–203–85447–1 (ebk)

Typeset in Times New Roman
by Swales & Willis Ltd, Exeter, Devon

Contents

List of figures   ix
List of tables   xi
List of boxes   xii

Introduction   1
     History and context 2
     Contemporary forces 3
     Understanding China's urbanization 5
     Using this book 7
     Bibliography 8

PART I
History and context   9

1 Geographical setting   11
     The physical environment 11
     Economic base 21
     Population and peoples 23
     Conclusion 26
     Bibliography 27

2 The historical urban system   28
     Thinking about the historical Chinese urban system 28
     How many Chinese cities were there? How large were they? 30
     The early traditional period (206 BC–AD 589) 35
     China's "medieval urban revolution" (589–1368) 38
     Cities of the late imperial period (1368–1911) 41
     Foreign empires expand into China: Treaty Port cities (1911–1949) 45
     Conclusion 46
     Bibliography 46

3 Traditional urban forms   48
Urban plan and layout 50
Monumental structures 57
Neighborhoods and markets in traditional Chinese cities 62
Treaty Port (1842–1949) and Republican (1911–1949) eras 64
Conclusion 67
Bibliography 67

PART II
Urbanization   69

4 The urban system since 1949   71
How big are China's cities? How many cities are there? 72
Shifting urban development in the Maoist period (1949–1979) 78
The urban system during the reform era 82
The Chinese urban system today 85
Conclusion 90
Bibliography 90

5 Urban–rural divide, socialist institutions, and migration   93
Urban–rural divide and its roots 94
Population mobility and migration 97
Migrants in the city 101
Conclusion 108
Bibliography 109

6 Cities in the global economy   111
Toward an open economy 111
Becoming "factory of the world" 116
Moving toward knowledge-based and consumer economies 122
Conclusion 128
Bibliography 128

PART III
Urban development   131

7 Urban restructuring and economic transformation   133
Revamping the urban economic landscape 133
Harnessing human resources 139
Redistributing economic growth 145
Conclusion 150
Bibliography 150

8 **Social–spatial transformation**   152
   Changing urban forms 152
   Work and accessibility 156
   Spatial reconfiguration and expansion 165
   Conclusion 171
   Bibliography 172

9 **Urban infrastructure**   176
   The state of urban infrastructure 177
   Financing urban infrastructure 182
   Challenges 187
   Conclusion 193
   Bibliography 193

10 **Urban land and housing**   195
   Legacies of state socialism 195
   Transition into land leasehold 198
   Housing reforms and implications 204
   Conclusion 214
   Bibliography 214

PART IV
Urban life   217

11 **Environmental quality and sustainability**   219
   State of the urban environment 219
   Strategies for environmental sustainability 227
   Environmental management 234
   Conclusion 238
   Bibliography 238

12 **Lifestyle and social change**   241
   Shopping and consumerism: changing landscapes and lifestyles 241
   Public spaces and entertainment 244
   Growing old in urban China 250
   Crime and urban neighborhoods 251
   Communities of difference: sexual preference, ethnicity, and art 252
   Conclusion 256
   Bibliography 256

13 **Urban governance and civil society**   259
   Reorganization of the state 259
   Developmental state in the making 264
   Evolving state–society relations 268

Conclusion 273
Bibliography 273

Conclusion: looking toward the future   275
Future urbanization 275
Challenges for China's cities 276
Implications for the world 278
Bibliography 280

Index   282

Figures

1.1	China's physical geography and cities	12
1.2	China's environments	14
1.3	Suzhou	18
1.4	Population density and provinces in China	23
2.1	"County-seat" cities established during the ancient (to 206 BC) and early imperial (206 BC–AD 589) periods	31
2.2	"County-seat" cities established in the medieval (589–1368) and late imperial (1368–1911) periods	32
2.3	Han and Tang Silk Routes	37
3.1	Continuity and change in elements of urban form	49
3.2	Changes in Chinese urban form	50
3.3	Reconstructed wall and gate at Jiayuguan, Gansu province, showing the rammed-earth wall	53
3.4	Top of the reconstructed Ming-dynasty wall at Xi'an	54
3.5	Wall forms for 233 traditional Chinese cities	55
3.6	Reconstructed courtyard structure in Xi'an	58
3.7	Qing-dynasty Beijing and its central axis	60
3.8	Shanghai as a Treaty Port	65
4.1	Changing distribution of cities, 1949–2009	74
4.2	Distribution of cities by size	75
4.3	Central Chongqing at night	77
5.1	Changing inequalities, measured by Gini coefficient	95
5.2	Net interprovincial flows of migration, 1995–2000	100
5.3	Migrant housing in a Shanghai factory	104
5.4	A migrant school in Beijing	107
6.1	Special Economic Zones (SEZs) and 14 coastal open cities	112
6.2	Main sources of foreign direct investment (FDI) in China, 1990–2009	117
6.3	Main urban destinations of foreign direct investment (FDI) in China, 2008	119
6.4	KFC in Shenzhen, 2008	123
7.1	Share of employment in three sectors, prefecture-level cities, 1999–2008	135

7.2	Urban employment by type of ownership, 1978–2009	137
7.3	Regional variation of Human Development Index (HDI), 2008	149
8.1	Distribution of local and migrant population in Shanghai, 2000	157
8.2	Plaza 66 on West Nanjing Road, Shanghai	160
8.3	Urban commercial landscape: street scene	161
8.4	Changing modes of travel in Shanghai, 1982–2004	162
8.5	Modes of travel in other cities	163
8.6	Heritage tourism in urban China	166
8.7	Urban expansion in Guangzhou	169
9.1	Index of per capita public investment in urban infrastructure by province, 1996 and 2006	189
9.2	Public investment in urban infrastructure by region, 2002 and 2005	190
9.3	Shanghai's maglev train serving Pudong Airport	190
10.1	Land use rights assignment cases, 2003–2007	199
10.2	Cases of land conveyance to commercial users, 2003–2007	201
10.3	Wu Ping's "nail house"	204
10.4	A typical commercial housing estate	209
10.5	Private housing for rental in an urban–rural transitional area	212
10.6	Growth rates of urban land and housing prices, 1999–2010	213
11.1	Air quality in downtown Shanghai, 2011	221
11.2	Solid-waste management containers fortified for high wind conditions, Shenzhen, 2009	225
12.1	An informal market in a temporarily vacant lot, Kunming, Yunnan province	245
12.2	Ribbon dancers at the Temple of Heaven, Beijing	247
12.3	New China Square, Höhhot, with fountains	250
13.1	China's administrative hierarchy	260
13.2	Resource allocations between urban and rural areas	262

Tables

1.1	Historical growth of China's population	24
2.1	Historical expansion of the Chinese urban system	36
4.1	Beijing and Shenzhen in 2000, by different measures (millions)	72
4.2	Changes in numbers of cities, 1949–2009	74
4.3	Changes in urban population, 1949–2009	76
5.1	Migrant access to urban housing	103
7.1	General indicators of regular higher-education institutions (RHEIs) and senior high-school education, 1978–2009	144
9.1	Average levels of infrastructure services in urban China, 1981–2009	180
9.2	Urban maintenance and construction revenues, 1990–2009 (billion yuan)	185
10.1	Types of urban housing tenure by income level, 2005 (percent of households)	208
13.1	Cities in the administrative hierarchy, 1949–2009	263

Boxes

1.1	Taiyuan: life at the edge of the Loess Plateau	15
1.2	Suzhou: a "land of fish and rice"	17
2.1	How important were cities in Chinese society?	29
2.2	Chang'an and its hinterland	33
2.3	The tea trade: a catalyst for growth	40
2.4	Höhhot and its hinterland	42
3.1	Beijing: forming an imperial capital	58
4.1	How big is Beijing?	72
4.2	Is China's largest city Shanghai or Chongqing?	77
4.3	The Pearl River Delta and Guangzhou: regional urban systems and airport development	87
5.1	Wenzhou migrants in Beijing's Zhejiang Village	104
5.2	Educating migrant children	108
6.1	"Factory of the world": manufacturing Dongguan and a new form of urbanism	121
6.2	Zhongguancun: China's "Silicon Valley" in the making	127
7.1	Breaking the "iron rice bowl" in Guangzhou	141
7.2	Shenyang: an industrial giant in decline	147
8.1	Catching the CBD fever: Harbin's new downtown	158
8.2	Gated communities on the rise	170
9.1	Upgrading Guangzhou's transport network to meet demands of growth	179
9.2	Foreign participation in urban infrastructure: Chengdu Water	191
10.1	Wu Ping's "nail house": one woman's quest for property rights in Chongqing	203
10.2	Equitable housing reform scheme in Guiyang	206
10.3	Bubble and burst: affording a home in urban China	211
11.1	What happened to the "bread loaf" taxis?	222
11.2	Dongtan: rise and fall of the "world's first eco-city"	229

11.3   Green Beijing, Green Olympics   232
12.1   Dancing in the streets: informal dance craze in Beijing   246
12.2   Riding tricycles in Höhhot: the evolution of public squares
       and recreation   248
13.1   Street democracy: protesting against proposed maglev line
       in Shanghai   272

Introduction

Looking toward the twenty-first century, a cover of the *New York Times Magazine* declared:

> The 21st Century Starts Here: China Booms. The World Holds its Breath.
>
> (Buruma 1996)

As China booms, its cities increasingly take center stage in the drama of explosive growth and unprecedented changes. Ten of the world's 15 fastest-growing cities are in China, and at current growth rates China could add about 350 million people – more than the population of the USA – to its cities over the next two decades. When the People's Republic of China was founded in 1949, only about 10 percent of the population lived in cities. Today more than half of the people live in cities that continue to grow at an unprecedented rate (Wines 2012).

While its historical roots date back to about 2000 BC, urbanization is really just taking off. Yet China's cities have lived through feudal dynasties, semicolonialism, and state socialism. Now, they display unprecedented dynamism and complexity under market transition and globalization forces. Economic reform and subsequent changes in national urban policies have brought about accelerated urbanization since 1979. The number of cities is growing, and the urban population is expanding. Studying urban China is important and exciting. It not only offers lenses into the character and complexity of the changing city, but also allows for a better understanding of the Chinese economy, society, and environment.

This book is anchored in the spatial sciences (e.g., geography, urban studies, urban planning, and environmental studies), targeting primarily upper-level undergraduate students. It offers a comprehensive survey of China's urban landscape, covering topics such as history and patterns of urbanization, spatial and regional context, traditional urban forms, social–spatial transformation, urbanism and cultural dynamics, housing and land development, environmental issues, and challenges of urban governance. Our intention is to provide a critical understanding of China's urbanization and cities that is grounded in history and geography. This distinguishes our book from many other recent studies on urban China. We also explore how the character and complexity of the Chinese city both conform to and defy conventional urban theories and the experiences of cities elsewhere around the world. The dominant ideas about the nature and development of cities are, at best, eurocentric. This is in fact drawn from the minority, as most urbanites now live outside the West, especially when considering cumulative historical instances. With one-fifth of the world's population, China has much to offer in its urban experience.

History and context

China is one of the largest countries in the world, with a diverse geography and wide range of climate regions. It has an area of about 9.6 million square kilometers which comprises about 6.5 percent of the world's total land area – fourth largest behind Russia, Canada, and the USA. As the most populous country, its current population of more than 1.3 billion accounts for 20 percent of the world's population. Situated in the eastern part of Asia, it is located on the west coast of the Pacific Ocean. The distance from the country's east to west measures about 5,200 kilometers, and from north to south over 5,500 kilometers – slightly smaller than the USA in territory. The two major rivers, the Yellow River (Huang He) and the Yangtze River (Chang Jiang or Yangzi), as well as the Pearl River (Zhu Jiang) in southeastern China, have provided the framework for agricultural development, population growth, and urbanization throughout China's history. The eastern region, with its rolling hills and wide plains, houses the bulk of the country's population. This is also where most of its cities can be found. The northern, western, and southwestern areas, which span a wide range of environments from deserts to subtropical forests, are more sparsely populated.

China has one of the world's oldest and most enduring urban systems. Ancient settlements first appeared along the Yellow River. The growth of the urban system proceeded in tandem with key events in history: the development of empires, extension of the Silk Road, and establishment of foreign Treaty Ports in the nineteenth and twentieth centuries, to name just a few. Overall, Chinese cities grew larger in size and more diverse in their economic functions, and the urban system became more integrated and mature. The country's long-standing cultures and societies are embodied in the evolution of elaborate and stylized urban forms. Cities were built not only as spatial expressions of the cultural and social divisions among different groups, but also as what Kevin Lynch (a renowned American urban scholar) has called "cosmic cities" that translate spiritual beliefs into the very layout of the city (Lynch 1984).

The course of urbanization saw a drastic turn when the Communist Party took power. Between 1949 and 1979, China remained a country with a large rural population (over two-thirds). The Soviet model of central planning shaped the structure of its economy. But, unlike the Soviets, the Chinese under Chairman Mao had a distinctly antiurban bias. The central authorities exercised direct administrative control over local governments and major enterprises, and distributed funds and supervised investment through a centralized budgetary allocation. Production was carried on almost entirely through state-owned enterprises and collectives, and the economy was closed to the outside world. Under the command economy, cities were assigned quite narrow roles. While important as producers of industrial goods and technology, they ceased to be the financial, trade, and business centers as well as the foci of regional economies. National policies discouraged the growth of large cities, encouraged the growth of industrial and urban centers away from coastal areas, and controlled rural–urban migration through food rationing and household registration.

Contemporary forces

Much of the book focuses on the post-1979 period, which fundamentally differs from the preceding 30 years of state socialism. After the death of Mao, China embarked on a new course of development. Monumental changes occurred in the domestic economic system (and, to a much lesser extent, the political system) and external relations. Shaping the urban landscape since then is a set of large forces transforming Chinese society: marketization, decentralization, industrialization, migration, and globalization. The consequences are no less drastic, in the pace of urbanization, scope of urban growth and development, and configuration of urban space and life.

Marketization

China's transition from a planned to a market economy has gone through a series of phased reforms. In general, these reforms were not the results of a grand strategy, but immediate responses to pressing problems. Between 1979 and 1994, in the early phase, the central government introduced economic incentives, aligned prices to supply and demand, and opened the economy up to the outside world. Since around the mid-1990s, there has been an emphasis on achieving the goal of transition to a market system, establishing market institutions, and restructuring state-owned enterprises. Throughout, two key features stand out. The first is pragmatism – criteria for success are determined by experiment rather than by ideology. The second is incrementalism – an idea is implemented locally or in a particular economic sector, and if successful it is gradually adopted piecemeal throughout the nation. Nonetheless, fundamental structural changes have been introduced to the economic system, with genuine competition and gradual reduction of state interference with economic activities. Reforms also have brought unprecedented improvements to Chinese people in terms of personal income and consumption of goods and services of all types.

Decentralization

Prior to the economic reform, the central government had direct control over local governments (both provincial and municipal governments) in three main areas: allocation of materials and resources, production planning for key industries, and budgetary control of revenues and expenditures. The fiscal system was characterized by centralized revenue collection and fiscal transfers. In tandem with market transition is the devolution of the centralized fiscal system that has given local governments increasing freedom. Central–local fiscal relations have been altered significantly. In addition, the central government has moved steadily to devolve responsibilities to lower levels – municipalities now have more economic power. The importance of local development and governance is further enhanced by the emergence of private interests and civil society organizations, albeit limited in their numbers and leverage over the local state. Much like the macroeconomic reforms, however, fiscal decentralization has been gradual and incremental, responding to immediate problems with short-term fixes. There continues to be a mismatch of expenditures and revenues at the local level, and municipal authorities have to cope with fund shortfalls through a variety of off-budget mechanisms.

Industrialization

Since 1979, China has gone through an industrial revolution. While development strategies under state socialism emphasized heavy industries, the country remained largely agrarian. In 1980, agriculture's share in employment was 69 percent (this share fell to 38 percent by 2009). The spread of market reform after 1983 brought industry to the forefront. Since then, it is the urban industrial economy that is largely responsible for China's phenomenal record of economic growth. Industrialization has proceeded on multiple fronts. Along the east coast at first and now extended into interior regions, an export-led manufacturing boom builds on China's true factor endowment – low-cost labor. China has become the world's factory. On the domestic front, rural industrialization took off in the early 1980s, as a way of generating employment opportunities for surplus rural labor. Township and village enterprises challenged the monopoly of state-owned enterprises and filled the void in consumer goods. To date, manufacturing continues to dominate the economies of many cities. In the meantime, the tertiary sector also has grown steadily after the central and local governments recognized more fully the contribution of services to growth and employment.

Migration

More than 100 million farmers have left the countryside for cities since 1983. There is little doubt that this is the largest tide of migration in human history. The transition from a planned to a market economy, combined with the partial relaxation of household registration, has paved the way for a drastic shift in the country's population mobility. This new mobility also is a reflection of the rapid process of industrialization and a product of persistent economic disparities between urban and rural areas. Much of the migratory flow involves circular movements of rural labor in search of work to augment agricultural income. Migrant workers now are indispensable for the full functioning of the urban economy, and thus for China's ability to compete strongly in the global marketplace. On the other hand, migrants are changing the face of urban China, challenging the long-standing rural–urban divide reinforced by state policies.

Globalization

Globalization, in a narrow sense, is the process by which markets and production in different countries are becoming increasingly interdependent through dynamics of trade in goods and services and flows of capital and technology. Intensified since the 1970s, it entails a network of worldwide production systems and supply chains favoring lost-cost development countries as manufacturing centers. Economic reforms introduced in 1979 entailed China's opening up to foreign investment and trade. The expansion of China's participation in the global economy ever since has been one of the most outstanding features of the country's transformation. China is moving toward a genuinely open economy, accented by its entry into the World Trade Organization in 2001. It has experienced a trade boom. Foreign direct investment has soared, making China the largest recipient of such investment among developing countries. The economic effects of openness and easier mobility have spurred urban economies, thereby reinforcing urbanization. However, China's participation is tempered by a strong state. Relative to many other countries, China remains a somewhat closed society. Central and local governments, for instance, continue to regulate and limit contacts with foreigners, control movement into and out of the country, and regulate internet traffic.

Understanding China's urbanization

"Urban" is a relative term. In general, urbanism refers to a way of life different from that in rural areas where livelihoods depend on agriculture. Aside from an anchor in non-agricultural activities, an urban area tends to have a higher density of people than rural areas. But countries have different definitions of how large a population is necessary to make a settlement "urban." Also, the terms "urban" and "city" are not equivalent, as some urban areas are not known as cities across countries. In the USA, for instance, there are urban places with names such as a county, town, or borough. In popular literature, "city" often is used as a generic term representing all things urban. As such, in studying urban trends and urbanization of any country, we must first understand the idiosyncrasies of the basic concepts and terms.

There are two types of urban places or settlements in China: cities and officially designated towns. Only the central government has the authority to designate such, with official criteria changing multiple times since 1949. In general, a city must have a population greater than 100,000. In some exceptions, provincial capitals, industrial bases, or major market centers also may be designated as cities even when their populations are smaller (Lin 2009). Designated towns include county seats that are not cities; seats of townships with a population over 20,000, of which at least 10 percent is non-agricultural; and seats of townships with a population of less than 20,000, of which at least 2,000 is non-agricultural. Since 1999, an urban place also must have a minimal population density of 1,500 people per square kilometer. There were 657 cities and 19,410 designated towns in 2010 according to official data. For the purpose of simplicity, we use the term "city" to refer to all urban places.

Counting urban population is no simple matter. First off, Chinese official statistics are collected on both the city proper (*shi qu*) and the entire designated urban place (*di qu*, including rural counties under the city's jurisdiction). Chinese cities, particularly the large ones, thus resemble metropolitan areas in the West with both a central city and surrounding areas (and less comparable to municipalities in the West). Many cities also embed considerable rural territory. Second, as a result of the long-standing rural–urban divide, residents are classified as non-agricultural or agricultural through a household registration system. Within a city's jurisdiction, there are both types of population. As such, the strictest definition of urban population would be non-agricultural population in the city proper while the broadest would be all population in the entire designated urban place, with two other versions in between. The classification of city size is based on non-agricultural population in the city proper (i.e., excluding counties).

Official data on urban population are known to be inconsistent and problematic (Chan 2007). For the period 1964–1982, the official measure of urban population was "city and town" population – the aggregate of all non-agricultural population in the designated cities and towns. The 1982 census used a different methodology, which defined urban population as all non-county population in all districts of cities, irrespective of agricultural or non-agricultural status. Because of growing concern over the large proportion of agricultural population entering the urban count, the 1990 census used a more complex system, similar to that used prior to 1982 in principle. In addition, population estimates from the national census source differ considerably from pre-census and post-census estimates from statistical yearbook sources. Migrants without local household registration were not always counted in urban population. The 2000 census was a turning point in that population was enumerated for the first time in their place of residence, not place of household registration. As such, data on the migrant segment of urban population became more readily available after 2000.

There is no doubt that the pace of urbanization has been rapid since 1979. According to official estimates, the urbanization level increased from 19.4 percent in 1980 to 51 percent in 2011. But such numbers are merely indications of the pace of change, and should be interpreted with caution, given the complexity in the accounting of urban population. Administrative changes also factor in the designation of cities. As the center of gravity shifts towards cities, the central government has made adjustments to the urban administrative system for the benefit of overall economic growth. For instance, at least three times (1983, 1986, and 1993), it reclassified a larger number of counties as cities and shifted more than 1,000 counties to under the jurisdiction of cities. All of these underscore the steady rise in the number of cities, size of the urban population, and ultimately urbanization level. On the other hand, the Chinese standards for classifying urban places are extraordinarily high. Official statistics likely underestimate urbanization levels. Most industrialized countries (except Japan) require only thousands of people for a settlement to qualify as urban (Kamal-Chaoui et al. 2009).

Using this book

The book is designed as a platform, covering essential information about Chinese cities and urbanization patterns. It focuses on mainland China; hence Hong Kong and Macao are not the subject of study, despite their storied development. There are four parts to the book, and a total of 13 chapters. Each chapter includes contextualizing discussions that place the Chinese city within broader urban theories as appropriate. The parts proceed from the broader urban landscape to the more specialized. Part I sets the context for the book, describing the geographical setting, China's historical urban system, and traditional urban forms. The next part covers the urban system since 1949, the rural–urban divide, and interactions with the global economy. Part III outlines the specific sectors of urban development, including economic restructuring, social–spatial transformation, urban infrastructure, and urban land and housing. The last part showcases urbanism through the lens of the urban environment, life and leisure, and urban governance.

All of the chapters can be used as stand-alone course materials, and the book makes frequent cross-references. In its whole, the book can serve as the primary textbook for courses on urban China in programs of geography, urban studies, urban affairs, city and regional planning, area studies, and environmental studies. Depending on the learning objectives, courses also can assemble the chapters in different sequences. A course using all or most of the chapters has a strong focus on urban China. Alternatively, select chapters can be used as specialized readings in comparative classes on urbanization and cities. For instance, a general course on urban morphology can draw from our chapters on traditional urban forms and social–spatial transformation. Students in a comparative course on urban development can benefit from reading the chapters on urban infrastructure and urban land and housing.

In addition, this book can help scholars with little prior expertise in urban China to obtain a quick but comprehensive grasp of the general area of study and specific topics. They can be China scholars who have not studied cities and urban scholars who have read little about China. It also appeals to the more educated general public who want a more systematic understanding of urban China. After each chapter we include a bibliography for further reading, including annotations on key books and relevant journal articles. There are case studies in each chapter to ground the discussion and demonstrate the diversity of cities and urban life in China. The book also contains many illustrative materials, in the form of tables, graphs, photos, and maps.

The majority of the data used in this book is based on official Chinese sources, including statistical yearbooks, census data, and official publications. Scholars have questioned the accuracy and reliability of such data (Naughton 2007). But there are no good alternatives. We interpret results based on such data with caution and point out the deficiencies whenever possible. The two main sources are the *Statistical Yearbook of China* and *China Statistical Yearbook for Cities*, both published by the National Bureau of Statistics annually. Detailed statistics for individual cities are contained in the latter and generally come out a year later than aggregate statistics. Another useful source is the *China Urban Construction Statistical Report*, published by the then Ministry of Construction (recently renamed as the Ministry of Housing and Urban–Rural Development), which includes individual city data too. All of the provincial-level units also publish their own statistical years, as well as a number of key cities. There are more specialized national statistical yearbooks, such as on land resources, population, science and technology, and education.

Bibliography

Buruma, Ian. 1996. "China: New York . . . or Singapore? The 21st Century Starts Here." *The New York Times Magazine*. February 18 (retrieved on 20 January 2012 from http://www.nytimes.com/1996/02/18/magazine/china-new-york-or-singapore-the-21st-century-starts-here.html).

Chan, Kam Wing. 2007. "Misconceptions and Complexities in the Study of China's Cities: Definitions, Statistics, and Implications." *Eurasian Geography and Economics* 48(4): 383–412.

Kamal-Chaoui, Lamia, Leman, Edward, and Zhang, Rufei. 2009. "Urban Trends and Policy in China." OECD Regional Development Working Papers, 2009/1. Paris: OECD Publishing.

Lin, George C. S. 2009. *Developing China: Land, Politics, and Social Conditions*. London: Routledge.
This is a systematic documentation of the pattern and processes of land development taking place in post-reform China. The author goes beyond the privatization debate to probe directly into the social and political origins of land development. The rural–urban interface is shown to be the most significant and contentious locus of land development where competition for land has been intensified and social conflicts frequently erupted.

Lynch, Kevin. 1984. *Good City Form*. Cambridge, MA: MIT Press.

Naughton, Barry. 2007. *The Chinese Economy: Transitions and Growth*. Cambridge, MA: MIT Press.
This is a comprehensive and accessible overview of the modern Chinese economy by a noted China expert. It first presents background material on the pre-1949 economy and the industrialization, reform, and market transitions that have taken place since. It then examines different aspects of the modern Chinese economy, including patterns of growth and development, the rural economy, industrial and technological development in urban areas, international trade and foreign investment, macroeconomic trends and cycles and the financial system, and the problems of environmental quality, and the sustainability of growth. The book also places China's economy in interesting comparative contexts, discussing it in relation to other transitional or developing economies and to such advanced industrial countries as the USA and Japan.

Wines, Michael. 2012. "Majority of Chinese Now Live in Cities." *The New York Times*, January 17 (retrieved on 20 January 2012 from http://www.nytimes.com/2012/01/18/world/asia/majority-of-chinese-now-live-in-cities.html).

Part I
History and context

1   Geographical setting

China is a vast and diverse country. Its citizens live in a wide range of localities – from the high plains of Tibet to the rocky Gobi Desert, from sandy deserts
to subtropical rainforests, and from nomadic herding settlements to some of the
world's largest cities. China's cities can best be understood in the context of these
widely differing landscapes and the people who inhabit them. The diversity of
environments, peoples, and subsistence systems is integral to understanding the
ways in which cities and systems of cities have developed in China, the current
inequalities between cities and between the regions in which they are located, and
the ways in which the Chinese government is approaching urban and regional
development today. This chapter provides a brief overview of China's physical
environment, its population, and its economic foundations as a context for understanding cities and urbanization. The following questions should help guide the
reading and discussion of the materials in this chapter:

- What are the basic patterns of China's topography, climate, and vegetation?
- What role have rivers played in the development of China?
- How does Chinese agriculture vary across different regions?
- What do Chinese industries produce?
- How has the distribution of the Chinese population changed over time?
- Where do the different peoples of China live?

The physical environment

China occupies about 6.5 percent of the world's land area, which makes it the
fourth largest country in the world. Its land area – 9,596,960 km^2 (3,704,426 mi^2)
– is about the same size as the USA and nearly as large as Europe. But China
covers a wider range of latitudes than Europe and thus has more climatic and
environmental variation. Palm trees line the streets of the city of Haikou, capital
of Hainan province in the southeast; elaborate ice sculptures have become the
hallmark of Harbin, capital of Heilongjiang province in the northeast; vast deserts
stretch toward the horizon near Urumqi, capital of the Xinjiang Uygur Autonomous region in the northwest, and glaciers carve the mountains near Lhasa, capital
of the Tibetan Autonomous region.

Topography

China is a mountainous country with about 65 percent of its land area covered in mountains (Figure 1.1). Mountains have a significant place in China's development and culture. Chinese cosmology is infused with sacred interpretations of this mountainous landscape. Daoists, for example, identify five sacred peaks, marking the compass directions, for pilgrimage: Hengshan Bei, in Shanxi province (north); Huashan, in Shaanxi province (west); Hengshan Nan, in Hunan province (south); Songshan, in Henan province (central); and Taishan, in Shandong province (east). Buddhists also have made pilgrimages to mountains in China, especially to four great peaks which they also associated with the cardinal directions: Wutaishan, Shanxi province (north); Emeishan, Sichuan province (west); Jiuhua Shan, Anhui province (south); and Putuoshan, Zhejiang province (east). Pilgrimage to these mountains was one of many factors that defined the ancient patterns of travel, trade, and settlement in China's core regions.

More prosaically, China's topography can be conceptualized as a series of three giant steps.

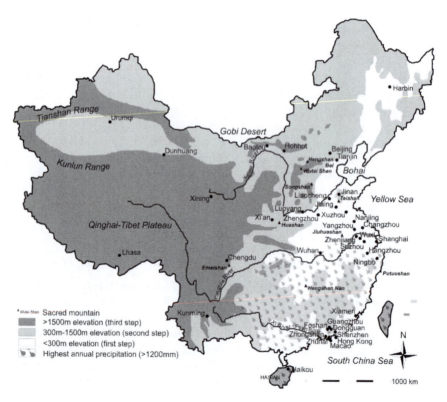

Figure 1.1 China's physical geography and cities.

Source: Adapted from Benewick and Donald 2009.

1. The highest of these topographic steps is western China's Qinghai-Tibet Plateau, sometimes called "the roof of the world," a vast high-altitude realm of interspersed plains and mountains which is bounded on the south by the Himalaya and on the north by the Kunlun Mountains. A number of the great rivers of Asia originate on the Qinghai-Tibet Plateau, including the Yellow River, the Yangtze River, the Indus River, and the Brahmaputra River. There are few cities in this cold, arid region. Those cities that have developed in this region, such as Lhasa and Xining, take on a functional and symbolic significance in their regional contexts beyond what their relatively small populations might indicate.

2. China's central topographic step comprises a complex series of basins and highlands that wrap around the Qinghai-Tibet Plateau from north to east. This region can be further divided into two subregions: the arid areas north of the Kunlun Mountains, which include the vast Tarim Basin and the Tianshan Mountains; and the basins and highlands of central China, to the east and southeast of the Qinghai-Tibet Plateau, which include the Sichuan Basin and the complex mountainous terrain of south and southeast China. The largest concentration of population in this region is in the Sichuan Basin, centered on the city of Chengdu. Other significant cities in this region include Chongqing and Wuhan along the course of the Yangtze River and Xi'an, the former imperial capital in Shaanxi province.

3. Most of China's cities and population can be found in the third topographic step, the rolling hills and plains of eastern China. All 25 of the most affluent metropolitan regions are on the east coast and 14 of the richest 20 metropolitan regions in China (measured by gross domestic product per capita) are in the two river delta mega-regions: the Pearl River Delta region, which includes cities such as Guangzhou and Shenzhen, and the Yangtze River Delta region, which includes cities such as Shanghai and Hangzhou (more in Chapter 4). The lower reaches of the Yellow River and the plains created by the river's course changes form a third concentration of population and urban development in north China, dominated by Beijing and Tianjin (Figure 1.2).

Rivers

Rivers occupy a special place in China's urban development. Not only are China's great rivers markers of vast differences in environmental systems, but they also have great significance culturally as fundamental elements in how the Chinese view space and territory, as each has come to define a large cultural and economic region. The three rivers mentioned above – the Yellow River in the north, the Yangtze River in Central China, and the Xi River/Pearl River system in southeastern China – are integrally tied to China's urban development. Each has served for centuries as the focus for concentrated and dense population growth; each has given rise to great cities which have survived China's long history.

Figure 1.2 China's environments.

Source: Piper Gaubatz.

Key:
1. Southwest: Yunnan Plateau.
2. Southeast: karst near Guilin, Guangxi Zhuang Autonomous Region.
3. Southeast: rice fields in Guangdong province.
4. Yangtze River Delta: canal in Suzhou, Jiangsu province.
5. North China Plain: farms in Hebei province.
6. Loess Plateau: cave housing in Gansu province.
7. Northern Grasslands: grasslands north of Höhhot, Inner Mongolia.
8. Sand dunes near Dunhuang, Gansu province.

The Yellow River (Huanghe)

The Yellow River has long been vital to agricultural production and urbanization in north China. A population map of China shows that the largest area of dense population and the region with the highest concentration of cities is found in the lower reaches of the Yellow River in the plains formed as the course of the river shifted over the centuries. At its southernmost extent, during the thirteenth century, the Yellow River reached the sea just north of the present-day outlet of the Yangtze River. Today, however, the river follows a northward course to its outlet into the Bohai Sea.

About 5,000 years ago, according to legend, the Chinese folk hero Yu the Great tamed the flooding of the Yellow River by, in legendary terms, lifting the rivers of China out of their beds and tying them into an orderly "net," or more prosaically, by dredging and re-routing a number of different water courses. In so doing, he is said to have defined the nine great regions that persist in fundamental conceptions of China's geography. Many archeologists and historians trace the origins of China's urban civilization to the region defined by the Yellow River and its dramatic course changes. The most prominent cities along the Yellow River include Lanzhou, capital of Gansu province; Baotou, Inner Mongolia; Luoyang, Henan province, which was repeatedly used as an imperial capital during six of China's dynasties between 771 BC and AD 936; Zhengzhou, capital of Henan province; and Jinan, capital of Shandong province.

The Yellow River stretches 5,464 km from its source on the Qinghai-Tibetan Plateau in Qinghai province to its outlet into the Bohai Sea. The muddy yellow coloring of the river develops as it passes through north China's loess plateau (Box 1.1). Loess is a lightweight, yellowish material which forms when glaciers

---

Box 1.1 Taiyuan: life at the edge of the Loess Plateau

Taiyuan, capital of Shanxi province, is an ancient city located near the hearth region of Chinese civilization. There has been some form of city or town on the site since about 476 BC. At various times it has served as an administrative center, a military outpost, and a center for trade and commerce. Taiyuan is located at the edge of north China's loess plateau. The city's name means "great plains," referring to the wide stretch of plains where the Fen River emerges from the mountains on its way to join the Yellow River. The deeply eroded loess hills in the city's western hinterland have, for centuries, supported one of China's unique rural landscapes – dry, yellow hills sculpted by centuries of agricultural terracing, and villages comprising elaborate cave houses built deep into the cliffs. Spring winds bring frequent dust storms to Taiyuan as the lightweight loess blows into the city. The challenges of this harsh environment and the indomitable spirit of its inhabitants became legendary in the 1960s as the efforts of the village of Dazhai, just east of

Taiyuan, were immortalized by Mao Zedong's 1964 declaration "[in] agriculture, learn from Dazhai." The message at that time was that any environmental challenge could be conquered through massive transformation of the natural environment (such as terracing), hard work, and patriotism. Although the loess is quite fertile, agriculture in Shanxi is limited by low levels of precipitation – the main crops in the area are wheat, corn, millet, potatoes, and beans. Thus local specialties include noodle dishes, porridges, and thick stews. Traditional housing in Taiyuan was formed of one-story courtyard-style complexes built with local materials – loess rammed into compact adobe-like bricks, fired bricks made in local kilns, and stone.

In the past, Taiyuan's residents had to cope with the ubiquitous yellow dust that blankets China's loess region. In recent years, coal dust has become an even greater issue. Today Shanxi is the largest coal-producing province in China; Taiyuan serves as the administrative center for a region of vast coal-mining complexes, and chemical and steel factories. Its traditional houses and temples have been overshadowed by factory complexes established during the latter half of the twentieth century and more recent high rises. As Kathleen McLaughlin (2009) explains, "[t]he residents of Taiyuan measure their air pollution in dirty clothes . . . when China's boom created endless demand for this area's coal, iron and steel, a white shirt stayed fresh only a few hours, turning black around the collar and sleeves before day's end."

scour rock. The loess in northern China developed in Europe during the last ice age, and was blown across the continent, to accumulate in layers averaging 50–80 meters thick in what is today northern China. Although the loess was once covered in forests that fixed the soil in place, centuries of deforestation have exposed this easily eroded and blown soil. As the river passes through the plateau, it carries the loess with it. The Yellow River in its lowest reaches has one of the highest silt loads of any river in the world, averaging more than 37 kg of silt per cubic meter (the silt load of the Amazon and Mississippi Rivers, by comparison, is less than 1 kg per cubic meter).

Despite the grand efforts of Yu the Great, the Yellow River has continued to periodically flood over the centuries. These floods have created great hardships for the people of north China, earning it the nickname "China's sorrow." One of the most consistent efforts of the Chinese empire and subsequent modern states has been to attempt to control the river. In Shandong province, a combination of these efforts and silt deposition has built up the banks of the river so far that it now flows in a channel nearly 10 meters above the plain. Ironically, however, the very floods that so devastated local settlements and agriculture also replenished them. Receding flood waters regularly deposited a layer of fresh silt over depleted soil, replenishing the land for intensive cultivation which supported the dense population of the region. Thus the Yellow River nurtured the very region that it regularly devastated.

The Yangtze River (Chang Jiang)

While the Yellow River was important primarily for its contributions to agricultural development, the Yangtze River served as a major transportation artery which contributed to the consolidation of both communications and trade as successive dynasties worked to control China's vast territory. Rice cultivation was well established in the region defined by the middle and lower reaches of the Yangtze River by about 6500 BC, and this region continues to serve as one of China's most productive agricultural regions. Over the centuries, the river has nurtured a number of China's most prominent cities – especially Chongqing, along the upper reaches of the river formerly in Sichuan province; Wuhan (Hubei province), traditionally a trio of cities at one of the river's confluences in its middle reaches; and the five major cities of the Yangtze Delta: Hangzhou, capital of Zhejiang province; Nanjing, which served as imperial/national capital for six different governments between 220 and 1949; Ningbo, Zhejiang province; Suzhou, Jiangsu province; and Shanghai, which is currently China's largest city (Box 1.2).

Box 1.2 Suzhou: a "land of fish and rice"

Suzhou is located in the lower Yangtze River Delta, one of the most resource-rich regions in China. The city's administrative area borders Taihu, the largest lake in the region, to the west and the Yangtze River on the north; the Grand Canal passes through its center. Shanghai, China's largest and perhaps most dynamic city lies about an hour's train ride to the east. Traditionally, this region was known as a "land of fish and rice" – an area of unsurpassed abundance. It combines a moderate climate with ample rainfall. Suzhou's climate, which is classified as humid subtropical, is characterized by hot, humid summers and cool, damp winters, but temperatures rarely fall below freezing. In this environment, paddy rice thrives, as well as the mulberry bushes that nourish silk worms. The region is also rich in copper and tin, which provided an early means of manufacturing in bronze. Thus Suzhou had both abundant natural resources and convenient natural and man-made water transport systems – a powerful combination in an empire that prized both rice and silk.

Although the first city on site was established as early as 514 BC, Suzhou was particularly prominent during the Han, Tang, Song, and Qing dynasties. Long before Shanghai became the economic powerhouse (after the mid-nineteenth century), popular opinion equated the health of the imperial economy with Suzhou. According to an adage dating from the Song dynasty, "when Suzhou and Huzhou ripen, the empire has enough" (Marmé 2005: 2). The city itself was laid out in a typical fashion with a more or less rectangular grid lined with courtyard houses. Canals threaded the grid and provided one of the main forms of transportation for the city, leading some to call Suzhou "the Venice of the east." During the late imperial period,

Suzhou served as the heart of China's textile industry. The period of prosperity Suzhou experienced during the Qing dynasty in particular attracted an unusually aesthetic nobility; Suzhou became famous for the elaborate private landscape gardens of its elites.

Today those natural advantages that made Suzhou an economic power in earlier centuries have been overshadowed by Shanghai, which controls the region's deepwater ocean ports. Yet Suzhou has benefited from its proximity to Shanghai's industrial base, most notably in the development of the Singapore-Suzhou Industrial Park, established in 1994 (288 km^2) , and the Suzhou New District, established in 1992 (52 km^2). Both new industrial areas have had mixed economic success but have certainly contributed to massive new industrial development in Suzhou. At the same time, extensive historic preservation efforts have restored Suzhou's traditional landscape gardens and opened them to the public. Suzhou also has benefited from Shanghai as an entrepôt for tourists who can easily day-trip to Suzhou from Shanghai or en route between Shanghai and Beijing. Suzhou and its gardens are one of the top tourist attractions in China.

Figure 1.3 Suzhou.

Source: Piper Gaubatz.

The southern part of the Yangtze River Delta was traditionally referred to as Jiangnan – a region known as one of the most productive in China, with its mild climate and fertile soils. The region earned a reputation as a "land of fish and rice," or great abundance and has remained significant throughout China's history. Today, led by Shanghai, it persists as one of the most economically vibrant regions in mainland China, and the locus of intense urban change and development (more in Chapter 4 and 6).

Grand Canal – the north–south link

Although the Grand Canal is not a natural watercourse, it is useful to think of it in relation to China's great river systems as it provided a functional linkage between the regions defined by the Yangtze River and the Yellow River. Construction began on the Grand Canal – a vast engineering work which combines natural watercourses with man-made canals – in the fifth century BC in order to bring the products of the Yangtze River region to northern China. The 1,776 km canal was completed toward the end of the sixth century AD, when all the lengths of the system were joined together to create a navigable waterway between Hangzhou and Beijing. As each segment was completed, this major transportation route generated urban development along its banks.

The southern Grand Canal in particular contributed to the development of noteworthy Chinese cities such as Suzhou, Wuxi, Changzhou, Zhenjiang, and Yangzhou. Farther north, Xuzhou, Jining, Liaocheng, and Tianjin marked the extension of the canal toward Beijing. In recent years, the Grand Canal has been a focus for the development of heavy industry.

The West (Xi) River and the Pearl River Delta

Three of China's most dynamic contemporary cities – Hong Kong, Shenzhen, and Guangzhou – are located in the Pearl River Delta. The Pearl River Delta is a complex region – there are well over 1,000 rivers in Guangdong province. The delta comprises a vast system of rivers, the most notable of which include the West (Xi) River, the North (Bei) River, and the East (Dong) River that join with the Pearl River itself as it flows from Guangzhou eastward to the sea. This system carries more water than the Yellow River, and defines a wet region well suited to rice agriculture. The Pearl River Delta itself is usually portrayed as a triangular region defined by Guangzhou, Hong Kong, and Macao. Hong Kong and Macao are about equidistant from Guangzhou (about 120 km); the distance from Hong Kong to Macao (the "short side" of the triangle) is about 60 km. This region is characterized by low-lying areas alternating between rice fields, ponds, and dikes and low hills covered with dense vegetation. Historically the rich natural resources of this environment were augmented by a sophisticated system of cultivation in which fish and ducks were raised in the flooded rice paddies, thereby both fertilizing the rice and producing much more food per field, and the dikes which contained the paddies were planted with a wide range of crops. (Similar cultivation systems were practiced throughout much of southern China.)

Although the physical characteristics of the Pearl River Delta have historical significance to the extent that the region was long able to support a high population density, the region has gained much greater significance during the past 50 years as the locus of urban, economic, and industrial innovation. Today the Pearl River Delta is one of China's wealthiest and most productive regions. The cities of Guangzhou, Shenzhen, Zhuhai, Dongguan, Zhongshan, and Foshan create a vibrant network of rapidly industrializing and expanding cities which maintain strong ties to the global economy (more in Chapter 6).

Climate

China's vast area crosses a wide range of climate regions. While Hainan Island in the southeast features a subtropical climate with little seasonal variation in temperature, portions of China's northernmost province, Heilongjiang, stand above 50° north in the subarctic realm. But much of China lies within the temperate zone and is characterized by seasonally cool winters and hot summers – the distinct seasonal changes characteristic of the temperate zones in the USA and Europe. An average winter's day in Beijing is similar to an average winter's day in Berlin or Chicago; an average summer's day in Shanghai is similar to an average summer's day in Athens or Atlanta.

Climate is important in understanding the ways in which urban construction styles, planning for natural hazards, and potential opportunities for and limitations on urban growth vary in China. Many cities in northern and western China face severe water shortages on a regular basis; cities in the southeast must plan for major seasonal flooding. As a general rule, the average annual temperature increases from north to south in eastern China. Precipitation is greater in the summer than in the winter throughout eastern China; the southeastern region experiences the most intense summer precipitation as monsoon conditions bring torrential summer rains and typhoons. Low-lying areas are often well suited to rice agriculture, which relies on seasonal flooding for irrigation, but poorly suited to village and city development; the villages and cities of southeastern China were often sited on higher ground as a result. The western region of China, however, especially Xinjiang and Tibet, is strongly affected by its position at the center of the Eurasian continent and its distance from the oceans. Rain clouds sweeping across India and Nepal during the monsoon season drop most of their moisture by the time they reach the Himalaya (the "rainshadow effect"), leaving the Qinghai-Tibet Plateau dry. Nor do the storms which bring rain to eastern China carry their moisture far enough to reach the desert regions of Xinjiang. Here the patterns of urban development followed the natural oases which provided access to groundwater.

In the most general sense, there are four basic climate zones in China:

1.   the cold high Qinghai-Tibet Plateau;
2.   the cold winter/hot summer deserts of the northwest;

3. the eastern temperate region of the northeast;
4. the eastern subtropical and tropical region of central and south China.

These zones are characterized by distinctive patterns of soil and vegetation, which have implications for China's agriculture, resources, and urban development.

Economic base

Although China's economy is becoming increasingly global, urban, and high-tech, it is important to understand the bases of China's economic strength as a generator of early and continuing urban development. China's vast area contains a wide range of natural resources enabling a diverse economy with a foundation in agriculture, mining, and other resources. China's development has been fueled by the exploitation of these resources to support both a growing domestic population and China's economic reach toward global markets (more in Chapters 6 and 7).

Agriculture

China has long been considered a country of farmers. Although the urban population is rising dramatically, and industrial production and the service economy are becoming increasingly important, close to 50 percent of China's people live in rural areas, and about 40 percent of China's workers are employed in agriculture. The Chinese revolution, which led to the establishment of the People's Republic of China in 1949, is often characterized as a rural movement, in contrast to the urban base of the Russian revolution of 1917.

Although China is about the same size as the USA, its potential to produce food and other agricultural products is quite different. Less than 11 percent of China's land area is arable (capable of being farmed) compared to about 20 percent of the USA. Another way to think of this is that China is home to about 20 percent of the world's population, but only 7 percent of the world's arable land.

Eastern China's agriculture can be thought of, in simple terms, as divided between the northern region, where wheat dominates, and the southern region, where rice dominates. The divide between these two regions lies between the Yangtze River and the Yellow River, a product of different growing conditions. Rice was domesticated in southern China at least as early as 4,000 BC and seems to have rapidly developed as a prized food crop throughout the Chinese world. One of the motivations for the construction of the Grand Canal in traditional China was to enable the movement of rice from south to north. Although rice production has gradually moved north with the development of improved cold-resistant and drought-tolerant varieties during the twentieth century, and recent development trends in southeast China have led to a decrease in rice production, this basic agricultural pattern persists.

In southern China's rice-growing regions, both rice and wheat are grown in the Yangtze region; the region south of the Yangtze is dominated by a mixture of rice production (especially in low-lying areas) and tea production on the surrounding slopes. In many areas of southeastern China, the climate is warm and wet enough that two crops of rice can be harvested each year. Southeastern China is also the center for the production of mulberry leaves and silk worms to fuel the silk industry. Pigs, water buffalo, chicken, and fish are raised in this region as well.

Northern China, by contrast, is relatively arid. Most northern Chinese agriculture is limited to a single harvest each year, although new crop varieties have extended the growing season in some areas. Soybeans, corn, *gaoliang* (a form of sorghum), millet, spring wheat, potatoes, oats, and rice are all cultivated in northeastern China. Pigs and chickens are raised as livestock. Northwestern China is more arid; wheat, cotton, corn, and melons are the main crops. But rice is also grown throughout northern China. Sheep, cattle, and horses graze on the northern Chinese grasslands and also contribute significantly to the agropastoral economy.

In China's western borderlands, the Xinjiang Uygur Autonomous region, which comprises much of the northwestern part of China, includes vast deserts bordered by high mountain ranges. Oases span the length of the desert close to the mountains. Urban development centers on these oases, whose underground water reserves support an agricultural economy centered on the production of grapes, melons, wheat, corn, cotton, and sugarbeets. The high-altitude grasslands of the Tibetan Plateau, by contrast, are best suited to nomadic pastoralism. Yet there are also cities and settled agriculture in Tibet and Qinghai – particularly in the high intermontane valleys, such as Lhasa, which formed along the mountainous rim of the Tibetan Plateau. Tibetan agriculture centers on barley, wheat, potatoes, beans, and turnips.

Industries and trade

China's traditional agropastoral, crafts, and industrial production all had a significant impact on the growth and development of cities and the urban system. Those products that relied on production in specific geographical regions – especially rice, salt, oil, tea, lacquer, porcelain, sugar, and indigo – were traded long distances early in China's history and contributed to the development of roads, inns, towns, and cities (the relationship between industry, trade, and urbanization in traditional China will be discussed in more detail in Chapter 2). Salt became so important in the economy that during the Ming and Qing periods it may have accounted for as much as half of the silver in the imperial revenues. While much of the economy of pre-modern China was centered on agricultural products, China's early industrial regionalization included concentrations of coal mining and iron production in the Yellow River and loess plateau regions of the north, and the silk, porcelain, and lacquerware production centers of the lower Yangtze and southeast coastal region.

China's industrial economy was transformed by the arrival of Europeans, Japanese, and Americans in the mid-nineteenth century. Although China had long resisted such outside influences, the "Opium Wars" of the nineteenth century forced China to open some cities for economic development by foreign firms. Foreign corporations fostered the development of trade and industry along the east coast and in some inland locations, along the Yangtze River (such as Wuhan). Most prominent among these rising economic centers was Shanghai. Shanghai's mid-nineteenth-century trade economy was gradually developed to include first, labor-intensive industries such as textile and food processing, and later, by the early twentieth century, a wide range of machine-based industrial ventures as well. By the 1930s, Shanghai may have accounted for as much as 40 percent of China's total industrial output, with most of the rest concentrated in other coastal cities such as Tianjin and Guangzhou (Canton). This uneven development was fueled by the emphasis on an export economy geared toward the eastern port cities. After the establishment of the People's Republic of China, the export economy was largely abandoned in favor of domestic production, which relocated industrial development closer to the sources of raw materials for industrial production. Since 1979, however, there has been a rejuvenation of the role of eastern cities, especially those with superior port facilities, such as Tianjin, Shanghai, Xiamen, and Guangzhou, in the national economy. All of China's regions have seen growth in industrial development which today draws not only on their natural resources, but on a skilled labor force increasingly engaged in high-technology and knowledge industries as well (more in Chapter 6).

Population and peoples

China is the most populous country in the world. Although it comprises 6.5 percent of the world's land area, it is home to more than 1.3 billion people – nearly 20 percent of the world's population. Most of these people live in the densely populated east and southeast regions (Figure 1.4).

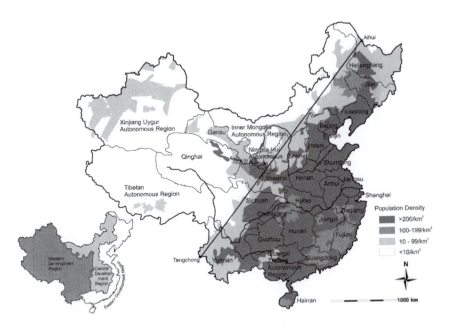

Figure 1.4 Population density and provinces in China.

Source: Adapted from Benewick and Donald 2009.

Population growth and distribution

When the People's Republic of China was founded in 1949, most people within its borders lived in the eastern third of the country's land area. The lower reaches of the three great river systems, the Yellow River, the Yangtze River, and the Pearl River, were home to dense concentrations of population living in both numerous cities and extensive rural areas. Chinese geographers often drew a diagonal line, from Aihui, in northern Heilongjiang province, to Tengchong, in western Yunnan, which illustrated this divide. The line divided the country's land area roughly in half. Most of the population lived east of the line. Much of the country to the west was sparsely populated. This pattern was long standing, and represented the historical development of settlement in China.

By the time of the Han dynasty (206 BC–AD 221), the Chinese population was about 60 million. For 1,000 years after that, the population fluctuated but did not grow. However, around AD 750 a shift began as Han Chinese migrants moved south. Estimates indicate that, while three out of every five Chinese lived in northern China during the Tang dynasty (618–907 AD), by some time during the Song dynasty (960–1279) fewer Chinese lived in the north than in the south, with about two out of every three Chinese living significantly south of the Yellow River region. This shift in distribution of the population was accompanied by a sharp rise in population, as the population doubled between 750 and 1250 to about 120 million with the introduction of new early-ripening and higher-yielding rice varieties, Chinese migration, and a growth in trade (Table 1.1). During this time urban populations also grew markedly, with many cities sprawling beyond their walls.

The Mongol Yuan dynasty was a period of population decline. But the Ming dynasty (1368–1644) ushered in an era of population growth which has yet to end. The population doubled during the Ming, from about 80 million to about 160 million. This era of growth was fueled by the introduction of new-world crops, such as peanuts and potatoes, to southern China in the sixteenth century. China's population further increased during the Qing dynasty (1644–1911), reaching about 430 million by 1850. China's population had reached more than

Table 1.1 Historical growth of China's population

Era	Year	Approximate population
Han dynasty	*c.* AD 2	60,000,000
Song dynasty	*c.* 1250	120,000,000
Ming dynasty	*c.* 1368	80,000,000
Ming dynasty	*c.* 1600	160,000,000
Qing dynasty	*c.* 1750	200,000,000
Qing dynasty	*c.* 1850	430,000,000
People's Republic of China	*c.* 1950	583,000,000
People's Republic of China	*c.* 1980	1,000,800,000
People's Republic of China	*c.* 2010	1,330,474,000

Sources: Buchanan 1970; Fairbank and Goldman 1998; NPFPC 2010.

500 million by the time of the establishment of the People's Republic of China in 1949. The population doubled by 1980 to more than 1 billion. Since 1980, the population growth rate has slowed through a combination of the one-child policy and the general trends of urbanization and economic development. See Chapter 2 for a discussion of the proliferation of cities as the population and the Chinese empire grew.

Peoples of China

China is a multicultural country with 56 distinct peoples (*minzu*) officially recognized by the Chinese government. Although 92 percent of China's population is classified as ethnic Chinese (Han *minzu*), the Han population is concentrated in the eastern provinces. The population of all of the east-coast provinces is at least 94 percent Han; eight eastern provinces have populations which are 99 percent Han. Interior China comprises the homelands of a number of different non-Han peoples, such as Tibetans, Mongols, and Uygur. The mountainous provinces of southwestern China – particularly the provinces of Yunnan, Guizhou, Guangxi, and Sichuan – are home to more than 25 different peoples; northeastern China is home to Mongols, Manchu, and Koreans.

The broadest generalization of China's cultural geography assigns the peoples into four groups: the Han, the Islamic peoples, the Lamaist peoples, and the peoples of southwestern China. The Han Chinese are united by many common traditions and beliefs, as well as a common written language and shared history. Traditional Chinese religious practice is a syncretic mix of traditional Chinese worship of ancestors and place-associated deities (such as the City God), Daoism, Confucianism, and Buddhism. There is also considerable regional variation among the Chinese in cultural practices such as settlement patterns, subsistence patterns, and spoken dialects. The common written language has been particularly important over the course of Chinese history in unifying peoples with otherwise quite different dialects, such as Mandarin (the dialect of north China) and Cantonese (the dialect of Guangdong and Hong Kong).

Origins of the Han Chinese are found among peoples who inhabited the region of the Great Bend in the Yellow River (the southern parts of today's Shanxi and Hebei provinces, and the northern part of today's Shaanxi and Henan provinces) 3,000–4,000 years ago. These peoples established proto-cities: ceremonial centers where priests and artisans lived in compact settlements along the riverbanks. The Chinese empire itself was first unified under the Qin (221–207 BC); the core area of Chinese settlement was firmly established in the subsequent Han (206 BC–AD 220) dynasty. This traditional core area extends from the Great Wall on the north to the south coast, and from the east coast to the eastern edge of Tibet in the west. The Chinese empire endured alternating periods of strength and relative weakness, expansion, and fragmentation over the centuries, with the great epochs of Chinese rule during the Han dynasty, the Tang dynasty (618–907), the Song dynasty (960–1280), and the Ming dynasty (1368–1644).

Islam entered China both from the east, where traders from Arabia and Persia introduced it to eastern port cities during the mid-seventh century, and from the west, when trade brought Islam along the Silk Routes (the multiple routes which made up the "Silk Road"). Today there are ten officially recognized Muslim nationalities in China; the largest of these include the Hui (Muslim Chinese), and the Uygur. "Hui" refers to a diverse group of people who are descended from marriages between Muslim traders and Han women. The Hui often have constituted the largest non-Han minority in eastern Chinese cities, where they tended historically to fill occupational and economic niches as traders, butchers, tanners, and innkeepers. As a result, most large eastern Chinese cities had a mosque and a small Muslim settlement. In northwestern China, the Hui are more broadly distributed among cities, towns, and villages. The Ningxia Hui Autonomous region is a small provincial-level unit with the largest concentration of Hui people in China. The Uygur are Turkic Central Asian peoples who traditionally occupied the oasis settlements of northwestern China. These peoples were converted to Islam between the tenth and sixteenth centuries. The Uygur primarily live in small villages and towns in a provincial-level unit, the Xinjiang Uygur Autonomous region, as well as in Kazakhstan and Kyrgyzstan.

Lamaism, or Tibetan Buddhism, links two distinct peoples in interior China: the Tibetans and the Mongols. Lamaism originally developed as a synthesis of traditional Tibetan religion (Bon) with Buddhist traditions imported to Tibet beginning in the seventh century from India, Nepal, and China proper. The militaristic and expansionist Tibetan empire was gradually transformed into a theocracy from the ninth century onward under the influence of Lamaism. Similarly, some years after the Mongol reign over China (the Yuan dynasty: 1279–1368) ended, the Mongolian prince Altan Khan (1507–1582) was converted to Lamaism and began what became a 100-year transformation of Mongol culture. This transformation was accompanied by increasing participation of Mongols in the establishment of cities and settlements in the northern Chinese frontier regions.

The diverse peoples of southwestern China inhabit a mountainous region adjacent to Vietnam, Laos, Burma, and Tibet. Some of their cultures share similarities with the hill tribes of mainland Southeast Asia, such as the Dai of Yunnan; others are relatively similar to Han Chinese culture, such as the Bai, while the region also has a significant population of Tibetan and Hui people. Although many of these peoples are rural dwellers, the Bai and the Naxi both had important and distinctive urban traditions in Yunnan province, and the Hui are a significant minority in larger cities such as Kunming.

Conclusion

China is one of the most geographically varied countries in the world. Regional variations in physical and human geography provide a context for a range of different patterns of urban life, from the densely packed cities of the east to the widely scattered cities of the west. In the most general sense, China's population and economic activity are heavily concentrated in the east, while the wide-open spaces of the interior, although more sparsely populated, house an interesting variety of peoples, environments, and resources. Understanding these patterns can contribute to a better understanding of the country's current urbanization and development.

Bibliography

Benewick, Robert and Donald, Stephanie Hemelryk. 2009. *The State of China Atlas: Mapping the World's Fastest-Growing Economy*. Berkeley, CA: University of California Press.

Buchanan, Keith. 1970. *The Transformation of the Chinese Earth: Aspects of the Evaluation of the Chinese Earth from Earliest Times to Mao Tse-tung*. New York, NY: Praeger.

Fairbank, John and Goldman, Merle. 1998. *China: A New History*. Cambridge, MA: Belknap Press.

Marmé, Michael. 2005. *Suzhou: Where the Goods of All Provinces Converge*. Stanford, CA: Stanford University Press.

McLaughlin, Kathleen. 2009. "Bad Economy, Better Lungs." *The Global Post* 2/26/09 (http://www.globalpost.com/dispatch/china-and-its-neighbors/090226/bad-economy-better-lungs).

National Population and Family Planning Commission of China (NPFPC). 2010. "Total Population by Urban and Rural Residence and Birth Rate, Death Rate, Natural Growth Rate by Region, 2009." Posted on December 22, 2010 (http://www.npfpc.gov.cn/data/201202/t20120228_382787.html).

Schinz, Alfred. 1989. *Cities in China*. Berlin, Germany: Gebrüder Bortraeger.

Veeck, Gregory, Pannell, Clifton, Smith, Christopher, and Huang, Youqin. 2011. *China's Geography: Globalization and the Dynamics of Political, Economic, and Social Change*. Lanham, MD: Rowman & Littlefield.

This compelling and comprehensive text covers the wide range of China's geography, from landforms to economy. Themed chapters offer straightforward summaries accompanied by rich illustrations in the form of maps, photographs, and tables. The book emphasizes the widely varying challenges which face China's distinctive regions as they encounter development and globalization.

Wang, Ling. 2005. *Tea in China's Culture*. San Francisco, CA: Long River Press.

2   The historical urban system

China has one of the world's oldest and most enduring urban systems. This chapter traces this history from the establishment of ancient settlements along the Yellow River, the development of empires, the extension of the Silk Route, to the establishment of foreign Treaty Ports in the nineteenth and early twentieth centuries. It emphasizes the growth of the urban system in tandem with key events in Chinese history. The following questions should help guide the reading and discussion of the materials in this chapter:

* How did the Chinese urban system change over time?
* What were the primary factors, in the different time periods, which contributed to the development and growth of Chinese cities?
* Why were certain commodities, such as silk, tea, and salt, particularly significant in the development of the urban system?
* What happened when Europeans, Japanese, and other foreigners developed Treaty Ports in China between the mid-nineteenth and mid-twentieth centuries?

Thinking about the historical Chinese urban system

There have been several different approaches to understanding the historical significance of cities in Chinese civilization and the ways in which the urban system grew and changed over time. Was the Chinese urban system primarily "top-down" (most cities established by administrative decree) or "bottom-up" (developed out of local needs for trade and cultural centers)? Was the growth of the urban system governed primarily by the needs of the empire to administer and control its territory, or did the system grow primarily as a result of the growth of local, regional, national, and international economic systems? Because the role of cities in society became one of the rallying cries of the Chinese revolution, as Chairman Mao sought to characterize the revolution as a movement of rural laborers against the overconsumption of urbanites, it is worth exploring the ways in which thinking about Chinese cities and the Chinese urban system changed over time.

Given the vast reach of the Chinese empires over widely differing regions and thousands of years of history, there are, of course, varying answers to questions about how the Chinese urban system evolved in different time periods and different places (see Box 2.1 for a discussion of how this scholarship evolved in the West during the twentieth century).

Box 2.1 How important were cities in Chinese society?

In 1915, sociologist Max Weber asserted that Chinese urban residents lacked the autonomy and rights of European urban residents and instead, due to their strong kinship ties, maintained their connections to their rural homelands. He believed that Chinese cities were so overwhelmingly involved in their bureaucratic and administrative functions that commerce and trade separate from these functions failed to develop. Thus, he argued, Chinese cities had not played the same role in shaping society that European cities did (Weber 1915). Weber's minimization of the role of cities in Chinese society contributed to Western historians' lack of attention to Chinese urbanization from the 1920s to the 1960s (Rowe 1984). The rhetoric of the Chinese revolution itself perpetuated this interpretation: the revolution was presented as resting on a fundamentally rural base, in contrast to the urban industrial base of the Russian revolution. When Chairman Mao announced the founding of the People's Republic of China in 1949, he ushered in an expressly antiurban approach to governance that persisted, despite various changes in the administration of China, until the 1978 announcement of the era of reform (see Chapter 4).

However, responses to the rural-based interpretation of Chinese society and the revolution brought new life to the study of Chinese cities after World War II. How could cities be thought to play such a minor, or even negative, role in society when so many great cities, and so many cities in general, had existed in China? As Frederick Mote, a historian, explained:

> Chinese civilization . . . has not granted the same importance to typically urban activities that other civilizations have. Thus Chinese values did not sustain a self-identifying and self-perpetuating urban elite as a component of the population. As a result, the Chinese have never felt the impulse to create one great city that would express and embody their urban ideals . . . Yet a large number of great cities have existed in China through the last 2,500 years.
>
> (Mote 1977: 102)

Studies of the Chinese urban system during the early years after the founding of the People's Republic of China identified linkages between the establishment of cities and the growth of the empires, thus asserting the integral role that cities played in Chinese society (Trewartha 1952; Chang 1963). G. William Skinner, an anthropologist, borrowed the "central place theory" (Christaller 1933) from geography and used it to present what was then a major reinterpretation of Chinese society, one in which cities served as nodes in a vast trade-based network that supported the empires (Skinner

1964/65). In this conceptualization, cities are at the core, rather than the periphery, of Chinese culture and society.

Skinner's work laid the groundwork for a reinvigoration of the study of Chinese cities. Skinner (1977) synthesized his work on defining and understanding China's regions with Chang's (1963) mapping of the establishment of new cities (Figures 2.1 and 2.2) in order to illustrate how processes of urbanization varied by region during different time periods. Extended discussions of the growth and development of the historical Chinese urban system can be found in more recent work (Mann 1984; Schinz 1989; Steinhardt 1990; Ning et al. 1994; Gaubatz 1996; Cartier 2002; Sit 2010).

Today, most scholars agree in general that there was a change in Chinese urbanization around the time of the Tang and Song dynasties (618–1280), during which there seemed to be a growth in the economic functions of cities. This change contributed to the development of a more integrated and maturing "urban system" that more closely resembled the urban systems found in Europe in pre-industrial times.

A simple scheme for understanding the development of Chinese cities in traditional times would identify three eras: an early traditional period that began to evolve more than 3,000 years ago, characterized by the establishment and development of cities primarily to serve administrative and military functions; a middle period that began during the Tang dynasty (618–907) and added trade-oriented cities to the urban hierarchy; and a late imperial period, which began with the significant fortification of cities throughout China during the Ming dynasty (1368–1911) and ended in the early twentieth century. Following these traditional periods, the Republican era (1911–1949) saw a transition from pre-modern urban forms toward modern cities heavily influenced by foreign development.

How many Chinese cities were there? How large were they?

China has an unusually extensive written historical record. Yet this record is not without its problems for modern-day scholars. Two issues are particularly challenging for understanding Chinese urban history: (1) despite China's rich historical records, the population of Chinese cities in historical times is very difficult to assess due to differences in reporting and definitions of cities and towns in different time periods and different places; and (2) official records regarding the nature of cities were sometimes altered in order to make the cities appear closer to the ideal. Of these, the first is particularly relevant to understanding the Chinese urban system, while the second is particularly relevant to understanding historical Chinese urban form and will be addressed in Chapter 3.

How many cities were there at any given period? This question requires deciding on definitions of cities and towns during different time periods and in different places. Many scholars have used officially designated counties as the best approximation of the number of cities in China during the imperial era, assuming that each county seat was designated as a town/city, and most cities up until about the twelfth century were fundamentally administrative in nature (Figures 2.1 and 2.2).

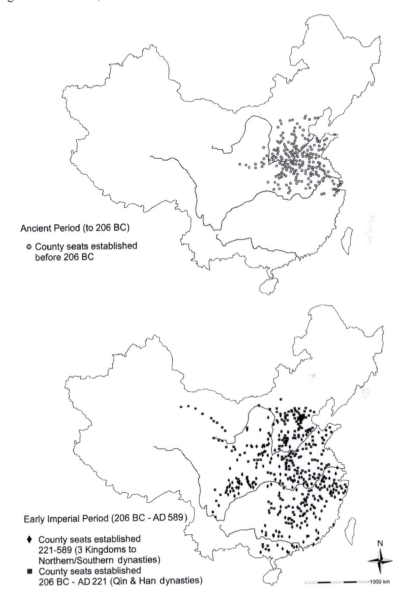

Ancient Period (to 206 BC)

⊚ County seats established
   before 206 BC

Early Imperial Period (206 BC - AD 589)

♦ County seats established
   221-589 (3 Kingdoms to
   Northern/Southern dynasties)
■ County seats established
   206 BC - AD 221 (Qin & Han dynasties)

Figure 2.1 "County-seat" cities established during the ancient (to 206 BC) and early imperial (206 BC–AD 589) periods.

Source: Based on data from Chang 1963.

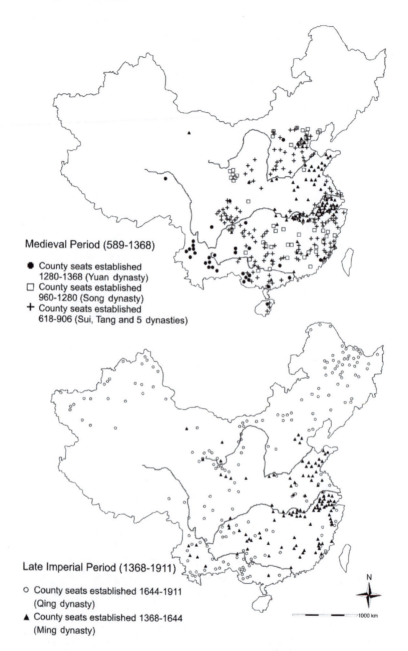

Medieval Period (589-1368)

● County seats established
   1280-1368 (Yuan dynasty)
□ County seats established
   960-1280 (Song dynasty)
+ County seats established
   618-906 (Sui, Tang and 5 dynasties)

Late Imperial Period (1368-1911)

○ County seats established 1644-1911
   (Qing dynasty)
▲ County seats established 1368-1644
   (Ming dynasty)

Figure 2.2 "County-seat" cities established in the medieval (589–1368) and late imperial
            (1368–1911) periods.

Source: Based on data from Chang 1963.

This becomes more difficult from the twelfth century onward, when there was an increase in the number of cities established for other purposes, such as trade. Recent efforts, such as Harvard University's China Historical GIS (2011), are experimenting with functional as well as administrative or population-based definitions of urban areas to produce more accurate interpretations of historical Chinese population change, including urbanization.

How large were Chinese cities? Traditional Chinese gazetteers – encyclopedic records of places – varied considerably in their approach to recording information. Not only were fewer records produced for lesser-known places, but the choice of what to include in a section on population was largely left to the scholars who wrote the records. Some reported only the population inside the city walls, while others included suburban and rural areas. Moreover, population was usually reported in terms of households rather than individuals, so all interpretations of historical Chinese population data include estimates of family size at different periods of time and in different situations.

Chinese cities did, in general, become larger over the course of history, especially over the course of the past 500 years as population grew dramatically. Large cities in the ancient era reached populations in the tens of thousands; large cities in the nineteenth century reached populations in the hundreds of thousands.

Did any traditional Chinese cities become "the largest city in the world?" In a comprehensive effort to estimate the sizes of cities worldwide during different time periods, Tertius Chandler theorized that a number of different Chinese cities have held the position of "the largest city in the world" during different time periods, such as Chang'an in 200 BC and again in AD 750 (Box 2.2), Kaifeng in

---

Box 2.2 Chang'an and its hinterland

Chang'an, literally, "the city of everlasting peace," served as the imperial capital for the Han, Sui, and Tang dynasties – the periods when the Silk Route was at its height. The valley of the Wei River, a tributary of the Yellow River in northwest China's Shaanxi province and the location of the present-day city of Xi'an ("western peace"), was one of the earliest regions of settlement in northern China. The region is frequently mentioned in the Chinese classics, and numerous archeological excavations have revealed a rich history of early cities, such as the royal palace complex of the Zhou dynasty (*c.* 771 BC). The Qin, China's first unifying empire, established their imperial capital in the vicinity of Chang'an in 350 BC. For the thousand years that followed, this area served almost continuously as either the seat of the imperial government or location of important palace complexes. The city of Chang'an itself was first built around 200 BC with the founding of the Western Han dynasty. It began as a palace complex, but the construction of an outer wall and moat eventually defined a complete walled city. Han dynasty

Chang'an was both the administrative center of the empire and an important regional and long-distance trade center. In the early years, the central walled area of the city was largely taken up by two large palaces. Later more palaces were established both inside and outside the city wall. Nine markets were built at the city – two within those walls, seven outside. Some of these markets would have served as regional centers; others were places for trade destined for the newly developing Silk Route (Steinhardt 1986).

According to the records of the Han dynasty, the population of Chang'an county was 80,800 households – 246,200 people. The walled city itself was located near the center of this county. It is likely that the entire county functioned, in an economic sense, as the city and its immediate hinterland. Certainly some portion of this total population would have been rural villagers. At the same time, it is likely that 246,200 is a significant underestimation. Chinese scholars estimate that households at that time averaged about five members, which would make the population of the county around 400,000 or more. Moreover, it is believed that this figure does not include soldiers, traders, sojourners, and other classes of "non-permanent" residents; thus scholars contend that the population may have been around 600,000 (Ning 1998).

Chang'an became less significant after the Han capital moved to the city of Luoyang in the first century AD, but it was revived and re-established as the imperial capital centuries later, first by the Sui dynasty (AD 582), who called it Da Xing, and subsequently by the Tang dynasty, who renamed it Chang'an during the seventh century and kept it as the imperial capital until the fall of the dynasty in 907. Tang Chang'an was more strongly linked to the "exotic" foreign influences of the Silk Route than Han Chang'an had been, and housed a significant, if fluid, foreign population. There were multiple palace complexes and more than 150 monasteries or temples (Buddhist, Daoist, Persian, Manichean, Nestorian, and Zoroastrian). By AD 640, the city's population included about 100,000 foreigners such as Syrians, Persians, and Japanese (Steinhardt 1990; Sit 2010). Some estimates place the suburban population at around 1 million (Sit 2010).

Although most discussions of the Silk Route describe Chang'an as the "starting-out point" on the Chinese end of the vast trade routes, in fact, Chang'an functioned more as the eastern collection and transshipment point. Goods produced throughout China – particularly silk and tea – were brought to Chang'an through a variety of means, traded in the city markets and loaded on to camel caravans which camped outside the city as they prepared for the long westward journey. In this sense, Chang'an was an important point within three nested economic systems – the local and regional economy, the imperial economy, and the vast international network that was the Silk Route.

1102, Hangzhou in 1348, Nanjing in 1358, and Beijing in 1800, but only Beijing in 1800 reached the 1 million population mark (Chandler 1987). Chinese scholars, however, long claimed that a number of Chinese cities became "million cities" over the years, including Chang'an and Luoyang in the eighth and ninth centuries, Kaifeng in the twelfth century, Nanjing in the fourteenth century, and Beijing in the nineteenth century. But more recent scholarship contends that many of these calculations were based on the population of the regions administered by the cities, rather than the cities themselves, so that only twelfth-century Kaifeng and nineteenth-century Beijing truly reached the million mark during the imperial era (Ning 1994).

While many cities in northern China were established during the earliest periods of Chinese history, hundreds of new cities were established during each of China's major historical epochs. As China's population expanded southward, new cities were increasingly established in the southern regions. The following pages characterize the expansion of the Chinese urban system from ancient times into the twentieth century. Table 2.1 provides a summary of the geographical expansion of the Chinese urban system to 1949, and Figures 2.1 and 2.2 illustrate the establishment of new cities in China during different time periods in China's core regions.

The early traditional period (206 BC–AD 589)

Urban origins

The earliest cities in China were established during the Shang period (from about 1600 BC onward) along the lower third of the Yellow River and its tributaries (in the modern provinces of Shaanxi, Shanxi, Hebei, Henan, and Shandong). Small urban systems developed as nucleated ceremonial and administrative complexes anchored to regions of agricultural and handicraft production which demonstrated a rudimentary level of specialization between villages (Wheatley 1971). Figure 2.1 maps these early cities. By the time of the first unification of China under the Qin dynasty (221–206 BC) this region was covered with a relatively dense network of walled cities that had expanded southward to the banks of the Yangtze River. A scatter of outlying Qin walled cities could be found to the south as far as the Pearl River Delta, west into Sichuan and Gansu, and north to the northern border of today's Hebei province (Chang 1963).

By the Qin dynasty (221–201 BC), cities had become the primary points for the complex and multilayered system of Chinese imperial administration. Cities were established by imperial decree in order to serve as the seats of county and provincial administrators who collected taxes, organized projects such as canal and road building, and arbitrated disputes. By the time the Qin emperor issued edicts defining the precise width of streets with different status within the imperial hierarchy, the Chinese empire was already dependent upon a network of small, medium, and large administrative cities to facilitate the running of a vast empire across rapidly expanding territories. About 800 cities, in the form of county administrative seats, were established by the first Qin emperor. This number increased to about 900 by

Table 2.1 Historical expansion of the Chinese urban system

Time period	Era	Main areas where new cities were established	Representative capital cities[a]
1111–221 BC	Zhou/Spring and Autumn/Warring States	Yellow River Valley	Fengyi (near modern Xi'an), Haoyang (near modern Luoyang)
221–206 BC	Qin	Sichuan Basin	Xianyang (near modern Xi'an)
206 BC–AD 221	Han	"River Empire" – Yellow River and Silk Route, Wei River, Yangtze River, and smaller rivers	Luoyang, Chang'an (near modern Xi'an)
221–265	Three Kingdoms	Hill country south of the Yangtze River; Sichuan	Luoyang, Chengdu, Jianye
265–589	Jin and Northern/Southern	Yangtze River valley	Luoyang, Chang'an, Jiankang, Pingcheng
589–960	Sui and Tang	"Tea Country" – hilly regions in southeast China, and southeast coast; cities along the Grand Canal; northwestern frontier region	Chang'an, Luoyang
961–1280	Song	"Northern Pastoral Region" – eastern Gansu, Hebei and northeast China (cities established by the Liao dynasty); Grand Canal area	Kaifeng, Hangzhou, Shangjing, Nanjing
1280–1368	Yuan	Yunnan	Shangdu, Dadu (Beijing)
1368–1644	Ming	Yangtze River Delta, southeast coast, Pearl River Delta	Nanjing, Beijing
1644–1911	Qing	Northeast (southern Manchuria), southwest, Yangtze River Delta, southeast coast, northwestern frontier	Beijing
1844–1911	Treaty Port era of the Qing dynasty	Eastern coast, then strategic points throughout China (more than 80 cities by the early twentieth century)	Beijing
1912–1949	Republican period	Frontier regions (northern Manchuria, southwest, northern Xinjiang)	Nanjing, Beijing

Sources: Chang 1963; Schinz 1989; Sit 2010.

Note:
a More cities served as capital cities for brief periods of time; those listed were the most long-lived capitals during each period of time.

the end of the Qin (206 BC), organized into nearly 50 prefectures, each of which accounted for a somewhat larger administrative seat (Sit 2010).

Growth of the early imperial urban system

Although the Qin dynasty brought the first unification of China as an empire, it was relatively short-lived. The empire first flourished in a sustained manner during the Han dynasty (206 BC–AD 221). The Han inherited the Qin urban system and added to it – as early as AD 2, there were nearly 1,500 counties, each with a county-seat town (Sit 2010: 123). As Figure 2.1 illustrates, the Qin and the Han witnessed the establishment of a vast network of cities. As the empire was consolidated an urban system was established which would serve as the backbone for subsequent urban development. For the first thousand years of the Chinese empires, cities were above all else established for administrative purposes and on the basis of administrative decrees. Yet this Chinese urban system eventually facilitated and protected vast trade networks. The Han dynasty (206 BC–AD 221) saw the first establishment of the Silk Routes (the multiple trade routes sometimes referred to as "the Silk Road"). Trade along this massive network (Figure 2.3) was enabled through the maintenance of regularly spaced garrison/trade towns along the great

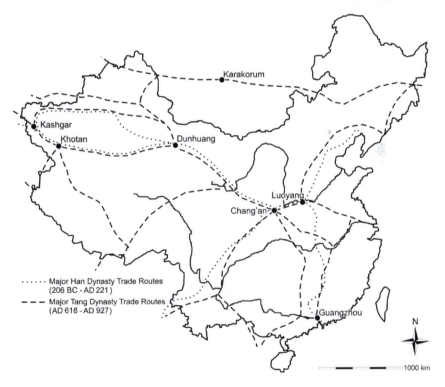

Figure 2.3 Han and Tang Silk Routes.

Source: Adapted from Tucker 2003.

arc of the route from the Han capital at Chang'an through Gansu province and westward across the deserts of Central Asia.

These settlements were vital both in their provision of military protection for trade and in their provision of centers for trade and resupply. Cities founded in frontier regions, in particular, served to protect these routes. Military governors had their headquarters in walled cities established to garrison troops; these cities also offered resupply services for caravans traveling the trade routes.

At the same time, silk itself was produced in the Yangtze region and the interior south. This led to an intensification of urban and administrative settlement, particularly in modern-day Hubei, Hunan, and Jiangsu provinces. A third new urban concentration developed in the Sichuan Basin as flood diversion and irrigation works established during the Qin provided a context for the development of new cities during the Han. Large cities of the Han dynasty were described in ancient texts as vibrant centers for trade in a vast array of products, from luxury items such as wine, silk, and lacquerware, to everyday objects such as straw, wooden implements, and vegetables (Trewartha 1952: 74). Nonetheless, cities remained primarily administrative and/or military in function. Trade was a controlled and subsidiary activity in these settlements, whose primary goal was to facilitate the maintenance of a vast empire.

This early urban system was also carefully structured to mirror the administrative hierarchy of the empire, with higher-ranked cities housing larger populations, and successively less important cities housing successively fewer people. There were three levels of cities during the Han period – the national capital at Chang'an, which eventually housed about 500,000 people within its walls (and possibly more outside the walls); about 100 commandery/prefectural seats, each with populations around 50,000; and more than 1,000 county seats, averaging populations of about 10,000 (Sit 2010: 124).

Thus, for much of the early history of imperial China, administrative and military planning governed the growth of China's urban system. While villages and towns grew in relation to the economic imperatives of an agrarian society, larger settlements were sited in relation to the administrative and military system.

China's "medieval urban revolution" (589–1368)

This pattern changed with the maturation of the Chinese economy that began during the Sui dynasty (589–607) and transformed cities during the Tang and Song dynasties (607–1280). There had been a gradual filling-in of the urban system over the centuries since the Han dynasty, but it was the dramatic increase in trade which began in the sixth and seventh centuries that led to new development in the system as a whole. Agricultural production and transportation were both improved during these centuries, enabling a significant growth in trade (Ma 1971). Tea (which had been produced in China as early as the third century AD) finally became widely consumed and traded in China and beyond. The tea-producing provinces of southeastern China, Anhui, Fujian, Zhejiang, and Jiangxi, saw the establishment of many new walled cities (Chang 1963). The Tang dynasty also revived the Silk

Routes, which had languished after the Han dynasty fell. The Silk Routes of the Tang dynasty were far more extensive than those of the Han dynasty, and with tea as well as silk as prime commodities, extended southward through Yunnan, overseas via the southeastern ports, and along the old northwestern desert routes. This massive commercial expansion generated new urban development along the southeast coast and southward into mainland Southeast Asia. The extension of the Grand Canal to the city of Hangzhou during this time period further integrated the southeastern cities and towns with a larger urban system.

At the same time local and regional marketing systems matured, and there was a corresponding commercialization of cities. Market towns grew into cities, and existing cities diversified their functions. Increasing numbers of cities were founded or expanded to serve the growing economy, rather than the needs of the administrative and military systems. This era has been described as China's "medieval urban revolution" (Elvin 1973). The development of a more robust urban system also can be seen in the growing diversity of city types and the increase in the size of cities during this period.

This expansion followed the general southward trend of the Chinese population over the years from the Han dynasty to the Tang dynasty. Whereas about 60 percent of Chinese cities were located in northern China in the region defined by the Yellow River at the end of the Han dynasty, by the middle of the Tang dynasty (around AD 740) the population had shifted – cities located in the large region from the Yangtze River to the Pearl River Basin (from the vicinity of modern-day Shanghai to the vicinity of modern-day Hong Kong and Guangzhou) accounted for 60 percent of all Chinese cities (Sit 2010). China's regional, trade-based economy continued to grow during the Song dynasty (961–1280). This contributed to urbanization patterns that supported regional economic specialization, such as tea and sugarcane from Fujian; paper from Sichuan and Zhejiang; lacquerware from Hubei, Hunan, and Zhejiang; and iron and steel from Hebei (Sit 2010). In this manner local and regional urbanization took place. As restrictions on market activities were relaxed, small and medium-sized towns proliferated, and existing cities expanded with new development outside the city walls devoted to trade (Elvin 1973).

Urban development was prompted not only by the typical infrastructural needs of local, regional, and long-distance trade, but also by a state monopoly and licensing system for key products. A state monopoly on the production and trade of iron and salt, for example, was established during the Han dynasty and generated a need for a system of urban-based offices to oversee the system. This, in turn, contributed to the development of an orderly and hierarchical system of towns and cities which served to facilitate the administration of the monopoly system. While salt was most consistently subject to state monopoly, other commodities also became subject to monopoly during some periods. Most notably, the Song dynasty saw monopolies over salt, tea, and liquor, and the imperial government also taxed and licensed other aspects of the economy such as foreign trade and trade in copper and jewelry. The Yuan and Ming dynasties also had monopolies on the production and trade of tea and salt; for the Qing, the imperial salt monopoly was one of the most important sources of imperial wealth.

The vast trade networks devoted to China's most prized products – silk, tea, and porcelain (Box 2.3) – during the times of the Silk Routes (Han and Tang dynasties) generated not only urban growth within China itself but also extensions of military outpost cities along the trade routes stretching from China proper toward Central Asia and Europe. The nomadic peoples, such as Mongols and Tibetans, who lived in regions bordering China, came to these outposts for trade more often than conquest, and a number of these Silk Route cities developed into large, multifunctional urban centers (see Figure 2.3 for a comparison of the Silk Route during the Han and Tang dynasties).

Box 2.3 The tea trade: a catalyst for growth

The tea trade, in particular, was a major driver of urban growth and expansion of the urban system. Originally domesticated for its medicinal properties in early China, over the centuries tea was refined into a ubiquitous drink and an important trade product. By the Han dynasty, tea had become a common drink for China's elite and priests. The tea-drinking habits of the nobility eventually filtered to the commoners; by the Tang dynasty, tea was in widespread use through China's regions and social classes. As China's far-flung population developed a taste for tea, tea became an increasingly important component of the domestic economy. The Chinese imperial state established a monopoly on tea production in 793 and began to develop an accompanying infrastructure – including urban markets, warehouses, and processing centers – to reap the tax benefits of China's growing tea consumption. As far north as the city of Kaifeng, in the northern province of Henan, archeologists have unearthed about 200 mills for grinding tea into powder (the preferred way to brew tea in north China during the eleventh and twelfth centuries).

The tea trade contributed further to urban expansion as China began to trade tea with the peoples of Tibet, Mongolia, and Central Asia at about the same time. By the Tang era, compressed bricks of tea were regularly traded in exchange for horses and other goods with the nomadic peoples along China's borderlands. Merchant organizations arose to manage this trade, adding their commercial institutions to the language of Chinese urban development. During the tenth and eleventh centuries, imperial offices to oversee and license the tea trade were established in Hangzhou, Ningbo, Shanghai, and Quanzhou, spurring the growth of the merchant economy in these cities. The proliferation of cities in southeastern China from the Sui period to the Song period can be seen in the growth in the number of counties (each of which required a county-seat town). For example, Fujian province had only five counties during the Sui period, but ended the Song period with 47 counties; Jiangxi province began with 23 counties, but ended the Song period with 70 counties (Ning et al. 1994).

Toward the end of the Song dynasty (981–1280), the Mongols began their drive to rule China. One of Kublai Khan's military strategies was to surround China with loyal forces. This involved establishing garrisons along the borders of China proper. In southern China, for example, the Mongols took over the Nan-zhao kingdom in Yunnan and moved thousands of soldiers, particularly Muslims, from northwestern China there in order to launch an attack on China. These men established new Muslim communities in southwestern China. After the Mongols ascended to power and established the Yuan dynasty (1280–1368), however, the urban system contracted. Both the number of cities and their sizes were relatively small compared to the preceding eras. There were perhaps about 1,500 administrative cities during this time. The urban system included the national capital at Dadu (Beijing) which at its height had a population of about 500,000; two major metropolitan centers – Hangzhou and Pengjiang (Suzhou), each with 200,000–300,000 inhabitants; a number of trade centers of about 50,000–150,000 people; smaller regional centers of 10,000–50,000, and the county seats, most of which had 5,000–10,000 inhabitants (Sit 2010: 193).

Cities of the late imperial period (1368–1911)

China's two last imperial dynasties – the Ming (1368–1644) and Qing (1644–1911) – are usually referred to as the "late imperial period." This period was one of dramatic changes and upheavals in Chinese society, as the empires expanded, came into contact with the West, and, ultimately, collapsed.

Ming dynasty urbanization: an empire of cities

Historian Si-yen Fei has described the Ming dynasty (1368–1644) as a period in which China experienced a "transition from an empire of villages to one of cities" (2009: 1). While the Tang and Song eras saw an increasing role for trade, the subsequent institutionalization and social and cultural change further expanded the roles and functions of cities in the Chinese world. In this manner, China's overall urban system was strengthened into a multifunctional and hierarchical web of economic, political, and social power.

In a geographical sense, the Ming dynasty saw the establishment of new cities along the coasts, in Yunnan and Guizhou, and in the Yangtze River Delta (Figure 2.2). Although the Tang dynasty was the first Chinese dynasty to establish substantial numbers of cities along the coasts, China's great seafaring age came during the Ming dynasty. The Ming, unlike both its predecessors and the subsequent Qing dynasty, developed an extensive fleet of ships engaged in exploration and trade with voyages throughout Asia and as far as the eastern coast of Africa. Cities were established during this period along the southeast coast, especially in Fujian province, which facilitated this seaward orientation. The development of port cities, in turn, led to (and was fueled by) development in the mountainous interior regions of the southeastern provinces. Tobacco-growing, for example, was introduced by way of Fujian's trade with the Philippines, and contributed to the growth of interior Fujian. New cities were established in Guizhou province as well, as migrants finally moved into this rugged region (Chang 1963).

The city of Nanjing, in the Yangtze Delta, served as the Ming imperial capital from 1368 to 1421. During this time, the entire Yangtze Delta region prospered, with new cities established and existing cities expanded to meet the growing needs of the empire. Although the imperial capital was moved to Beijing in 1421, the lower Yangtze region continued to play a vital role in the urban system, with Hangzhou, Nanjing, Suzhou, and, eventually, Shanghai all functioning near the top of China's urban hierarchy. Cotton became a major Yangtze Delta crop for the first time, eventually accounting for 70 percent of the region's farmland and newly developed cottage industries. At the same time, the economy as a whole was increasingly monetarized, which led to a substitution of taxes for official monopolies and thus encouraged the development of private, diversified industries (Sit 2010). This diversification of the economy contributed to the development of a larger and more stratified urban system throughout China. An elaborate urban system emerged with four levels in the Yangtze River Delta: seven main cities (Suzhou, Songjiang, Changzhou, Zhenjiang, Hangzhou, Jiaxing, and Huzhou), 112 county seats, 166 market towns, and 205 towns. The economy of Ming and Qing was able to support towns and cities at a number of different scales in nested local, regional, and national trade networks (Skinner 1977). New cities were established, or grew out of existing villages, which were centered around specialization in single industries or handicrafts, such as the porcelain-manufacturing center of Jingdezhen.

But the establishment of new cities and towns is not the only reason for Ming China to be described as an empire of cities. Not only did the Ming dynasty solidify the existing system of cities, but indeed they solidified the cities themselves. The walls of many Chinese cities, which had been formed of rammed earth, were expanded and faced with stone and brick during this time period as systems of domestic trade and transportation matured. Thus cities that had existed for centuries were restored, strengthened, and expanded (more in Chapter 3). Particularly well-fortified cities could be found along the rebuilt Great Wall, where nine new military garrison cities were founded during the Ming period in order to protect the empire and trade with its neighbors. Other cities, such as Tianshui (Gansu) were established during the Ming dynasty to support the "tea-and-horse" trade with China's northern neighbors. See Box 2.4 for a discussion of the development of Höhhot, an important trade center along China's northern frontier.

Box 2.4 **Höhhot and its hinterland**

During the Ming and Qing periods Mongol nomads and their steppe, mountain, and gobi homelands, in what today are central Inner Mongolia and Mongolia, were drawn into imperial China in new ways. The development of the frontier city of Höhhot, in what is today Inner Mongolia, facilitated that territorial, social, and economic restructuring of the expanded empire.

Höhhot became the center of a vast trade network that reached from south-east China to far western Xinjiang, and from northern Mongolia to China's southwestern provinces. The city was first founded by the Mongol leader Altan Khan in 1572 as part of a concerted effort to establish trade with the Chinese along their northern border. Altan Khan had ordered the construction of two large Buddhist monasteries there, and five more major Buddhist complexes were built there soon after his death. As a result, this Mongol city served continuously as an administrative and religious center through successive redevelopment and multiple political and military regimes. The Chinese imperial administration eventually established an administrative outpost and built a new city wall in 1582, but the Mongols continued to administer the city until it was largely destroyed in a battle between Manchu and Mongol troops in 1632. In 1634, the city was re-established by the Manchu, who rebuilt the city and refurbished the temples after the Qing dynasty was established in 1644. The Qing city was the headquarters for a large Qing military territory. A new city was built adjacent to the existing city in order to house the large military establishment which was garrisoned there. This new city, called Suiyuan, was a substantial square-walled city with four massive gates and a moat. But the earlier city, called Guihua, continued to serve as the region's population and marketing center. These twin cities formed the basis for today's city of Höhhot (Gaubatz 1996; Gaubatz and Stevens 2006).

From the time of its first establishment, Höhhot was a major center for trade between the nomads of the northern frontiers and the Chinese. In 1572, about 11,000 head of cattle were traded at Höhhot. By the late nineteenth century, Höhhot served as a trade center for more than 1 million sheep and about 200,000 other livestock (especially cattle, horses, and camels). Tea from Hunan and other locations on the southeast coast of China, as well as cotton and other Chinese manufacturing, supported this massive trading system. The trade was monopolized by a small number of trading companies that were granted licenses by the Qing government in Beijing. The largest of these, Da Sheng Kui, employed about 7,000 people and controlled much of the economy of Mongolia and the northern Chinese frontier during the nineteenth and early twentieth centuries from its two headquarters in Shanxi province and Höhhot. Thus Höhhot served as a major node in a vast network of cities and markets which stretched across much of the Chinese and Mongol worlds (Gaubatz 1996; Gaubatz and Stevens 2006).

Qing dynasty urbanization: the Chinese empire expands across Asia

There were more Chinese cities established during China's last dynasty, the Qing (1644–1911), than in any previous era since the Han. Although this was to some extent an intensification of China's medieval urban revolution, new generators of urban growth also were introduced. These included the vast extent of the territory administered by the Qing, unprecedented rapid population growth, diversification of the economy and urban functions, and the impact of foreign economic and urban development. Cities were established both on the northern and southern borderlands and in the interior as the Chinese population tripled between 1740 and 1850. There were three major regions of new urban development during this time period: (1) the interior mountainous regions south of the Yangtze river (especially in today's Hunan and Guizhou provinces); (2) the Silk Route oases of Xinjiang; and (3) the northeastern Manchurian region (Figure 2.2). The Manchurian urbanization, in particular, did not occur until the final years of the Qing toward the end of the nineteenth century. The peripheral regions continued to be the locus of new urbanization even after the fall of the Qing dynasty in 1911. At the same time, existing cities expanded beyond their walls to accommodate population growth and a continuing increase in the commercial functions of cities. There may have been more than 3,000 places that demonstrated urban functions by the end of the nineteenth century, ranging from small market centers with populations as low as 1,000 to vast metropolises, such as Shanghai, with populations in the hundreds of thousands (Skinner 1977).

There is some disagreement among scholars over the nature of the urban system that resulted. While Skinner found large regional variations in levels of urbanization, he also found, in many parts of China, a relatively even distribution of cities of differing sizes according to the rank size rule (e.g., a few large cities, a medium number of medium-sized cities/towns, and many small towns). Gilbert Rozman (1973), in contrast, concluded that the Qing urban systems lacked a sufficient number of medium-sized cities and therefore failed to attain a "regular" distribution (Mann 1984; Cartier 2002).

Another way to view the growth of urban trade networks in late imperial China can be found in the work of the Chinese historian Chengming Wu, who observed that, while long-distance trade during the Ming period was primarily in luxury goods traded north–south, Qing trade, which included a larger share of basic commodities, was largely east–west. In this analysis, the overlay of the Qing east–west trade system on the Ming north–south trade system helped to generate a relatively comprehensive network of trade-based urban centers (Cartier 2002).

In addition to new trade-based cities, the Qing also introduced an elaborate system of military garrison cities designed not only to defend their far-flung empire but also to facilitate communication and trade with and along the border regions. These cities were established especially within five different linear networks, or chains, which crossed the northern and coastal regions of China (but not the far south). For example, the Grand Canal Chain covered the territory between Beijing and Hangzhou; the Great Wall Chain stretched from Beijing to Liangzhou (Gansu province). But there were also garrison cities in far-flung locations such as Kashgar, in far western Xinjiang, and Aihui, along the Amur River. The placement of these garrisons contributed to development along their access routes (Elliott 2001).

Foreign empires expand into China: Treaty Port cities (1911–1949)

The period from 1842 into the 1940s saw the growth of a number of cities in eastern China in response to new economic development spurred by the opening of the Chinese Treaty Ports. There were two "Opium Wars" (1840–1842 and 1856–1860) fought during the nineteenth century between China and Great Britain as the British sought to trade with China, and China resisted. When the British negotiated access to five of China's port cities – Guangzhou (Canton), Xiamen (Amoy), Fuzhou, Ningbo, and Shanghai – to establish trade following their victory over the Chinese in the first Opium War, they created opportunities for other foreign countries to negotiate for trade access as well. A French–Chinese treaty in 1844 offered the French the same rights as the British in the port cities. These Treaty Port cities granted extraterritorial rights to the Europeans. This means that the British, and later the French, were permitted to operate their assigned sections of these cities without being subject to Chinese law. By the end of the nineteenth century, British, French, German, Japanese, American, and other foreign companies maintained offices in about 80 cities, most of which were in eastern or central China. Cities such as Shanghai, Tianjin, Xiamen (Amoy), Guangzhou (Canton), Ningbo, Fuzhou, Hankou, and Harbin grew dramatically.

This new development had several impacts on China's local, regional, and national urban systems, including the rapid growth of coastal cities, an increasing disconnection from rural China, and urban growth in those places in the hinterland which had direct linkages to the port cities through either traditional or newly built infrastructure. Concentrations of foreign trade and settlement in the Treaty Port cities themselves generated urban growth as rural migrants flocked to the cities seeking economic opportunities. Newly arriving migrants may have accounted for as much as 10–20 percent of the total population annually in early twentieth-century Shanghai (Elvin 1974). Many of these cities, particularly those along the coast, grew large and more economically significant in striking disproportion to their previous status. Cities such as Tianjin and Shanghai were elevated from regional ports to world ports, handling increasingly large volumes of cargo and cash. Industrial development also became concentrated; by 1949, about 70 percent of China's industrial capacity was located in the coastal Treaty Ports.

The introduction of outwardly focused trading cities arguably generated "disruption and conflict" in systems which had evolved gradually over the course of centuries (Murphey 1977). Some scholars argue that Chinese cities became increasingly disconnected from their rural hinterlands as their economies turned from serving as administrative and trade centers which supported those hinterlands toward global trade, a more monetarized economy, and industrial production (Mann 1984).

Economic activities within the Treaty Port cities also generated economic development, and, in some cases, urban development, in hinterland regions. New infrastructure was developed to facilitate the extraction of resources from the hinterland to the Treaty Ports. This new connectivity particularly benefited the cities of Manchuria – the largely land-locked region of northeastern China which today comprises Heilongjiang province, Jilin province, and Liaoning province – which were increasingly dependent upon the railroads for their development. Two different railroad systems – the South Manchuria Railway Company and the Chinese Far East Railway – were established in this region. Both were eventually owned and operated by the Japanese, who established numerous industries in northeastern China during the early twentieth century. As a result, this region developed large industrial cities, such as Changchun and Shenyang. On the eve of World War II, about half of the railroad trackage in China was located in this rapidly developing region (Trewartha 1951).

Conclusion

The urban system in China expanded from a few proto-cities which emerged along the Yellow River from about 1600 BC onward to a massive, continent-spanning network capable of administering to, supporting, and defending some of the world's largest empires. The growth in this system was spurred by imperial expansion, long-distance and regional trade, and the need to defend far-flung borders. During the nineteenth century, the establishment of Treaty Ports along the east coast generated enough new development to pull the system toward the nascent coastal mega-cities – a pattern that was to re-emerge in the twenty-first century (see Chapters 4 and 6).

Bibliography

Cartier, Carolyn. 2002. "Origins and Evolution of a Geographical Idea: The Macroregion in China." *Modern China* 28(1): 79–142.

Chandler, Tertius. 1987. *Four Thousand Years of Urban Growth: An Historical Census.* Lewiston, NY: Edwin Mellen Press.

Chang, Sen-dou. 1963. "The Historical Trend of Chinese Urbanization." *Annals of the Association of American Geographers* 53(2): 109–143.

Christaller, Walter. 1933. *Die zentralen Orte in Suddeutschland.* Jena, Germany: Gustav Fischer. (Translated (in part), by Charlisle W. Baskin, as *Central Places in Southern Germany.* Englewood Cliffs, NJ: Prentice Hall, 1966.)

Elliott, Mark. 2001. *The Manchu Way: The Eight Banners and Ethnic Identity in Late Imperial China.* Stanford, CA: Stanford University Press.

Elvin, Mark. 1973. *The Pattern of the Chinese Past.* Stanford, CA: Stanford University Press.

Elvin, Mark. 1974. "Introduction," in Skinner, G. William (ed.) *The Chinese City Between Two Worlds.* Stanford, CA: Stanford University Press, pp. 1–16.

Fei, Si-yen. 2009. *Negotiating Urban Space: Urbanization and late Ming Nanjing.* Cambridge, MA: Harvard University Press.

Gaubatz, Piper. 1996. *Beyond the Great Wall: Urban Form and Transformation on the Chinese Frontiers.* Stanford, CA: Stanford University Press.

Gaubatz, Piper and Stevens, Stan. 2006. "Transforming a 'Sea of Grass': Urbanization, Nomadic Pastoralism, and Agricultural Colonization on the Sino/Manchu-Mongolian Frontier, 1550–1937." Paper presented at the Yale Agrarian Studies Colloquium. http://www.yale.edu/agrarianstudies/colloqpapers/23seaofgrass.pdf

Harvard University. 2011. China Historical GIS. http://www.fas.harvard.edu/~chgis.

Ma, Laurence J.C. 1971. *Commercial Change and Urban Development in Sung China*. Ann Arbor, MI: Michigan Geographical Publications.

Ma, Zhenglin (ed.) 1998. "China's Urban Historical Geography [*zhongguo chengshi lishi dili*]." Jinan: Shandong Education Press.

Mann, Suzanne. 1984. "Urbanization and Historical Change in China." *Modern China* 10(1): 79–113.

Mote, Frederick. 1977. "The Transformation of Nanking 1350–1400," in Skinner, G. William (ed.) *The City in Late Imperial China*. Stanford, CA: Stanford University Press.

Murphey, Rhoads. 1977. "The Treaty Ports and China's Modernization," in Skinner, G. William (ed.) *The Chinese City Between Two Worlds*. Stanford, CA: Stanford University Press, pp. 17–74.

Ning, Yuemin. 1998. "City Planning and Urban Construction in the Shanghai Metropolitan Area," in Foster, Harold D., David Chuen-yan Lai, and Naisheng Zhou (eds) *The Dragon's Head: Shanghai, China's Emerging Megacity*. Victoria, BC: University of Victoria Press.

Ning, Yuemin, Wudong Zhang, and Jinxi Qian et al. 1994. *The History of Urban Development in China [zhongguo chengshi fazhanshi]*. Hefei: Anhui Science Press.

Rowe, William T. 1984. *Hankow: Commerce and Society in a Chinese City 1796–1889*. Stanford, CA: Stanford University Press.

Rozman, Gilbert. 1973. *Urban Networks in Ch'ing China and Tokugawa Japan*. Princeton, NJ: Princeton University Press.

Schinz, Alfred. 1989. *Cities in China*. Berlin, Germany: Gebrüder Bortraeger.

Sit, Victor. 2010. *Chinese City and Urbanism: Evolution and Development*. Hong Kong: World Scientific Press.

Skinner, G. William. 1964/5. "Marketing and Social Structure in Rural China, Parts 1, 2, 3." *Journal of Asian Studies* 24(1): 3–44; 24(2): 195–228; 24(3): 363–399.

Skinner, G. William, (ed.) 1977. *The City in Late Imperial China*. Stanford, CA: Stanford University Press.
Many of the papers from key conferences about Chinese cities organized by G. William Skinner were published in three volumes in the 1970s: *The City in Late Imperial China, The Chinese City Between Two Worlds*, and *The City in Communist China. The City in Late Imperial China*, in particular, established an important framework for subsequent work – most Western scholarship on the historical Chinese urban system over the past 30 years has refined or reacted to this book.

Steinhardt, Nancy Schatzman. 1986. "Why were Chang'an and Beijing so Different?" *Journal of the Society of Architectural Historians* 45(4): 339–357.

Steinhardt, Nancy Schatzman. 1990. *Chinese Imperial City Planning*. Honolulu: University of Hawaii Press.

Trewartha, Glenn. 1951. "Chinese Cities: Numbers and Distribution." *Annals of the Association of American Geographers* 41(4): 331–347.

Trewartha, Glenn. 1952. "Chinese Cities: Origins and Functions." *Annals of the Association of American Geographers* 42(1): 69–93.

Tucker, Jonathan. 2003. *The Silk Road: Art and History*. Chicago, IL: Art Media Resources.

Weber, Max. 1915. *The Religion of China: Confucianism and Taoism*, transl. by Hans Gerth, 1951. Chicago: The Free Press.

Wheatley, Paul. 1971. *The Pivot of Four Quarters: A Preliminary Enquiry into the Origins and Character of the Ancient Chinese City*. Chicago, IL: Aldine.

3 Traditional urban forms

China's long-standing culture and society are embedded within the evolution of elaborate and stylized urban forms established perhaps as early as the age of Confucius, 2,500 years ago, and have survived to the present. Cities were built not only as spatial expressions of the cultural and social divisions among different groups, but also as what Kevin Lynch (1984) has called "cosmic cities" that translate spiritual beliefs into the very layout of the city. How did the internal form and structure of China's walled cities develop? This chapter explores the evolution of China's urban walls, markets, districts, and institutions. It emphasizes the persistence of ancient forms and ideals alongside the changes and upheavals that took place over more than two millennia.

There have been five major epochs of fundamental transformation in China that have given rise to distinct urban forms: an early traditional form that began to evolve more than 3,000 years ago; a late traditional form dating from the Tang dynasty (618–907); transitional forms developed during the years of foreign influence between 1842 and 1949; the socialist city (1949–1978); and the contemporary city that is developing since China reopened to the world of global trade at the end of 1978 (Gaubatz 1999a). This chapter addresses the first three of these eras; the more recent eras are discussed in Chapter 8. Key changes in urban form during these eras are summarized in Figures 3.1 and 3.2. These figures illustrate the ways in which key elements of Chinese urban form – from city walls and gates to monumental structures – have changed over time. While some fundamental aspects of Chinese urban form have demonstrated a marked consistency, others were lost or extensively transformed by the early twentieth century (Gaubatz 1996, 1999a, 1999b). The following questions should help guide the reading and discussion of the materials in this chapter:

- What basic principles guided the design of Chinese urban form in ancient times?
- What were the basic components that made up a traditional Chinese city?
- Why was traditional Chinese urban form so consistent across time and space?
- How did traditional Chinese cities accommodate expansion?
- What were the functions of Chinese city walls?

- Who lived inside the city walls? Who lived outside?
- What role did guilds come to play in Chinese cities?
- What happened to Chinese urban form after the Europeans arrived in large numbers during the nineteenth century?

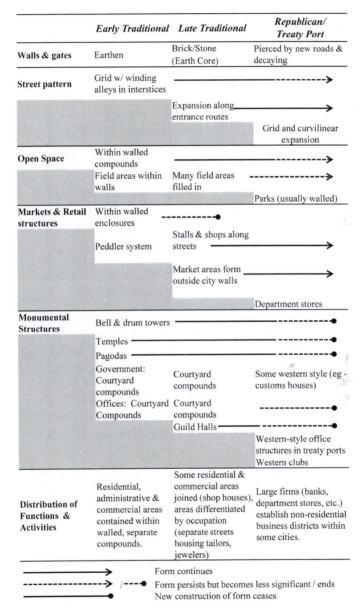

	Early Traditional	*Late Traditional*	*Republican/ Treaty Port*
Walls & gates	Earthen	Brick/Stone (Earth Core)	Pierced by new roads & decaying
Street pattern	Grid w/ winding alleys in interstices	─────────────────────────────>	
		Expansion along entrance routes ────>	
			Grid and curvilinear expansion
Open Space	Within walled compounds	─────────────────────────────>	
	Field areas within walls	Many field areas filled in ──────────────>	
			Parks (usually walled)
Markets & Retail structures	Within walled enclosures	──────────●	
	Peddler system	Stalls & shops along streets ────────────>	
		Market areas form outside city walls ───>	
			Department stores
Monumental Structures	Bell & drum towers ──────────────────────────●		
	Temples ─────────────────────────────────────●		
	Pagodas ─────────────────────────────────────●		
	Government: Courtyard compounds	Courtyard compounds	Some western style (eg - customs houses)
	Offices: Courtyard Compounds	Courtyard compounds ──────────────●	
		Guild Halls ─────────────────────●	
			Western-style office structures in treaty ports Western clubs
Distribution of Functions & Activities	Residential, administrative & commercial areas contained within walled, separate compounds.	Some residential & commercial areas joined (shop houses), areas differentiated by occupation (separate streets housing tailors, jewelers)	Large firms (banks, department stores, etc.) establish non-residential business districts within some cities.

────────────> Form continues

-------------> /----● Form persists but becomes less significant / ends

──────────● New construction of form ceases

Figure 3.1 Continuity and change in elements of urban form.

Source: Adapted from Gaubatz 1999a.

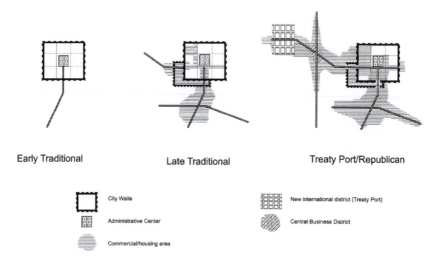

Early Traditional   Late Traditional   Treaty Port/Republican

☐ City Walls   ▦ New international district (Treaty Port)

▦ Administrative Center   ◍ Central Business District

▦ Commercial/housing area

Figure 3.2 Changes in Chinese urban form.

Source: Adapted from Gaubatz 1999a.

Urban plan and layout

"Traditional" Chinese urban form evolved gradually over the space of centuries, but always remained anchored in ancient principles designed to find harmony between the natural world and the constructs of humankind. From the most basic ancient traditions calling for grid pattern, symmetrical cities oriented toward the cardinal directions, Chinese cities developed into highly sophisticated, preconceived constructions, which served as a physical manifestation of cosmological beliefs, bureaucratic hierarchies, and the practicalities of daily life. By the late imperial period, Chinese cities were massive walled constructs, with major streets and architecture oriented in the cardinal directions, all conforming to a grid pattern. Most structures, except the most monumental, were low to the ground (one to two stories in the north, one to three stories in the south); many were surrounded by high walls of their own. The most important structures were usually clustered near the center of the city.

Principles

In the traditional Chinese world, ethics and ideals were as important as, and sometimes more important than, mundane everyday practice in the design and layout of cities. Cities were designed to achieve an ideal deemed so important that official histories and maps were sometimes altered in order to appear more true to the ideal (Steinhardt 1986, 1990; Gaubatz 1996). The Chinese urban ideal incorporated aspects of both the process and practice of urban design, planning, and construction and the form of the city that results from that process. The elaborately ritualized form of Chinese cities was already legendary by the time of the Qin dynasty. Classic Chinese literary works said to be written versions of long-standing oral traditions, such as the *Kaogongji* – the book of trades, including architecture and urban design – which appeared some time during the second century BC as part of the annals of the Zhou dynasty (eleventh century BC – 221 BC) described and idealized the careful layout, for example, of the Royal Zhou capital as a square with sides nine measures long, each side with three gateways defining nine east–west and nine north–south roads. Each of these roads was nine chariot-widths wide. The square shape of the Zhou capital was particularly significant. Traditional Chinese design ideals were tied to cosmological representation. Shape, rather than size, was critical in these ideals, which applied to shape, proportion, and placement for all scales of design and building projects. Similar organizing principles applied to furniture placement, architecture, and urban design.

These principles, which reflect Chinese beliefs in geomancy, have had a strong influence on the establishment, planning, and administration of cities since ancient times. The use of geomancy in urban design is by no means unique to the Chinese – ancient Etruscan and Roman cities, for example, used aspects of geomancy in their design. Chinese geomancy, or *fengshui* (literally, "wind and water"), assigns significance to the ways in which structures built by people interact with the natural world. Three fundamental aspects of *fengshui* affect Chinese urban design: (1) orientation with the cardinal directions; (2) shape (square, rectangular, rounded, irregular) and symmetry; and (3) relative location.

According to *fengshui,* lines of cosmic energy flow through the landscape like wind and water. Depending on its placement, a wall or building can block or collect and augment these forces as it interrupts or redirects the natural energy flows (*qi*). Because these forces can be benevolent or evil, the way in which structures are placed and built is critical to the well-being of those who will occupy them. A badly placed house can bring misfortune to the family who occupies it; a carefully placed temple can bring good fortune to the populace of an entire city. The ideal Chinese city is a map of the Chinese cosmos, in which the walls represent the world. The principal streets served as an idealized organizing force to tame the chaos, much in the way the legendary folk hero Yu the Great rerouted rivers to control flooding in ancient China. This use of the layout of the city as a representation of the cosmos underscores the idea that layout and ground plan were fundamental to the Chinese conceptualization of the city. The paths of walls and streets and the siting of monumental structures were executed with great attention to the urban ideal.

Two aspects of the natural landscape are particularly important for choosing an auspicious site for a city on the basis of *fengshui*. The first is the shape of the landscape; the second is its *yin* and *yang* characteristics. Landscape morphology is a key aspect of *fengshui*, as particular shapes of landscape elements are associated with various mythical beasts or objects with varying degrees of impact on the good fortune of the city. The most auspicious mountain forms, for example, are those with sharp ridges resembling a dragon's back, or undulations in a form associated with tigers.

The male (*yang*) and female (*yin*) aspects of landscape elements also need to be carefully considered in order to ensure an auspicious city site. The balance of *yang* and *yin* forces is critical to fostering advantageous circulation of *qi* through the landscape. *Yang* sites tend to be open toward the south, or to have water on the south side – the direction associated with the sun and "male" energy. *Yin* sites, which open toward the north (mountains on the south side, plains on the north, for example) often are considered inauspicious and required mitigation, such as the construction of a temple, in order to be suitable. Most Chinese cities with names ending in "-yang," such as Luoyang, Hanyang, or Guiyang, are located on the northern banks of rivers or the southern slopes or bases of mountains.

Analysis of a potential city site's *fengshui* was a critical first step in Chinese urban design. In eastern China, authorities intending to establish a city would first engage a *fengshui* practictioner for a site analysis. This practitioner, or geomancer, made use of an astrological compass and a wealth of knowledge to analyze the landscape. In one famous illustration of city site analysis, the geomancer stood at the top of a hill above a site bordered by hills (where he stood) and water (in the foreground). His assistants carefully tested and recorded various measurements to aid him in preparing his report as noblemen, possibly those who commissioned the study, looked on. The best site would permit the *qi* to flow freely and would collect good *qi* for the benefit of its inhabitants. Flowing water was usually considered to carry and store good *qi*, unless it flowed away from the site in a manner which would drain the city of its good *qi*. Mountains and hills on the north were effective in blocking the ill winds associated with the north.

The same principles of geomancy applied to other scales of construction, from the internal layout of the city to the placement of furniture within the home. Careful manipulation of *fengshui* within the city was considered beneficial to its inhabitants. Positive manipulations included orienting all the buildings and streets within the city along the cardinal directions with structures oriented due south to receive *yang* energy, strategic additions of artificial pools and streams to enhance the flow of *qi*, and tree planting. The most auspicious sites in the city were usually located in the center and "back" (north) part of the city, where the *qi* would collect within the walls, thus the most important structures were placed there. *Fengshui* also was manipulated in order to ward off the ill will that often came from the north. "Spirit walls," long brick walls placed about a meter in front of openings, such as doorways, were important in blocking any evil that otherwise might travel into a courtyard structure. Mirrors often were placed on the outsides of structures to deflect ill forces away. Artificial hills were constructed north of important structures or complexes to protect them. For example, "coal hill" in Beijing, due north of the large imperial palace complex called the Forbidden City, was constructed out of dirt dug from the palace moat in order to defend the emperor, and thus the empire as a whole, from evil.

Superstructure

The spatial conception of the ideal Chinese city consisted of several basic elements – walls and gates, a street grid, and, most importantly, monumental structures (administrative compounds, temples, and bell and drum towers). The walls, gates, and street grid served as a superstructure to contain and organize the remainder of the city. The rudiments of this superstructure nearly always were planned, and often completed, before the remainder of development took place. Thus all subsequent construction was carried out in relation to the frame defined by the superstructure. This superstructure also defined the cardinal directions and helped to organize the city as it grew. Many ancient diagrams and plans of Chinese cities showed only the superstructure, underscoring its significance. In fact, the adherence of the superstructure to ancient urban ideals was so important that city plans were sometimes altered for the official histories to make cities appear more in compliance with these ideals than they really were (Steinhardt 1986; Gaubatz 1996).

Walls defined Chinese cities from the earliest times. In fact, the Chinese term for city, *cheng shi*, means "wall and market." The ancient city walls in northern China were made of loess tamped with wooden tools into a dense mass. Ancient poems describe the rhythmic "klak-klak" sound as the city walls were tamped and gradually built up. Most city walls were made of rammed earth alone until the Ming dynasty (1368–1644) when there was a concerted effort to face them in brick and stone. Dynastic records report as many as 564 walls receiving fresh brick facings during the early Ming period (Sit 2010) (Figure 3.3). Most city walls were tall

Figure 3.3 Reconstructed wall and gate at Jiayuguan, Gansu province, showing the rammed-earth wall.

Source: Piper Gaubatz.

Figure 3.4 Top of the reconstructed Ming-dynasty wall at Xi'an.

Source: Piper Gaubatz.

(at least 3 meters – much taller than a basketball hoop) and quite thick at the base, with a trapezoidal profile. The tops of city walls were always wide enough to walk on; the most important cities, such as imperial or provincial capitals, had walls several chariot-widths wide. The wall of Xi'an, for example, was about 12–14 meters wide, and about 12 meters tall in its Ming dynasty form (Figure 3.4).

While the ancient ideal called for walls following a square, or at least rectangular, path around the city, in reality many city walls did not meet this ideal. In a survey of 233 ancient city walls, only 61 percent followed square or rectangular paths; most of the rest were quite irregular in shape. Nearly all of the irregular or round-path city walls were in eastern China (Figure 3.5). This may be due, in part, to a combination of complex terrain (hills and water) in eastern China and the careful adjustments made by *fengshui* masters to adjust to that terrain. Cities in the more arid regions of western China were more likely to have truly square walls, either as a result of their more open sites or the lack of consultation with a *fengshui* master for subtle realignments of city fortresses erected by the military on the basis of a standard design (Gaubatz 1996).

City walls both defined the highest order of the urban superstructure and fulfilled three primary functions: symbolic and physical definition of the city, defense, and urban space. In symbolic (and literal) terms, city walls separated urban residents from the rest of the world. Urban residence during many eras was exclusive either to clans with specific status within the administrative hierarchy or those who could afford the higher costs of urban life. The symbolism of walls was social in this sense, but it was also cultural in several different ways. For example, in multicultural settlements in China's interior frontier zones, city walls marked the boundary

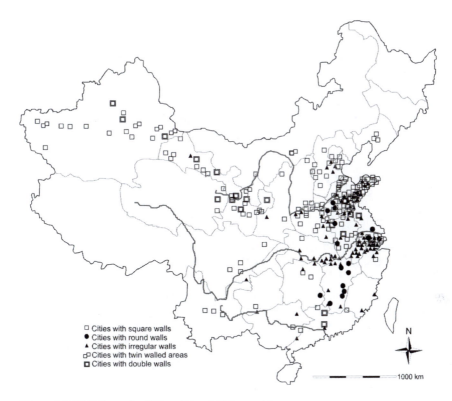

Figure 3.5 Wall forms for 233 traditional Chinese cities.

Source: Adapted from Gaubatz 1996.

between the Han Chinese and non-Chinese realms. In double-walled cities (Figure 3.5) the inner wall surrounded the Chinese core of the city, an outer wall surrounded privileged others (such as Muslim merchants), while other aspects of urban life (such as bazaars) might be carried out beyond the walls altogether.

At the same time, city walls were defensive structures. By the fifteenth century most had crenellations at the top to protect archers, as well as "horse faces" (*ma mian*) – outward rectangular protrusions at regular intervals designed to enable the city's defenders to stand above and on three sides of enemies who attempted to scale the walls. Many walls also had watchtowers on top which extended their height; these towers sometimes served as landmarks that could be seen from great distances. Other defensive features included moats, which surrounded most cities just outside the wall, guard towers at the corners of city walls, and the city gates themselves. City gates were placed for both defensive purposes and to be consistent with *fengshui*. At a minimum, there was one gate on each side of the city. But it was common for all but one of these to be closed much of the time, to limit and control access to the city. The principal gate of the city was often built as an enceinte, that is, as a double gate with an enclosed court-yard between them. This enabled defenders to trap invaders in the courtyard between the two gates.

Walls also served as a kind of exclusive urban space. Most were wide enough for several soldiers to walk abreast, and the tops of larger walls held a variety of structures, usually associated with their military functions, such as watchtowers, barracks, or temples dedicated to patron spirits of the military. Soldiers could walk or ride horses quickly around the walls without needing to pass through the crowded cities.

Within the city walls, typical Chinese cities were laid out along a grid pattern of streets. These streets were planned in hierarchical fashion and served as tangible markers of the city's social, political, and cosmological status geographies. Together with the envelope provided by the city walls themselves, the street grid thus served as an organizational device for the city's power structure.

Status and power in the city were confirmed by location in relation to the city walls and streets. The principal streets were usually laid out in an orthogonal grid aligned with the cardinal directions, thus reaffirming the cosmological layout of the city. The principal axes, which usually bisected the city north–south and east–west, were wider than the rest of the streets in the grid. The widest of these streets was limited to imperial cities and was decreed in terms of chariot-widths as early as the Qin.

Because the most powerful position in the city, in terms of geomancy, was at the center and toward the "back" (north) of the city, it was not unusual for the principal structure of the city – such as a temple, palace, or administrative compound, to be placed directly on the north–south axis. This means that the north–south axis was interrupted at that point, as was the case in the imperial capital of the Tang dynasty at Chang'an (modern Xi'an). Bell and drum towers sometimes interrupted the principal streets as well. Secondary streets refined the basic grid into large square or rectangular blocks. Like the primary streets, these formed an orthogonal grid pattern and led from city gates across the city. Such secondary streets remained true to the cosmological layout of the city.

Chinese cities were preconceived; it was common, particularly in the case of small and medium-sized cities, for the walls and principal grid-patterned streets to be laid out long before there was enough settlement to fill them out. In such cases it was common for the actual settled area to cluster near the center of the grid and along the principal streets, with large, yet undeveloped areas within the walls used for agriculture or grazing.

Streets in early Chinese cities served primarily as the sinew of the grid. Markets and housing remained, for the most part, locked behind the gateways of the city's wards (see below). The commercial transition that began in the Tang and Song periods, however, ultimately transformed the nature of the streets. Streets became spaces in and of themselves, with shops and houses opening directly into them.

Over the years leading to the twentieth century, many Chinese cities outgrew their original walls, as trade and commerce contributed more to population growth than the bureaucratic functions locked behind city walls. At times, additional walls were built to protect these new suburbs, but more often than not, the population simply sprawled into the surrounding countryside, where land values and controls were often cheaper and less strict than for the area within the walls. The distribution of "double" city walls can be seen in Figure 3.5. The higher incidence of double-walled cities in the interior frontier regions most likely is correlated with greater cultural diversity in those areas.

Monumental structures

The city walls and main street system established a framework influencing the way in which the rest of the city developed. The city's primary monumental structures – administrative compounds and government bureaus (sometimes called *yamen*), temples, and bell and drum towers – were all sited within the grid laid out by the superstructure. Moreover, the basic architecture of the monumental structures – especially the administrative compounds and temples – was very similar regardless of the specific functions of these institutions.

Administrative compounds

There were often multiple administrative compounds in a city – not only the *yamen* of the highest-level administration, but also the government offices housing a wide range of officials. The higher the status of the city in the imperial hierarchy, the more of these there would be, such as tax offices, salt administration offices, or official archives. These were all more or less built in the same architectural layout. The basic unit was a courtyard surrounded by four single-story (occasionally two-story) rectangular structures containing one, two, or three rooms each. Each rectangle faced the courtyard. Walls on the courtyard sides of these structures usually had folding or sliding panels with full or half-height windows or latticework, which could be used both as doorways and windows. Thus each square courtyard was defined by four structures and the covered (but often unwalled) corridors connecting them at the corners. These corridors often included gateways arranged so that each courtyard could be closed off from the others (Figure 3.6).

The physical layout of these compounds reflected the careful spatial differentiation and hierarchical ordering of public and private places in traditional China. The same principles apply to all scales of traditional Chinese architecture – from simple single-courtyard houses to massive imperial complexes, thus creating a distinct cellular structure in the city as a whole. Courtyards were arranged along axes (usually south to north) with communication between the different courtyards limited to the gateways connecting them. In a *yamen* compound serving as an office for an administrator or government bureau, the public rarely entered past the first courtyard, and the processing of business related to the public took place in the second courtyard. The third courtyard was devoted to the staff offices of the governing official, who did not deal directly with the public. The innermost courtyard housed the private residence of the administrator himself (Watt 1977). This structure, from public to private, can be seen on a grand scale in the Forbidden City – the Qing dynasty imperial palace – in Beijing. The massive,

Figure 3.6 Reconstructed courtyard structure in Xi'an.

Source: Piper Gaubatz.

monumental courtyards along the southern end of the great south–north axis that bisects the palace compound were used for interactions with foreign emissaries and provincial officials; progression along the axis from south to north led to ever more exclusive (and smaller) courtyards and structures, culminating in the private gardens of the emperor at the far north of the entire assemblage (Box 3.1).

Box 3.1 Beijing: forming an imperial capital

Beijing, or Peking, as the Westerners called it, long symbolized the mystique of the Orient to the Western world. The inaccessible center of a vast, opulent empire, rumors of the city's splendor reached Europe as early as 1601, when Matteo Ricci, a Jesuit priest who had established the first mission in China in 1582, was first permitted to visit the imperial capital. But for China, Beijing served as the most long-lived and perfected representation of the ideal capital city described in the ancient texts. Beijing – or at least the general site of the present-day city – served as a capital city for many different dynasties – first during the Warring States period (475–221 BC) and, more recently, for at least part of every dynasty since the Liao – a dynasty ruled by the Qidan people of northern China – established it as their southern capital early in the tenth century. Each city built on the site was walled and gated in the traditional Chinese fashion, yet there were significant variations in the form and size of the city over time. During the Tang dynasty, it served as a modest-sized border trade outpost inhabited primarily by military personnel and their families, with rectangular walls and ten gates (Sit 1995: 48). When Beijing was re-established as the southern capital of the Liao dynasty, its walls were built along a square path, each side about 1 km long. Its eight gates defined an orthogonal and symmetrical grid of four principal streets, and the palace area was walled to create an inner palace city. But in the Liao city, the inner walled palace was in the southwest corner of the city (Steinhardt 1990). When the Jin displaced the Liao in 1115, they made the city one of their capitals, and named it Zhongdu or Zhongjing (central capital) (Sit 1995: 46–48). The Jin capital was planned by Han officials and designed as a replica of the Han dynasty capital at Kaifeng. It was a square city, completed in the year 1153, with walls nearly 4 km on a side which sat on top of the site of the much smaller Liao city (Sit 1995: 49; Wu 1999: 4).

In 1279, the Mongols established their own imperial capital – the center of the newly established Yuan dynasty – just northeast of the site of the Jin capital at Beijing. Dadu (Great Capital), as it was called, was laid out in accordance with the ancient principles for an ideal imperial capital suggested by the classic text *Kaogongji*, on a much grander scale than the previous cities at Beijing. Building the quintessential Chinese city was one visual means of expressing and legitimating Mongolian power in the Chinese context. Dadu was surrounded by three concentric walls – the outer wall, the palace city, and the palace. Each side of the outer wall was about 5 km long. There were 11 gates – three on each side, except the north, which had two gates (Sit 1995: 51; Wu 1999: 4). The major streets of Dadu connected the outer gates and were 37 meters wide. Secondary streets, which averaged about 18 meters wide, defined 50 wards. The strict symmetry and grid of this city plan were interrupted by watercourses, both natural and manmade, which subdivided the imperial city.

The Ming dynasty completed the main part of their imperial city at Beijing in 1420 after a massive construction project that by some estimates involved 2 percent of the Chinese population (Steinhardt 1990). This was a rectangular city, about 6 km east–west, 5 km north–south, which enclosed an imperial city (about 2.5 × 3 km) and a palace city near its center (Figure 3.7). The city was reconstructed on the same site as Yuan Dadu. However, in 1553, another walled area was completed – a large rectangle (about 8 km east–west, 3 km north–south) attached to the southern wall of the city, which enclosed residential areas and the massive temple complexes of the Temple of Heaven and the altar of agriculture.

When the Manchu founded the Qing dynasty in 1644, they maintained the Ming city as their own, rather than destroying or rebuilding it (which had been the norm in previous dynasties). Thus Beijing's form was largely set in

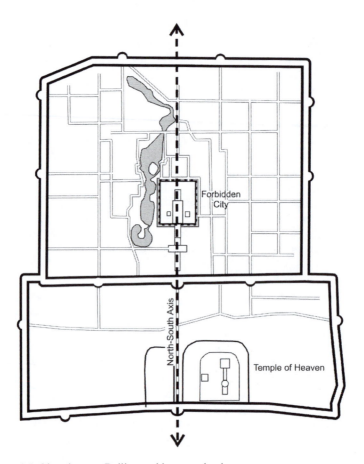

Figure 3.7 Qing-dynasty Beijing and its central axis.

Source: Adapted from Schinz 1989.

1553. It comprised four massive walled spaces: the palace (Forbidden City), the imperial city, the capital city (the main northern walled area), and the outer city (the walled southern extension, completed in 1553). The whole of the city was symmetrical, with the exception of the watercourses that interrupted the grid in the western half of the city. Its social form, especially in the early years of the Qing, was rigidly defined: the Forbidden City was for the exclusive use of the emperor and his servants; the imperial city, also inaccessible to the public, housed the private offices of the imperial administration. The capital city housed the public offices of the empire and many of the day-to-day urban functions necessary for a large city. In the early Qing period, only Manchu could live within its walls; ethnic Chinese lived in the outer, or southern city. The residential division between Chinese and Manchu broke down later as the Chinese moved up the ranks in the Manchu imperial hierarchy (Zhu 2004).

Temples

Temples, monasteries, and mosques also fit within the city's superstructure. These reflected China's syncretic religious culture, which varied in different time periods but often led to cities housing separate temple compounds devoted to Chinese folk religions, Buddhism, Daoism, and Confucianism. Mosques were relatively common as well (see Chapter 12). The basic structure and layout of most temple compounds followed similar principles to that of the administrative *yamen*. Temples, like *yamen*, had a spatial progression from the entrance (usually on the south) toward the "back." The first gateway often housed statues of guardian spirits. Subsequently, each successive courtyard/structural unit was devoted to a more venerated spirit or set of spirits. For example, in a Buddhist temple the most precious statue was located toward the rear of the complex.

Temples varied in size from small local shrines to vast monastic complexes housing hundreds or thousands of monks. Temples were located throughout the city, but the largest and most important temples were often in good *fengshui* locations, such as the center or northern half of the city. Of special note are the temples to the City God – most cities had a patron saint or spirit and incorporated a temple devoted to that spirit. City Gods – part of the folk pantheon – were often the spirits of notable ancient officials or generals and were usually unique to the specific city. Ceremonies and festivals at the City God's temple helped to ensure the good fortune of the city as a whole.

Temples served multiple functions in urban life. As urban spaces, they served as a limited form of urban public space by offering open spaces within their courtyards that were open to all. Children might play in the open courtyards, while adults met and gossiped in the temples' shady arcades. The forecourt or an enlarged space just outside the front gate was often used for periodic temple fairs – elaborate markets that brought goods and entertainment into the city. Many of these grew into permanent markets during the late imperial period. Some temples, especially Confucian temples, also served as schools to educate young boys who might eventually end up in the civil service or as monks. Temples also might be used for political organizing or charitable activities (Naquin 2000).

Bell and drum towers

Bell and drum towers (in larger cities, one of each; smaller cities might have only one or the other) stood near the center of the cities. These structures were usually taller than the surrounding structures, and might resemble large pagodas. The Ming dynasty drum tower in the city of Xi'an, for example, was about 34 meters high. Both bell and drum towers were used for time keeping. Not only was a bell or drum sounded at regular intervals to announce the time, but, more importantly, drums in particular might be beaten for about an hour each evening before the city gates were closed for the night, in order to give residents who had gone outside the walls time to get inside before the gates closed. Bells and drums also would be sounded as an alarm in case of an impending attack.

Neighborhoods and markets in traditional Chinese cities

In 1378, the Ming emperor decided to establish his imperial capital at Nanjing. He ordered an urban atlas. The resulting *Hongwu Atlas* was thematically organized with remarkable attention to topics rarely included in official Chinese maps. In addition to the usual palaces, city walls and gates, topography, shrines and temples, schools, government offices, bridges and streets – the basic elements of the superstructure – this atlas depicted entertainment quarters, warehouses, pasturelands, and gardens. This mapping project followed upon an evolving tradition in literature and the arts that paid growing attention to the everyday urban life of Chinese cities. The most notable example of the new tradition, a scroll painting called *Qingming Shanghe Tu* ("Spring Festival on the river" or "A city of Cathay") brought the streets, markets, and entertainments of twelfth-century Kaifeng, the Song capital, into vivid motion (Fei 2009). Yet these high-profile illustrations of everyday life were relatively uncommon, and much of what we know of the vernacular life of traditional Chinese cities must be inferred from tax records, literary sources, and archeology.

In ancient Chinese cities from at least the Han dynasty until some time during the Tang and Song periods, "neighborhoods" took the form of walled and gated wards which, like the city itself, could be closed off at night. Most of the streets of the city thus would have appeared as corridors between high walls punctuated by gateways; street-fronting shops and stalls had not yet developed. The wards themselves were rarely shown on urban maps and diagrams, which tended to leave blank spaces between the monumental structures (Steinhardt 1990). Written accounts suggest the wards were built within the uniform grid pattern. Each ward contained a number of courtyard dwellings whose residents had clan or native-place ties. These areas were granted to the clans by the imperial or local administration of the city. Larger wards – associated with high-level officials – might extend for the distance of several modern "blocks" in each direction. Chinese social hierarchies tended to place those who engaged in trade in low social ranks. Throughout much of China's imperial history, small merchants and artisans often were required to live outside the city walls, retaining the favored urban wards for those with high status in the imperial bureaucracy (Trewartha 1952: 75). A few places, however, such as the imperial Silk Route capital at Chang'an, assigned spaces inwards near the markets for foreign traders, perhaps so that officials could keep a close watch on them. In this manner, the wards contributed a cellular form to the interior fabric of the city not unlike that of the large rectangular cells containing monumental structures (Gaubatz 1996).

Trade took place primarily in designated (walled) markets within the city (or sometimes outside the walls) or within the wards, where itinerant peddlers might be invited in to display their wares. Walled markets within cities were carefully controlled, with set opening and closing hours. Bureaucrats regulated the weights and measures used within the markets and charged taxes and fees on all trade. For example, during the Han dynasty, when the Silk Route trade was controlled and taxed in walled imperial markets within cities, the largest market of all was the western market in the imperial capital of Chang'an where there were more than 3,000 shops or stalls and traders from Central Asia and China offered wares ranging from silk to vegetables. The market also offered a wide range of regulated entertainment, from restaurants and juggling shows to brothels and storytellers.

Trade activities began to develop outside the ward-designated market system during the latter half of the Tang dynasty. Permanent shops seem to have appeared first within the residential wards (but outside the designated markets) and eventually along the streets themselves. A land market developed as land within the wards became available for purchase or rent for commercial purposes, and land prices became increasingly differentiated according to commercial access. Eventually shops were built along the outer walls of residential wards. This significantly changed the character of urban street life in China, as streets were transformed from circulatory to commercial space. In this manner there was increased differentiation between the relatively quiet walled courtyard housing areas and the bustling public life of the streets outside their walls (Gaubatz 1996).

Just as the walled wards had been organized by clans and/or place of origin (which were often one and the same), the emerging commercial areas within the city also became organized by trade and/or place of origin. It was common for clan-based villages in rural China to specialize in certain trades, such as metalwork or tailoring. These clan and trade connections were replicated within the city, as migrants tended to congregate within cities along clan and trade lines. This led to both the organizational development of trade guilds within the cities based on not only trades but also native place association, and to the development of spatially distinct specialized markets and districts within the city where one or more guilds controlled and organized commerce within the defined set of trades they were associated with (Rowe 1984).

Chinese place-of-origin guilds have their roots in Chinese folk religion, in which the veneration of clan ancestors associates people strongly with their clan homes. Proper veneration of the ancestors requires access to clan shrines, which often formed the core of these associations. In China, urban residents often continued to identify their home place as the place of their clan headquarters, even though they as individuals may never have lived in that place. Because it was not uncommon for most of the migrants to a city from a given place to ply the same trade, guilds became associated with both place of origin and occupation. The place of origin in some cases also represented regional dialects, so that migrants might be speaking local dialects with other members of the guild in the city, further strengthening the interpersonal bonds within these institutions. It was not uncommon for guilds to control tracts of property within the cities, and to steer their trades toward those neighborhoods they controlled (Rowe 1984).

As trade- or native place-associated guilds developed, they often came to control tracts of land within the city. Each guild established a guild hall – a temple devoted to its patron "saint," a meeting area, and a place for sojourning crafts- or kinsmen to stay. Guilds often controlled housing which they would assign to their members, and organized and performed a wide range of local urban administrative functions, from cleaning and maintaining the streets in their part of the city to operating fire brigades. It was also common for guilds to erect large arched entranceways, called *pailou*, to the streets they controlled, to mark and advertise their position within the city.

Treaty Port (1842–1949) and Republican (1911–1949) eras

In the coastal port cities of eastern China, a transitional form developed during the late nineteenth and early twentieth centuries – that of the Treaty Port. This development came as the volume of China's international trade more than doubled between the mid-nineteenth century and the early twentieth century. When the Treaty of Nanking was signed by China and Britain in 1842, it opened selected cities for enclave developments of British (and subsequently other European, American, and Japanese) residents. Foreign merchants, missionaries, and entrepreneurs remade the extraterritorial sections of these cities along European lines, often known as "concession areas." These areas had substantial populations of both foreigners and Chinese, and were to a large extent governed and policed by foreign emissaries. They were built in European style, with "modern" infrastructure and social institutions such as churches and clubs. While China was not colonized by foreign powers, parts of these cities were quite similar in form and function to colonial cities in other parts of Asia, such as Vietnam or India. Subsequent treaties and agreements between the Chinese and foreign governments during the nineteenth century opened yet more cities to foreign residents and business concerns.

By the early twentieth century, more than 30 cities incorporated foreign "treaty"-based settlements, and numerous ports and small towns became small outposts of foreign economic and religious concerns. Thus a European order was imposed upon these cities which hastened, to some extent, the demise of their traditional forms. The economic activities generated by the Treaty Ports also instigated labor migration from the countryside to these cities. The cities grew rapidly in terms of economic functions, population, and areal extent. City walls and some temples and administrative structures fell into disrepair and were either abandoned or used for other purposes. City walls in particular were subject to being dismantled in the early twentieth century to make way for roads and tramways. In Guangzhou, for example, the walls were dismantled early in the twentieth century, first in a piecemeal manner and later by the Canton Tramways Syndicate, which was given a municipal contract to demolish the walls and redesign the streets to accommodate omnibuses and trams (Gaubatz 1999a; Schinz 1989).

Nonetheless, the European districts more often than not were constructed as additions to the existing cities, creating new urban development without completely destroying the older forms. Each of the coastal port cities, including Shanghai, Xiamen, and Guangzhou, entered the twentieth century as dual cities, with one or more European settlements fully functional and adjacent to a Chinese settlement. The cities that arose in this context had complex geographies which both compartmentalized and overlaid widely varying approaches to urban spatial order. In Shanghai, for example, the European settlements were better serviced

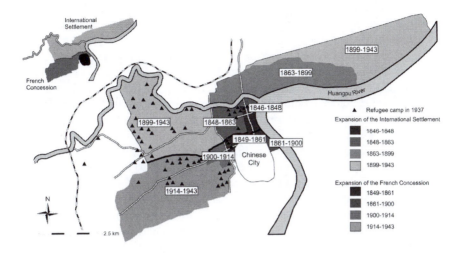

Figure 3.8 Shanghai as a Treaty Port.

Source: Adapted from Schinz 1989.

with electricity, street paving, and other amenities. This physical difference had deep social significance. Electric light was available later into the evening in the European settlements than in the Chinese city. This mundane fact contributed to the development of a disproportionate share of the city's nightlife in the European areas of Shanghai (Yeh 2008). As is evident in Figure 3.8, Shanghai had developed a complex social, economic, and political layout. Much like colonial cities elsewhere around the world, Treaty Port cities also showed a pattern of residential differentiation. There were so-called "upper corners" for well-to-do residents and "lower corners" for the lower classes in the cities, a legacy that seems to have reappeared in these cities since market and housing reforms took hold in the late twentieth century (more in Chapter 8).

Cities that were not Treaty Ports were far less affected by the arrival of the Europeans, but they, too, experienced change during the early twentieth century. The arrival of the automobile and streetcars meant that city gates were sometimes widened to accommodate traffic, railway stations generated new economic development in their vicinities, and in some cities modernized industries were established which not only introduced new landscapes into the cities but also generated labor migration and contributed to changing economies. In Lanzhou, Gansu province, for example, not only were new factories established for paper, electric power, machine tools, woolens, and automobile parts, despite the distance from the coastal Treaty Ports, but the city's administrators embraced modernity as well by repairing roads, opening new public bathrooms, establishing clean water supply systems, and installing streetlamps (Gaubatz 1996; Strand 2000). By 1918, most provincial capitals in China had electricity, match and soap factories, telegraph and telephone services, modern hospitals and hotels, and Christian churches. Two-thirds of them had libraries. Asphalt street pavement was common (Strand 2000).

These changes reflect new approaches to urban planning and administration. Early twentieth-century China was witness to a growing interest in modernity. With the collapse of the Qing empire in 1911, the Chinese experimented with a wide range of new social, political, and economic institutions. In the urban realm, the 1920s saw the rise of experimentation with urban administration and planning. The late Qing government itself had laid the groundwork for urban reform by establishing national guidelines for chambers of commerce, police, school systems, and local self-government between 1889 and 1911. During subsequent years, there were a number of experiments with new forms of urban administration, from the establishment of short-lived City Government Civic Associations to the 1928 publication of the *Complete Book of Urban Administration*, which laid out guidelines for urban planning. Yet the near-continuous political unrest, civil warfare, and the concentrated military incursions of the Japanese generated near-impossible challenges to ordered urban development during the first half of the twentieth century. Moreover, widespread unrest in the countryside generated migrations of refugees into cities, which contributed to the growth of slums and shantytowns at the urban periphery (Stapleton 2007). Figure 3.8 shows the locations of refugee camps within the international settlement areas of Shanghai in 1937.

Thus cities expanded from the mid-nineteenth to mid-twentieth centuries in terms of population, physical extent, trade, and industrial production. Western styles of architecture, planning, and transportation became relatively common. But there were serious problems with life in these cities as well. While European and elite Chinese businessmen built lavish mansions and dined at the best restaurants, slums and informal housing districts formed on the outskirts of the foreign concession areas to house rural migrants who had come to the cities looking for work and refugees from war and unrest. On the outskirts of Shanghai, for example, such slums began to form at the end of the nineteenth century. In 1936, a Shanghai municipal government survey estimated that there were about 100,000 people living in such slums. In the late 1940s, another survey estimated at least 300,000 Shanghai slum-dwellers – about 10 percent of the urban population. Both of these surveys probably underestimated the extent of the slums. By the early 1950s, when the newly established People's Republic of China conducted far more careful and systematic urban surveys, about 1 million people – one-sixth to one-fifth of the city's population – lived in crowded neighborhoods of straw huts on the outskirts of Shanghai, without access to running water or electricity, in makeshift shacks built on stilts over the polluted streams which ran through the city, or in boats in the streams themselves (Lu 1995). These large, poorly serviced slum areas, combined with the damage inflicted on many cities by the ravages of the Sino-Japanese War (World War II), the civil war that followed, and associated economic upheavals, left many Chinese cities in a state of collapse by the time the People's Republic of China was declared in 1949.

Conclusion

Traditional Chinese cities were remarkable artifacts which reflected the ideals of Chinese culture and society. Their sites and layout were carefully chosen and preconceived to align with beliefs in cosmology, to serve as formidable defensive structures, and to maintain the ordered social structures of Confucian society. Despite centuries of political upheaval, cultural and social developments, and natural disasters, these cities were built and rebuilt much along the same lines from ancient times to the nineteenth century. Change was accommodated through accretion. Either new cellular structures of courtyards and walls were added to the outside of existing cities, or newly built or walled areas were conveniently ignored with the assumption that if the central walled city was consistent with the ideal, the subsequent sprawl was irrelevant.

This consistent tradition was changed, however, when Europeans brought their own ideas about the formation of urban space to the nineteenth-century Treaty Ports. The more open and, to some extent, haphazard European approach challenged these traditional ideals. In the early twentieth century, city gates were enlarged or removed to facilitate traffic flow, neighborhoods of elite European villas were built along curving boulevards at odds with the Chinese grid, and in some cases, city walls were torn down to make way for new development. Yet on the eve of the declaration of the People's Republic of China in 1949, the vast majority of Chinese cities retained their distinctive urban form.

Bibliography

Fei, Si-yen. 2009. *Negotiating Urban Space: Urbanization and Late Ming Nanjing.* Cambridge, MA: Harvard University Press.

Gaubatz, Piper. 1996. *Beyond the Great Wall: Urban Form and Transformation on the Chinese Frontiers.* Stanford: Stanford University Press.

Gaubatz, Piper. 1999a. "Understanding Chinese Urban Form: Contexts for Interpreting Continuity and Change." *Built Environment* 24(4): 251–270.

Gaubatz, Piper. 1999b. "China's Urban Transformation: Patterns and Processes of Morphological Change in Beijing, Shanghai, and Guangzhou." *Urban Studies* 36(9): 1495–1521.

Lu, Hanchao. 1995. "Creating Urban Outcasts: Shantytowns in Shanghai, 1920–1950." *Journal of Urban History* 21(5): 563–596.

Lynch, Kevin. 1984. *Good City Form.* Cambridge, MA: MIT Press.

Mote, Frederick. 1977. "The Transformation of Nanking 1350–1400," in Skinner, G. William (ed.) *The City in Late Imperial China.* Stanford, CA: Stanford University Press.

Naquin, Susan. 2000. *Peking: Temples and City Life, 1400–1900.* Berkeley, CA: University of California Press.

Rowe, William T. 1984. *Hankow: Commerce and Society in a Chinese City, 1796–1889.* Stanford, CA: Stanford University Press.

Schinz, Alfred. 1989. *Cities in China.* Berlin, Germany: Gebrüder Bortraeger.

Sit, Victor. 1995. *Beijing, the Nature and Planning of a Chinese Capital City.* New York, NY: John Wiley.

Sit, Victor. 2010. *Chinese City and Urbanism: Evolution and Development.* Hong Kong: World Scientific Press.

Stapleton, Kristin. 2007. "Warfare and Modern Administration," in Cochran, Sherman and Strand, David (eds) *Cities in Motion: Interior, Coast and Diaspora in Transnational China*. Monograph #62. Berkeley, CA: Institute of East Asian Studies China Research.

Steinhardt, Nancy Schatzman. 1986. "Why were Chang'an and Beijing so Different?" *Journal of the Society of Architectural Historians* 45(4): 339–357.

Steinhardt, Nancy Schatzman. 1990. *Chinese Imperial City Planning*. Honolulu: University of Hawaii Press.

Strand, David. 2000. "'A High Place is Better than a Low Place': The City in the Making of Modern China," in Yeh, Wen-Hsin (ed.) *Becoming Chinese: Passages to Modernity and Beyond*. Berkeley, CA: University of California Press.

Trewartha, Glenn. 1952. "Chinese Cities: Origins and Functions." *Annals of the Association of American Geographers* 42(1): 69–93.

Watt, John. 1977. "The Yamen and Urban Administration," in Skinner, G. William (ed.) *The City in Late Imperial China*. Stanford, CA: Stanford University.

Wu, Liangyong. 1999. *Rehabilitating the Old City of Beijing: A Project in the Ju'er Hutong Neighborhood*. Vancouver: UBC Press.

Yeh, Wen-Hsin. 2008. *Shanghai Splendor: Economic Sentiments and the Making of Modern China, 1843–1949*. Berkeley, CA: University of California Press.

Zhu, Jianfei. 2004. *Chinese Spatial Strategies: Imperial Beijing 1420–1911*. London: Routledge.

Part II
Urbanization

4   The urban system since 1949

The place of cities in Chinese politics, society, and economy has changed markedly since the mid-twentieth century. Chairman Mao reviled cities as symbols of unbridled consumption; today cities are perceived as the primary mechanism for new economic growth and development. These differing visions have had a fundamental influence on the development of China's urban system. Although there were hundreds of cities distributed throughout China (see Chapter 2), nonetheless, in 1949, only about 11 percent of China's 542 million people lived in cities. During the first decade of the People's Republic of China, the urban share of the population nearly doubled. Yet between 1960 and 1980, there was little change in this share, as it hovered between 18 and 20 percent of the total population. By 1980, about 19 percent of China's 985 million people lived in cities. Today, however, the urban share of China's 1.4 billion population is about 51 percent, and is expected to continue rising well into the twenty-first century. These drastic shifts are inextricably linked to dramatic changes in development policies over the past 60 years, population movements, and changing definitions of what is urban.

This chapter presents the fundamental changes which have taken place since the founding of the People's Republic of China in 1949, including urbanization patterns and policies under Mao and the shifts since the start of market reform. It also situates the study of urban China in the context of regional and territorial development, as geographic and environmental conditions have rendered Chinese urbanization a highly uneven process both across and within provinces and large regions. Despite all of these changes, there are three aspects of the Chinese urban system which have remained remarkably consistent: (1) China has never developed a single "primate" city or urban region, as have many other countries during the process of development and increasing globalization; (2) national economic and urban policies have played an important role in shaping the Chinese urban system; and (3) there continue to be marked regional imbalances in China's urbanization. The following questions should help guide the reading and discussion of the materials in this chapter:

- How big are Chinese cities? What does "big" mean in this context?
- Why are China's cities distributed unevenly across the landscape?
- How has the Chinese urban system grown and changed since the 1949 establishment of the People's Republic of China?

- How have the post-1979 reforms changed China's urban system?
- How might inter-regional and inter-city economic competition change the urban system?

How big are China's cities? How many cities are there?

Geographical approaches to analyzing urban systems have shifted in recent years from studies of rank-size distributions to analyses of the processes of urban growth, changing functions of cities, and inter-urban connectivity within political, economic, and historical contexts. Urban growth can be interpreted as horizontal (growth in the number of cities) and/or vertical (growth in urban population) (Fan 1999). Analyses of urban growth in China, however, can be challenging due to the coexistence of multiple and changing definitions of urban areas and multiple ways of counting the urban population. How many people live in China's cities today? How many cities are there? These questions are surprisingly difficult to answer, and depend largely upon how "urban" is defined, as outlined in the Introduction. As Box 4.1 and Table 4.1 demonstrate, there are great variations in how China's urban populations are counted and reported.

Box 4.1 How big is Beijing?

Chinese urban populations are notoriously difficult to interpret because there are many different ways in which urban population is reported. For example, Table 4.1 presents the population of Beijing and Shenzhen in 2000 using different official figures.

Does Beijing have 9.74 million or 13.57 million inhabitants? Does Shenzhen have 1.25 million or 7.01 million inhabitants? Can these counts truly reflect the massive, often unregistered, temporary migrant populations of these cities? The staggering variations in these figures illustrate the difficulties in understanding the size and distribution of the Chinese urban population.

These differences reflect different ways of counting urban population. The smallest figures, 9.74 million for Beijing and 1.25 million for Shenzhen, represent counts of people who are officially registered as "urban" inhabitants of the city (see Chapter 5 for more on the household registration system).

Table 4.1 Beijing and Shenzhen in 2000, by different measures (millions)

	Beijing	Shenzhen
Registered inhabitants of the city	9.74	1.25
Census (population of urbanized (built-up) area)	9.88	6.48
Census regional (area administered by city, including rural counties)	13.57	7.01

Source: MGI 2009.

Beijing appears slightly larger, with 9.88 million inhabitants, when all population (registered or not) is counted in the built-up urban area. Shenzhen,
however, appears vastly larger (6.48 million instead of 1.25 million) when all
the people living in the built-up area, rather than the registered population, is
used as a basis for the count. This is due, in part, to the fact that Shenzhen's
population includes a large proportion of migrants who are not officially registered. The 2000 census counted rural-to-urban migrants for the first time.
Thus, especially in a city like Shenzhen, which experienced staggering levels
of migration, there is a huge difference between the size of the registered
("permanent") population and the actual number of people living in the city.

Another level of complexity involves the administrative area of the city.
Most Chinese cities administer not only the built-up central city, but vast
areas of rural hinterland as well. The much larger population figure of 13.57
million for Beijing, reported as the "census regional" population, includes
the population of the entire area administered by the city, which encompasses many smaller towns within the rural counties surrounding Beijing.

Although the official population figures for cities such as Beijing and
Shenzhen can appear remarkably different, their message remains clear:
many Chinese cities are quite large, and have grown dramatically over the
past several decades as China has experienced rapid globalization and economic development. In 2011, the population of Beijing's administrative
area was about 20 million, while Shenzhen had grown to about 11 million.

The United Nations, for example, uses figures provided by the Chinese National
Bureau of Statistics. The Chinese national census uses different counts (*de facto*
population) such as population living in the area administered by the municipality,
or population of the urbanized (built-up) area, while many Chinese government
bureaus use the number of urban residents officially registered (*de jure* population) (Chan 2007; MGI 2009). (See Chapter 5 for further discussion of urban
household registration in China.)

There are two main components to understanding the changing counts of urban
population in China: what constitutes a "city" or "town" in an administrative or
spatial sense, and what constitutes an urban "person." The official definitions of
"towns" and "cities" by the Chinese government have changed five times since
the establishment of the People's Republic of China in 1949, in 1955, 1963, 1984,
1986, and 1993. These changes make accurate analyses of the changing Chinese
urban system difficult. In 1955, cities needed a clustered population of more than
100,000; towns required a clustered, non-agricultural population of 2,500. This definition remained unchanged for cities until 1986, but the definition of "towns" was
changed in 1963 and 1984. In 1986, the definition of a "city," which had remained
unchanged since 1955, was significantly modified to include economic as well as
population considerations. A settlement could be counted as a city with a population of only 60,000 people if its gross national product (GNP) was more than 200

Table 4.2 Changes in numbers of cities, 1949–2009

Cities by population size	1949		1978		2009	
	# of cities	% of total	# of cities	% of total	# of cities	% of total
Super-large cities (>2 million)	—	—	—	—	42	6
Extra-large cities (1–2 million)	5	4	13	7	82	12
Large cities (0.5–1 million)	7	5	27	14	110	17
Small and medium cities (<0.5 million)	120	91	153	79	423	64
Total	132	100	193	100	657	99

Source: National Bureau of Statistics of China 2010.

— Data not available.

million yuan (about US$60 million at that time), or if it functioned as a significant local center and was located in a border or remote area. Counties with more than 100,000 non-agricultural residents and a GNP of at least 300 million yuan could also be classified as cities (Chan and Xu 1985; Liu et al. 2003). Thus the increase in the numbers of cities (Table 4.2 and Figure 4.1) is due, in part, to changing definitions of cities as they are officially designated by the government. In 1997, however, the Chinese government announced a halt to new designations of cities, and the number of cities has remained around 660 since then (Figure 4.2).

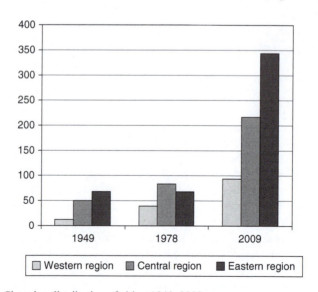

Figure 4.1 Changing distribution of cities, 1949–2009.

Source: National Bureau of Statistics of China 2010.

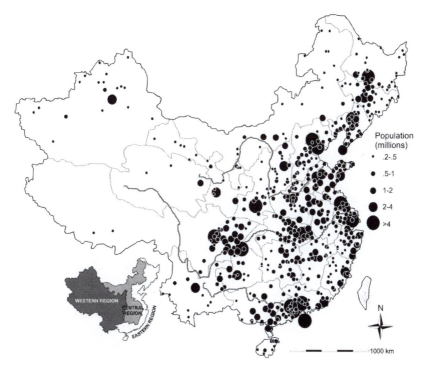

Figure 4.2 Distribution of cities by size.

Source: Adapted from National Geographic 2007 and National Bureau of Statistics of China 2010.

At the same time, definitions of who gets counted as "urban" in the census also have varied greatly over time. For example, the national level of urbanization in 1999 would have been 23.91 percent on the basis of the pre-1982 definition of urban population, 72.99 percent on the basis of the 1982 definition, or 30.90 percent on the basis of the 1990 definition (Liu et al. 2003). The Chinese census made major changes to the definition of urban population in 1964, 1982, 1990, and 2000. From 1963 to 1964, urban population was counted as the total population of places that were designated as cities and towns. In 1964, this was changed to distinguish between "agricultural" and "non-agricultural" population (official designations based on occupation), so that from 1964 to 1981 urban population was counted in terms of non-agricultural population of places. In 1982, the census returned to reporting urban population in terms of the total population in cities and towns, in order to produce more internationally comparable figures (Chan and Xu 1985). The 1990 census refined this system by counting urban population as people living in "urban" areas who had local household registration (see Chapter 5), and non-registered residents who had lived in the city for at least 1 year. The 2000 census counted non-registered migrants for the first time (Chan and Hu 2003). This dramatically increased the official counts of China's urban population. The

Table 4.3 Changes in urban population, 1949–2009

Year	Urban population (millions)	Urban population as % of total population
1950	58	11
1960	131	20
1970	101	12
1980	191	14
1990	302	26
2000	459	36
2010	666	50

Source: National Bureau of Statistics of China 2010.

most consistent counts are probably those of the inner urban districts. Table 4.3 illustrates changes in the urban population over time within these inner districts.

The quantification and analysis of China's urban system have been further complicated by the conversion of rural areas to urban areas. This conversion has both actual and administrative components. In terms of the actual conversion of rural areas to urban land uses and functions, during the reform era, thousands of townships and villages established township and village enterprises – industrial and commercial enterprises that took advantage of local government's ability to control land. These activities eventually led to urban development in rural areas, sometimes known as "rural urbanization" or "urbanization from the bottom up" in China. Reform-era rural urbanization has to some extent followed the *desakota* model, like cities in Southeast Asia, in which the rural hinterlands of cities develop scattered urban functions to create an "urbanized countryside" (McGee 1991). But ultimately the Chinese version of rural urbanization is distinctive. Rather than the *ad hoc* process through which the *desakota* regions in Southeast Asia are developed, to a large extent, in a piecemeal fashion on the basis of their transport connections to an urban center, rural village governments in China have made decisions to turn relatively unprofitable agricultural lands to industrial use. This change eventually results in urban-type development patterns in formerly rural areas, although the formal designation of the land may remain agricultural. Only local governments, equivalent to municipality or county, can officially convert agricultural land into urban land (see Chapter 10).

The administrative aspect of the conversion of rural areas to urban areas involves both the redefinition of formerly rural areas to "urban," and the annexation of rural areas by cities (Box 4.2). In 1949, most Chinese cities' administrative boundaries ended where the urbanized (developed) area ended. But since the 1950s, there has been a major increase in the area administered by cities. Most Chinese cities today administer regions that include both an urbanized core area and surrounding rural counties. These rural counties may include both farming areas and towns or even small cities. Other rural areas became urban when they were annexed by neighboring cities (Chan 2007; Zhang 2008). All of these involve the coexistence of rural and urban administration within cities (see Chapter 13). As cities grow and develop, villages within their administrative areas take on increasing urban functions. In some cases, a "village," in administrative terms, may develop a completely "urban" land use pattern. Much of the urban growth in China between 1990 and 2005 can be accounted for by such administrative changes. Recently, however, migration has become (and will likely continue to be) a more important factor in urban growth (MGI 2009).

***Box 4.2* Is China's largest city Shanghai or Chongqing?**

Most people think of Shanghai as China's largest city. Despite the many different ways to count the urban population, Shanghai comes out on top of all urban population counts, with the exception of the crudest count of *de facto* regional population. That is, the "urban" population of Shanghai, whether counted by the census or by administrative reporting, is much larger than any other city in China. Chongqing is the city with the largest administrative area (82,000 km², compared to Beijing's 16,578 km², Guangzhou's 7,435 km², or Shanghai's 6,341 km²). When the entire population of this massive area is counted as Chongqing's population, Chongqing also appears to be the most populous city in China (and in the world, with more than 30 million inhabitants: Chan 2007). However, this is truly a regional population rather than a "city" population. Chongqing, formerly located in Sichuan province on the upper reaches of the Yangtze River, was long one of the most significant cities in southwestern China (Figure 4.3). During the 1940s

Figure 4.3 Central Chongqing at night.

Source: Wikimedia Commons, retrieved 13 January 2012 from http://commons.wikimedia.org/wiki/File:Chongqing_Night_Yuzhong.jpg# (donated by Jonipoon).

it served as the national capital, and during the mid-1960s it was developed as a key inland center for defense industries by the central government.

During the mid-1980s, it became the headquarters for economic development coordination between China's southwestern provinces (Sichuan, Yunnan, Guizhou, Guangxi, and Tibet). But it was its strategic location on the Yangtze River, in preparation for the Three Gorges Dam project, and the central government's concern with addressing the growing economic inequality between eastern and western China that led to the elevation of Chongqing from an "ordinary" city to the fourth "municipality" with the political status of a province (along with Beijing, Shanghai, and Tianjin; see Chapter 13) in 1997. Along with this changed designation, Chongqing was merged with more than 40 surrounding counties. This expansion accounts for its large size – the city's population grew through this administrative change, from 15 million to more than 30 million; its land area increased from about 23,000 km^2 to about 82,000 km^2 (Han and Wang 2001). For the sake of comparison, New York City has about 8 million residents; the New York metropolitan region as defined in the US census has about 21 million people. Chongqing's redefinition is the equivalent of what would happen if, for example, "New York City" was redefined to include all of New York state (19 million), Connecticut (3.5 million), and New Jersey (8.8 million). If the count is limited to Chongqing's urban districts, the population is about 6 million, compared to 13.5 million in Shanghai's urban districts (Chan 2007).

Given the many alternative ways of quantifying China's urbanization, understanding the growth and change of China's urban system requires understanding patterns and trends more than precise figures. It is more important to understand that there has clearly been massive growth in the share of the Chinese population which lives urban lifestyles, and a complementary decrease in the share of the population living as farmers, than to know precisely how many people fall into each category. Similarly, while it is complex to define, at some levels, exactly how many "cities," or how many urban residents, there are in China, it is clear that the numbers of places that function as cities have grown. Moreover, many cities in China face increasing challenges with administering larger areas and larger populations than they did in the past.

Shifting urban development in the Maoist period (1949–1979)

The period from 1949 to 1979, under state socialism, saw a doubling of the urban share of China's population. In short, national development and urbanization policies played a large role: first, discouraging the growth of very large cities and promoting medium and small cities; second, developing industrial and urban centers away from coastal areas; and third, using a variety of administrative measures to control the general growth and distribution of cities. There was rapid growth, in particular, in provincial capitals. Ten of 25 provincial capitals more than doubled their populations between 1953 and 1970; five of these tripled their populations. This is to say that, in general, large cities became even larger. Provincial capitals in the northwest and southwest grew more rapidly than those in eastern China. Four of the five cities that tripled their populations – Urumqi, Höhhot, Lanzhou, and Xining – were located in China's inland frontier regions. By the mid-1970s, 22 of 26 provincial capitals were the largest city in their province (Chang 1976). The eastern dominance of the urban system, in which eastern cities accounted for 52 percent of all cities in 1949, had been diluted. The total number of cities had grown by about one-third, and about 44 percent of cities were located in the central region. On the eve of the reform era, after a number of initiatives aimed at generating urban development in underdeveloped regions and limiting development of the eastern cities, the regional imbalance had been addressed somewhat, as the percentage of cities in the eastern region had dropped to 36 percent (Figure 4.1). The eastern-centered system of the early twentieth century had been substantially realigned toward the interior (Lin 2002). In 1949, about two-thirds of the urban population was concentrated in the eastern region, but by 1980, only about half of them resided in that region.

Basic generators of urban expansion: rail transport, coal, petroleum, and hydropower

The sections below describe the impact of national development policies on the urban system in terms of three distinct periods between 1949 and 1979. However, there were fundamental infrastructural developments that contributed to the expansion of the urban system that were less subject to policy changes during those three periods: the development of railroad infrastructure, coal mining, exploitation of petroleum reserves, and hydroelectric development. All of these were linked to a steady drive to increase China's industrial production, which has remained a constant priority throughout the history of the People's Republic of China (Chan 1992).

Before the 1949 founding of the People's Republic of China, most railroad development had taken a colonial form. Through joint projects of foreign governments and corporations and the Chinese government, most railroad lines were constructed in the northeast and along the eastern seaboard, and the system was skewed toward movement of goods in and out of port cities. After 1949, there was massive new rail development – 30,000 of the 50,000 km of rail line in operation by 1979 had been built since then. Seventy-five percent of these new lines were located west of the Beijing–Guangzhou (east-coast) line. The railroads had a significant impact on urbanization. Whereas traditional urbanization had, for the most part, been concentrated along waterways and coasts, the railroads linked the drainage basins and thus provided opportunities for new patterns of development and urbanization. Cities such as Baoji in Shaanxi province, Liuzhou in Guangxi Autonomous region, and Bengbu in Anhui province developed as a result of becoming regional rail centers (Chang 1981).

Coal, the primary fuel of China's industrialization, also has been significant in post-1949 urbanization. Most of the major coal fields are in the north and northeast. In particular, six new coal mines developed since 1949 generated new urban development: Huolinhe in Inner Mongolia, Huainan-Huaibei in Anhui, Yanzhou in Shandong, Xuzhou in Jiangsu, Liupanshui in Guizhou, and Pingdingshan in Henan. Due to the cost of transporting coal, each of these areas, along with the traditional coal-mining centers at Datong and Kailuan, generated significant regional industrial and urban development (Chang 1981).

The Chinese began to exploit large petroleum reserves with the opening of the Daqing oil field in 1959 in Heilongjiang province. This led to the development of three new cities. The development of oil production complexes and pipelines in Xinjiang and Gansu and the establishment of refineries in port cities such as Luda, Tianjin, and Nanjing generated further urban development prior to 1979 (Chang 1981).

Hydroelectric systems were important long before the completion of the now-famous Three Gorges project on the Yangtze River. There were about 100 major power stations constructed on Chinese rivers between 1949 and 1979, which helped fuel expansion and development in China's river-oriented urban system (Chang 1981).

Industrialization and urbanization: 1949–1957

When the People's Republic of China was established in 1949, it faced the challenges of years of wartime devastation and weak governance, massive unemployment, and extreme economic inequalities. Decades of poverty and civil unrest had left cities ill-equipped to serve the needs of their residents. Grand monuments of the imperial past were decaying and neglected, slums ringed the cities, and urban problems mounted as hopeful rural residents flocked to the large cities (particularly Beijing) after the revolution, seeking a new life under the new regime. In the northwestern corner of Beijing's walled city, for example, near today's Xizhimen subway station, new migrants from the northeastern provinces built humble shacks in what had been a defensive clear perimeter between the settled area and the wall. In fact, the entire system of cities was in need of rapid change to stimulate the economy and bring new development to the country. The new government worked quickly to transform both individual cities and the urban system as a whole.

During the 1950s, the fate of Chinese cities remained controversial. Chairman Mao's ideological approach celebrated a rural ideal and vilified cities as centers of consumption and excess. The Chinese Communist revolution had, unlike the Soviet revolution, relied upon a rural base. Cities were portrayed as centers of consumption, which had fueled the rising inequality between rich and poor regions and people in China. Soviet advisors, who did not share this stigmatized vision of urban life, urged the Chinese to remake cities as centers of production. The Chinese desire to change cities from centers of consumption to centers of production combined with Soviet advisors' orientation toward urban industrialization to produce a development strategy centered on heavy industries and urban industrial development. Much of this industrialization focused on key inland cities, rather than the coastal trade cities that had grown during the Treaty Port era. The larger eastern cities, such as Shanghai and Guangzhou, were perceived as having played an unduly large role in the national economy during the Treaty Port era. In fact, in 1949 about 70 percent of China's industrial capacity was located in the large coastal cities (Fan 1995). To counter this, growth was encouraged in inland cities, through allocation of central government investment in heavy industries there. Large coastal cities, in contrast, experienced substantial disinvestment and were used as net revenue contributors.

The overall pace of urbanization was rapid as the total number of officially designated "cities" grew from 120 in 1949 to 176 by 1958. The urban share of the population rose from about 11 percent to nearly 20 percent by the end of this first decade of the People's Republic of China, both through migration and through a 1950s "baby boom" as Chairman Mao encouraged Chinese families to have more children in order to strengthen China (Tables 4.2 and 4.3 and Figures 4.1 and 4.2) (Chang 1976; Fan 1995; Pannell 2002; Kamal-Chaoui et al. 2009).

The Great Leap Forward: 1958–1960

Dissatisfaction with the pace of change in the first decade of the People's Republic of China led to a yet more ambitious industrial development program as Mao turned away from Soviet advisors and advocated a new approach based on small-scale industry. The "Great Leap Forward" aimed to increase the self-sufficiency of regions and localities by widely distributing industries, such as steel production, throughout both urban and rural areas. Although the consequence was unintended, this program initially contributed to further urban growth through rural–urban migration as new urban industrial ventures attracted workers. In response, the central government introduced a household registration system in 1958 in order to restrict most people to a residential status (either rural or urban) assigned at birth (see Chapter 5 for an extended discussion of the household registration system and its implications). This measure, combined with economic decline generated through the failure of many of the small industries developed under the Great Leap Forward and a series of natural disasters, halted the growth of the urban population in the early 1960s, which remained at 18–20 percent throughout the 1960s and 1970s (Chang 1976; Chan and Xu 1985; Chan 1994; Chan and Zhang 1999; Kamal-Chaoui et al. 2009).

The "Third Front" and Cultural Revolution: 1964–1978

The People's Republic of China's conceptualization of its urban economy in the early 1960s developed, largely on the basis of defense-based thinking, into a "First Front (or Line)" (coastal cities), a "Second Front" (central interior cities), and a "Third Front," the mountainous western interior regions. In 1964, a new campaign was initiated to reorient industrial investment toward the Third Front region in order to reallocate resources away from the militarily vulnerable coastal region. This resulted not only in industrial development, including factory relocations from the east to China's west, but also in the relocation of workers and their families to cities such as Lanzhou, Chengdu, and Xi'an (Kamal-Chaoui et al. 2009; Fan 1995).

Another factor that had the potential to transform the urban system was the "Rustication of Youth." Building upon the rural ideals of the Chinese revolution, a movement had begun in 1957 to send people (primarily teens and young adults) from cities to the countryside and less-developed areas to generate more even development and to infuse the youth with a greater sense of China's rural base. This movement was interrupted by the Great Leap Forward in 1958, but resumed in 1962 and continued throughout the 1960s. Some 14–17 million youth were sent away from cities to the countryside during this time period. The cities that sent the most youth out under this program were Beijing, Tianjin, Shanghai, Guangzhou, and Wuhan. Consequently, the urban share of the population dipped to about 15–16 percent of the national population by the late 1960s (Chang 1976; Kamal-Chaoui et al. 2009).

Despite the efforts of both the Third Front strategy and the Rustication of Youth, however, economic and urban development remained relatively uneven. By the late 1970s, the coastal region still retained about 60 percent of China's industrial output and remained the locus of the most developed urban areas (Fan 1995). Some of the population that had been relocated to the countryside returned during the 1970s; by the end of the decade the urban share of the population had returned to somewhere between 18 and 20 percent.

The urban system during the reform era

The decade of the 1970s was a transitional time period for China. Diplomatic relations were normalized with the USA and other countries early in the decade, opening the possibility of increased international trade. Chairman Mao's death in 1976 initiated a far-reaching discussion among China's leaders about the future of the country, which resulted in the announcement of the "Four Modernizations" program at the end of 1978. This sweeping program, that the country should concentrate development in four areas – industry, agriculture, the military, and science and technology – ultimately led to the complete physical, economic, and social transformation of nearly every city in China. The new era is usually referred to as the "reform era," or the era of "reform and opening up (to the world)."

During the reform era, the growth rate of the urban population exceeded that of the population of China as a whole. That is, the population became significantly more "urban." While the population of the country as a whole has risen from just under 1 billion in 1980 to just under 1.4 billion today, the urban share of the population has risen from just under 20 percent to about 51 percent. Although the largest cities have grown particularly quickly during this era, this growth also has been spread across the urban scale, from the largest mega-cities to the smallest towns

(Figure 4.2). New cities have been created both through the establishment of new settlements and through redesignation of previously "rural" areas to urban. There is also unprecedented rural–urban migration, after the relaxation of government control on mobility in 1983 (see Chapter 5 for further exploration of the processes of rural–urban migration). The McKinsey Global Institute estimates that, by the year 2030, 1 billion people will live in China's cities, and 221 Chinese cities will have populations of more than 1 million (MGI 2009).

The early reform years: 1979–1992

The implementation of the Four Modernizations program was centered on a dramatic and aggressive effort to invigorate the Chinese economy through increasing levels of engagement with the global economy. During the 1980s national economic development policy was guided by a "three economic belts" approach. The three belts, which corresponded to the eastern, central, and western regions, were first delineated in the Sixth Five-Year Plan (1981–1985) and formally adopted in the Seventh Five-Year Plan (1986–1990). In this macroscale scheme, export-oriented industries and foreign trade would be concentrated in the eastern region, agriculture and energy would concentrate in the central region, and animal husbandry and mining would concentrate in the western region. Theoretically, economic growth in the eastern region would eventually spread to the interior (Fan 1995). This was a complete reversal of the Third-Front strategy which had governed the previous two decades of development, and shifted urban investment back to the eastern region. The coastal region was recognized as the most promising growth area.

During the 1980s the Chinese government exercised caution in its opening of the economy to increased foreign trade. The initial strategy was to place strict geographic limitations on foreign economic activities by permitting foreign trade only within designated development zones. This began with the establishment of four Special Economic Zones (SEZs) by the State Council (the chief executive body of the central government) in 1980 (for details, see Chapter 6): Shenzhen, Zhuhai, Xiamen, and Shantou (Hainan Island became the fifth SEZ after it was designated a province in 1988). Modeled after the successful export-processing zone established in Shannon, Ireland, during the 1960s and other similar zones in Asia, SEZs offered financial incentives to simultaneously encourage and control foreign investment. Within only a few years, it became evident that not only were the zones successful in economic terms, but there was also demand by forces both within and outside China to provide more opportunities for foreign investment on Chinese soil.

In response to this new demand, the Chinese government embarked on a geographically based development strategy. It designated specific, clearly demarcated areas within and adjacent to cities as zones open to foreign investment in production, services, and infrastructure. This development zone strategy offered a means for cities to invite foreign investment and establish external economic ties. To some extent, it encouraged cities to develop an entrepreneurial approach to economic and spatial development. This ultimately contributed to changes in the urban system by fostering growth in those cities able to take advantage of the program. The first manifestation of this new policy was the State Council's 1984 designation of 14 coastal cities as "open" cities (see Chapter 6). Another program designated three Open Coastal Economic Areas in 1985 – the Pearl River Delta, the Yangtze River Delta, and the Min River Delta – and a fourth Open Coastal Economic Area – Shandong and Liaoning – in 1988.

Since the establishment of these early development zones, there have been a number of expansions of the state-sponsored programs. In 1986 the State Council began a formal program to develop high-technology development zones, focusing on Beijing, Jiangsu province, Shanghai, and Guangdong province. In 1988, the State Council extended its support for development zones to nearly all of China's provinces by establishing 52 High and New Technology Industries Development Zones through the Torch Program (aimed at technology diffusion). The central government was not, however, the sole player in development zone planning in China. The ability of local governments to establish development zones on their own was enhanced by the 1987 land reform, which began to create land markets by allowing the paid transfer of land use rights and land leasing (see Chapter 10).

By the early 1990s, China had redefined its urban network from a largely closed domestic system, to a system increasingly linked to the more open-ended global system of cities, by fostering foreign direct investment (FDI) and joint-venture industries, by engaging in trade requiring economic linkages both to markets beyond China's borders and suppliers of raw materials throughout China, and through numerous other changes, such as allowing Kentucky Fried Chicken, McDonald's, and other global consumer-oriented firms to open stores in eastern China's cities.

The 1990s and beyond

In 1992, Deng Xiaoping, who had led China for most of the reform period in a variety of high-level offices, made a trip from Beijing to southern China to visit exemplars of the new economy. Deng's "Southern Tour" was perceived as a "green light" for the newly emerging market economy. China's urbanization and economic development were growing at unprecedented rates, and the goal of becoming a world economic superpower seemed increasingly realistic. While Deng's tour inspired urban development along the southeast coast, Beijing pumped new development resources into urban infrastructure to prepare its 1991 bid to host the 2000 Olympics – a first official entry of a Chinese city into the arena of world cities (although this bid failed, Beijing entered again in 2000 and won the bid to host the 2008 Games).

The 1990s sometimes has been referred to as an era of "unanticipated urbanization" for China (Kamal-Chaoui et al. 2009). Central government fiscal reforms in 1988 and 1994 had allowed local governments to lease lands under their jurisdiction for real estate development, to retain most of the profits from such economic activities, and to make their own decisions about urban development (see Chapters

9 and 10). Real estate development became the main source of income for many local governments, and a massive real estate boom resulted.

During this decade there was also growing diversification of urban economic bases and international investment sources from the more "traditional" sources sought out by the former Treaty Ports (see Chapter 7). The drive to diversify was intensified by the Asian financial crisis of 1997–1998, which led to slow-downs and defaults in projects with Japanese, Taiwanese, Indonesian, and South Korean investors.

But by the early 1990s there was a rise in concerns about overdevelopment, irrational investment, and rising inequality. The phrase "zone fever" came into use to describe the overbuilding of economic development zones as cities competed to attract foreign investment (Cartier 2001). Cities beyond the rapidly developing east coast often had difficulty attracting investment to these zones, many of which remained largely empty or underutilized after construction. Even in the rapidly growing eastern cities, construction sometimes was carried out more on the basis of blind investment than on projections of demand. On the outskirts of large cities, such as Shanghai and Beijing, some urban projects – especially large tracts of free-standing, ultra-expensive "villas" – stood empty as construction outpaced demand. The 1989 National Urban Planning Law offered specific directives meant to shape the urban system by advocating "controlling the big cities, moderating the development of medium-sized cities, encouraging the growth of small cities." This approach continued in the urban policy put forward in the Eighth Five-Year Plan (1991–1995) and the Ninth Five-Year Plan (1996–2000). In these plans, the strictures against the growth of large cities were made even stronger ("strictly control the growth of big cities") (Kamal-Chaoui et al. 2009).

Large cities, however, continued to proliferate. The number of cities with populations over 2 million grew from nine in 1990 to 13 in 2000 and to 42 by 2009; the number of cities with populations from 1 to 2 million grew from 22 to 27 and then to 82 by 2009. Mid-sized cities proliferated even faster – the number of cities with populations from 500,000–1 million grew from 117 in 1990 to 218 in 2000 (although this slowed down subsequently, to 110 in 2009: see Table 4.2). The urban population lived primarily in large cities. In 1990, 53.1 percent of the urban population lived in cities of more than 1 million residents. In 1999, the share of urban population in million+ cities had only marginally dropped, to 52.5 percent. Global firms were attracted to the largest cities in order to take advantage of superior infrastructure and labor force skills. In turn, residents of million+ cities enjoyed a higher standard of living and higher wages than urban dwellers in smaller cities. Moreover, the large urban agglomerations were located in the east, especially in four agglomerations – the Pearl River Delta, the Northeast, Beijing-Tianjin-Qingdao, and Shanghai-Suzhou-Nanjing (Zhao et al. 2003).

The Chinese urban system today

More than 400 cities were added to the Chinese urban system since 1979; the number of "towns" had exploded from 2,968 (1983) to 19,410 (2010), and by the dawn of the twenty-first century, the urban system had matured well beyond its shape at the start of the reform era. The Chinese economy as a whole made a substantial transition to a market system and the level of urbanization had doubled since the start of the reform era. Although the numbers of cities had become more even across China's regions, urban population remained concentrated in the east, with more than half the urban population living in 37 percent of cities located in that region.

The contemporary urban system is driven not only by the continuing development of natural and infrastructure resources, but by added reform-era urban stimuli such as foreign investment, state investment programs (such as the Go West policy), state development projects (such as the Three Gorges Dam), and industrialization. Many of these recent stimuli have disproportionately favored cities over rural areas, eastern cities over central and western cities, and a few preeminent cities above the rest. In addition, as their economies grow and diversify, the eastern region's cities have received large numbers of rural–urban migrants (see Chapter 5).

China is not dominated by a single "primate" city, but rather by several massive "city regions" in the eastern region. These city regions, whose growth is inextricably linked to China's participation in the global economy and major inter-city transportation infrastructure development, include the Pearl River Delta (centered on Guangzhou, Shenzhen, Zhuhai, and Hong Kong), the Yangtze River Delta (centered on Shanghai, Hangzhou, Nanjing, and Suzhou), and the Bohai Bay Region (centered on Beijing, Tianjin, and Shijiazhuang). By 2005, these three regions accounted for 54 percent of China's national gross domestic product (GDP), 79 percent of China's foreign trade, and received 85 percent of all the FDI received by China (Zhang 2008). The McKinsey Global Institute, in its forecast for the growth of the Chinese urban system, predicts that a large share of China's future urban growth will be centered on these regions, but that new giant city regions might arise inland, around Chengdu and Chongqing, as the Chinese economy continues to grow (MGI 2009). The economic structure of these city regions, on the other hand, is quite different. It ranges from the spatially dispersed patterns of growth in the Pearl River Delta that are based on small-scale industries and investment from Hong Kong and Taiwan, to the more concentrated pattern of urban growth in the Beijing and Shanghai city regions. The growth of the city region centered around Shanghai in the Lower Yangtze Delta is especially impressive, and it now rivals the Pearl River Delta region as the country's leading economic area (Ma 2002).

Regional and inter-regional competition

One of the most significant trends in the Chinese urban system today is the development of competition among urbanized areas for investment and development funds. City regions, such as the Pearl River Delta area and the Shanghai region, are competing with each other to lure investors, and competition for investment is increasing among communities within those regions as well (Box 4.3).

Box 4.3 **The Pearl River Delta and Guangzhou: regional urban systems and airport development**

Guangzhou has long stood at the political and economic heart of a larger economic area and regional urban system commonly referred to as the Pearl River Delta. Guangzhou was China's primary economic gateway to the rest of the world prior to the reform era. China's international trade fair – the Canton (Guangzhou) Fair – has been held in Guangzhou nearly every year since 1957. The Pearl River Delta as a whole has had the fastest-growing regional economy, registering a 17.9 percent annual growth rate in per capita GDP between 1978 and 2000 (Yeung 2010). The Pearl River Delta region includes not only Guangzhou, a former Treaty Port and traditional regional center, but also Shenzhen, the fastest-growing of the new cities generated by the SEZ designation in the early reform era, and Hong Kong and Macao. Hong Kong (a former British colony handed over to China in 1997) and Macao (a former Portuguese colony handed over to China in 1999) are both subject to the "one country, two systems" policy, which allows them to maintain a degree of independence in their economies and governance for the first 50 years (until 2047 and 2049, respectively) that they are part of the People's Republic of China.

During the reform era, it has become clear that Guangzhou itself, while still a center for many economic, political, and cultural activities within the region, must compete with other regional economic powerhouses for investment, production contracts, and other resources. Guangzhou's primary local competitors are not only Hong Kong, Macao, and the eight other cities of the Pearl River Delta (Shenzhen, Zhuhai, Foshan, Jiangmen, Dongguan, Zhongshan, and part of Huizhou and Zhaoqing), but also numerous villages and towns that have been the locus of rapid industrialization during the reform era as the Pearl River Delta communities have served as the source of cheap labor initially for Hong Kong corporations, and more recently for a wider range of global firms.

Who wins and who loses in regional inter-urban competition for investment and development resources? Development has not been well coordinated between the Pearl River Delta and Hong Kong. The most high-profile example of this is the overbuilding of international airports in the region. At the start of the reform era, only Hong Kong and Guangzhou had international airports. In the early 1990s, Beijing authorized Shenzhen and Zhuhai to build new airports. At that time, Guangzhou also had requested permission to rebuild its airport, but such permission was not immediately granted. Eventually, new international airports were opened in Shenzhen in 1991 and both Macao and Zhuhai in 1995. Hong Kong opened a new, expanded international airport in 1998, while Guangzhou finally opened a new international airport in 2004. All five international airports are within 100 km of each other. Hong Kong's new airport retained and expanded its global prominence, and today is one of the world's top five air cargo hubs. But the Zhuhai international "aerotropolis," an expensive mega-project, essentially failed as an international airport. It was unable to generate sufficient international business, and Beijing ultimately banned international flights to it in favor of Guangzhou. The management of this airport was reassigned to the Hong Kong International Airport authority in 2007. This costly failure illustrates the problems of uncontrolled development and competition within a regional urban system. In this case, Guangzhou was able to maintain a strong regional position despite a lack of state support early on, and in the face of strong competition from other cities.

In recent years planning for the Pearl River Delta has centered on increased regional integration in order to reduce intra-regional competition and increase the Pearl River Delta's ability to compete with the other urban mega-regions (Yangtze River Delta and Bohai Bay). In 2003, Zhang Dejiang, the Communist Party Chief for Guangdong province, proposed a Pan-Pearl River Delta Economic Zone. Efforts are underway in many different sectors, from transportation to environmental management, for greater regional integration. Meanwhile, the airspace over the Pearl River Delta has become the busiest in all of China. The five airports are now managed in coordination; the most recent plan (2009) assigns domestic and East Asian flights to Guangzhou. Long-haul international flights will be led by Hong Kong but also assigned to Shenzhen, Macao, and Zhuhai. But the transition from competition to cooperation may not be enough to save the now-faltering economy of the Pearl River Delta and Hong Kong, as increasing numbers of manufacturing operations are being relocated inland in search of more economic production costs (Lam 2011; Shih 2011).

Shanghai, for example, caught up with and eventually surpassed Guangzhou in economic performance during the reform era. Between 1979 and 1993 Shanghai averaged a GDP growth rate of 7.9 percent a year, well below the 11.3 percent achieved by Guangzhou. But the growth of Shanghai's GDP accelerated significantly and averaged somewhat more than 14 percent a year during 1993–96, raising Shanghai's performance above the national average, approximately on par with that of Guangzhou (Yusuf and Wu 1997). Since then, both cities have maintained a double-digit growth rate.

The linkage between inter-urban competition and urban development has been strengthened by national government controls on land pricing in development zones. In the absence of much latitude in offering price breaks on land to investors, cities have turned instead to competitive incentive programs based on infrastructure development, local tax holidays, and workforce training. Today "development zones" have largely been replaced by "industrial parks" and other large-scale entrepreneurial ventures. China's entry into the World Trade Organization in 2001 limited or eliminated many of the preferential policies that had been offered to foreign investors within the development zones. Development zones or industrial parks now offer tax and other incentives to all tenants without preference for foreign firms.

The successes and failures of such programs, to the extent they are linked with investment patterns, are changing the structure of China's national urban system. The system continues to favor cities in the eastern coastal region, but cities in the central and western regions also are participating in the new investment-driven urban development. The redistribution of investment and development which has not been effective via state policy may be happening, to some extent, as a result of the global financial crisis and the maturing or development of China's domestic economies. China's massive eastern urban regions are, in some cases, outpricing themselves; investors may increasingly look toward relocating to less costly interior cities. Southeast China in particular has been losing industry to the less costly interior regions.

Should the Chinese urban system be dispersed or concentrated?

The persistence of massive eastern agglomerations reflects a continuing imbalance in the Chinese urban system among the eastern, central, and western regions. Uneven foreign investment and domestic development have turned the Chinese urban system "top-heavy" in recent years. For example, a disproportionate share of foreign research and development investment is concentrated in Beijing and Shanghai (more in Chapter 6). Several different studies have all concluded that more than 75 percent of all such establishments in China have been located in Beijing and Shanghai. This concentration seems based more in historical precedent than in actual competitive factors, such as infrastructure, costs of operations, or labor force. In fact many of these factors are more favorable in second-tier cities, such as Nanjing, Chengdu, or Wuhan (Sun and Wen 2007). Does this imbalance matter? If so, to whom?

When the Ninth Five-Year Plan (1996–2000) was in preparation, it was clear that the rapid economic and urban development of eastern China was not generating significant development in the interior. The Ninth Five-Year Plan included discussions of support for the inland areas and greater coordination among regions, but did not operationalize these ideas. The Tenth Five-Year Plan (2000–2005), however, emerged with a major new focus on the reduction of economic disparities through an aggressive and targeted approach to development in western China. The Western Development Strategy, also known as the Go West program, was designed to direct domestic investment toward western provinces that had not received as much FDI as the cities in the east. The western region for this program includes Sichuan, Guizhou, Yunnan, Shaanxi, Gansu and Qinghai provinces; Ningxia, Xinjiang, Tibet, Inner Mongolia and Guangxi Autonomous regions; and the municipality of Chongqing. The old industrial base in northeastern China – the largely heavy industrial region encompassed by Liaoning, Jilin, and Heilongjiang provinces – also was targeted with similar investment policies during the Tenth Five-Year Plan. Ironically, those cities which have received heightened levels of domestic investment during the Go West policy have experienced increased difficulties, as investment in the real estate sector, for example, has driven up housing prices without raising wages.

In a major study of China's urban system, the McKinsey Global Institute suggests that, if current trends continue, the Chinese urban system will become increasingly dispersed – that is, distributed in a relatively even fashion across the regions. This would bring the national urban system closer to what has become a goal of the central government: to lessen the inequality between regions. But the McKinsey analysis offers alternative scenarios that suggest economic, energy efficiency, and other advantages to several forms of agglomeration, which might be achieved through planning and controls on development (MGI 2009). These scenarios would lead to a concentrated, rather than dispersed, urban system.

Conclusion

In 1949, China's cities and urban population were concentrated in the country's coastal region as a legacy, in part, of the Treaty Port-oriented development of the nineteenth and early twentieth centuries. The concerted efforts to develop the country's infrastructure and antiurban ideology resulted in a somewhat more even distribution of urban places and resources by the 1970s. During the early reform era, however, there was a sharp reversal of this trend, as a few key cities along the east coast were designated to lead the rest of the country in economic development and connectivity with the global economy. The meteoric rise of massive urban regions such as the lower Yangtze River Delta and the Pearl River Delta was fueled by growing economies and labor migration. By the mid-1990s, the central government perceived growing disparities between the eastern regions and the rest of the country. There have been a number of efforts, such as the Go West policy, to address these disparities. But the eastern region continues to dominate the urban system in a number of different ways, from housing the largest share of China's urban citizens to generating a significantly disproportionate share of the country's wealth.

Bibliography

Cartier, Carolyn. 2001. "'Zone Fever,' the Arable Land Debate, and Real Estate Speculation: China's Evolving Land Use Regime and its Geographical Contradictions." *Journal of Contemporary China* 10(28): 445–469.

Chan, Kam Wing. 1992. "Economic Growth Strategy and Urbanization Policies in China, 1949–1982." *International Journal of Urban and Regional Research* 16(2): 275–305.

Chan, Kam Wing. 1994. "Urbanization and Rural–Urban Migration in China since 1982: A New Baseline." *Modern China* 20(3): 243–281.

Chan, Kam Wing. 2007. "Misconceptions and Complexities in the Study of China's Cities: Definitions, Statistics, and Implications." *Eurasian Geography and Economics* 48(4): 383–412.

Chan, Kam Wing and Hu, Ying. 2003. "Urbanization in China in the 1990s: New Definition, Different Series, and Revised Trends." *The China Review* 3(2): 49–71.

Chan, Kam Wing and Xu, Xueqiang. 1985. "Urban Population Growth and Urbanization in China since 1949: Reconstructing a Baseline." *The China Quarterly* 104: 583–613.

Chan, Kam Wing and Zhang, Li. 1999. "The Hukou System and Rural–Urban Migration in China: Processes and Changes." *The China Quarterly* 160: 818–855.

Chang, Sen-dou. 1976. "The Changing System of Chinese Cities." *Annals of the Association of American Geographers* 66(3): 398–415.

Chang, Sen-dou. 1981. "Modernization and China's Urban Development." *Annals of the Association of American Geographers* 71(2): 202–219.

Fan, C. Cindy. 1995. "Of Belts and Ladders: State Policy and Uneven Regional Development in Post-Mao China." *Annals of the Association of American Geographers* 85(3): 421–449.

Fan, C. Cindy. 1999. "The Vertical and Horizontal Expansions of China's City System." *Urban Geography* 20(6): 493–515

Gaubatz, Piper. 2005. "Globalization and the Development of New Central Business Districts in Beijing, Shanghai, and Guangzhou," in Wu, Fulong and Ma, Laurence (eds) *Restructuring the Chinese City: Changing Society, Economy and Space*. New York: Routledge Press, pp. 98–121.

Han, Sun Sheng and Wang, Yong. 2001. "City Profile: Chongqing." *Cities* 18(2): 115–125.

He, Canfei. 2002. "Location of Foreign Manufacturers in China: Agglomeration Economies and Country of Origin Effects." *Papers in Regional Science* 82: 351–372.

Kamal-Chaoui, Lamia, Leman Edward, and Zhang, Rufei. 2009. *Urban Trends and Policy in China*. OECD Regional Development Working Papers, 2009/1. Paris: OECD Publishing.

Kasarda, John. 2009. "Aviation Infrastructure, Competitiveness and Aerotropolis Development in the Global Economy: Making Shanghai China's True Gateway City," in Chen, Xiangming (ed.) *Shanghai Rising: State Power and Local Transformations in a Global Megacity*. Minneapolis, MN: University of Minnesota Press, pp. 49–72.

Kirkby, R.J.R. 1985. *Urbanization in China: Town and Country in a Developing Economy, 1949–2000*. New York, NY: Columbia University Press.

Lall, Somik and Wang, Hyoung Gun. 2011. "China Urbanization Review: Balancing Urban Transformation and Spatial Inclusion." World Bank *An Eye on East Asia and Pacific* series. www.worldbank.org/eapeye.

Lam, Anita. 2011. "Congested Skies a Barrier to Growth – and Flight Safety: Delta Airports Say Surging Traffic Needs Careful Planning." *South China Morning Post* March 26.

Lin, George C.S. 2002. "The Growth and Structural Change of Chinese Cities: a Contextual and Geographical Analysis." *Cities* 19(5): 299–316.

Liu, Shenghe, Li, Xiubin, and Zhang, Ming. 2003. *Scenario Analysis on Urbanization and Rural–Urban Migration in China*. Interim Report IR-03-036. Laxenburg, Austria: International Institute for Applied Systems Analysis.

Ma, Laurence J.C. 2002. "Urban Transformation in China, 1949–2000: A Review and Research Agenda." *Environment and Planning A* 34(9): 1545–1569.

McGee, Terry. 1991. "The Emergence of Desakota Regions in Asia: Expanding a Hypothesis," in Ginsburg, Norton S., Koppel, Bruce and McGee, Terry (eds) *The Extended Metropolis: Settlement Transition in Asia*. Honolulu: University of Hawaii Press.

McKinsey Global Institute (MGI). 2009. *Preparing for China's Urban Billion*. Retrieved on 20 October 2010 from www.mckinsey.com/mgi.

This comprehensive report was developed by a large team of economists and urban scholars who made one of the most thorough analyses yet of the changing Chinese urban system. The report is well illustrated, and contains 14 case studies of individual cities in addition to China-wide analyses. The 540-page pdf version is available for free download from the McKinsey website (see above).

National Bureau of Statistics of China. 2010. *China Statistical Yearbook 2010.* Beijing: China Statistics Press.

National Geographic. 2007. *Atlas of China*. Washington, DC: National Geographic.

Pannell, Clifton. 2002. "China's continuing urban transition." *Environment and Planning A* 34: 1571–1589.

Shih, Toh Han. 2011. "Hong Kong, Shenzhen Losing Export Role as Factories Move Inland: Shipping Data Points to Rise of New Manufacturing Hubs." *South China Morning Post* April 19.

Sun, Yifei and Wen, Ke. 2007. "Uncertainties, Imitative Behaviors and Foreign R&D Location: Explaining the Over-concentration of Foreign R&D in Beijing and Shanghai within China." *Asia Pacific Business Review* 13(3): 405–424.

Sung, Yun-wing. 1996. "'Dragon-head' of China's Economy?" in Yeung, Y. M. and Sung, Yun-wing (eds) *Shanghai: Transformation and Modernization under China's Open Policy*. Hong Kong: The Chinese University Press.

United Nations Center for Human Settlements (UN-HABITAT), with China Science Center for International Eurasian Academy of Sciences and China Association of Mayors. 2010. *State of China's Cities 2010/2011: Better City, Better Life*. Beijing: Foreign Languages Press.

Yeh, Anthony G. O. and Fulong Wu. 1995. "Internal Structure of Chinese Cities in the midst of Economic Reform." *Urban Geography* 16(1): 521–554.

Yeung, Yue-man. 2010. "The Further Integration of the Pearl River Delta: A New Beginning of Reform." *Environment and Urbanization Asia* 1: 13–26.

Yusuf, Shahid and Wu, Weiping. 1997. *The Dynamics of Urban Growth in Three Chinese Cities*. New York: Oxford University Press, for the World Bank.

Zhang, Li. 2008. "Conceptualizing China's Urbanization under Reforms." *Habitat International* 32: 452–470.

Zhao, Simon X.B., Chan, Roger C.K., and Sit, Kelvin T.O. 2003. "Globalization and the Dominance of Large Cities in Contemporary China." *Cities* 20(4): 265–278.

Zhou, Yixing. 1991. "The Metropolitan Interlocking Regions in China," in Ginsburg, Norton S., Koppel, Bruce, and McGee, Terry (eds) *The Extended Metropolis: Settlement Transition in Asia*. Honolulu, HI: University of Hawaii Press.

5 Urban–rural divide, socialist institutions, and migration

Notwithstanding China's long urban history, the country remained largely an agrarian society until very recently. But urban superiority has taken hold since the turn of the twentieth century. Despite efforts to reduce the distinction between city and countryside after the Communist Party took power in 1949, an urban–rural divide forms the basis of the broadest kind of social inequality. Rural areas continue to have the poorest of the poor and lag behind in health status, nutrition, education, life expectancy, and overall living standards. Under market transition and globalizing forces, however, population mobility has grown drastically. Close to 200 million migrants have left the Chinese countryside for cities since 1983. This recent migratory flow is perhaps the largest tide of migration in human history. It has become a prominent feature of China's economic transition and is changing the face of the country (Fan 2008).

This chapter outlines how the persistent urban–rural divide has formed historically and then has been reinforced by a set of socialist institutions. Particularly critical is the household registration system (*hukou*). The chapter also shows how a confluence of rising agricultural productivity and globalizing forces in urban manufactures opened the flood gate of migration in the early 1980s. Since then, migrant workers and entrepreneurs have provided substantial human impetus for the rapid modernization of cities. But most of them continue to face barriers to settle there permanently and exhibit a temporary or circular pattern of mobility. The following questions should help guide the reading and discussion of the materials in this chapter:

- What were the historical roots of China's urban–rural divide?
- What was the trinity of institutions that enforced this divide under state socialism (1949–1979)?
- How did the Chinese government use *hukou* to control the distribution of social welfare and the migration of peasants to cities?
- What are the characteristics of a typical migrant worker in urban China today?
- In what ways is the access to urban amenities by a migrant worker different from that by an urban resident?
- What are the common patterns of housing and settlement by migrants in Chinese cities? How are they different from and similar to patterns seen in other developing countries?

Urban–rural divide and its roots

The gulf between city and countryside is long standing, across countries and throughout history. In the global south, urban areas are places where people can find more employment opportunities, access better public services, and enjoy higher living standards. Rural areas, on the other hand, receive less public investment and experience constant exodus of the young and brightest. The chief culprit lies in an urban bias inherent in development strategies adopted by government authorities, particularly as countries embark on industrialization. China is no exception. While egalitarian ideology was a key in lifting millions of people out of extreme poverty during 30 years of state socialism (1949–1979), the urban–rural divide failed to subside. Aside from large gaps in income and living standards, the central government established different systems of property rights, health care, and welfare provision in urban and rural areas. Two types of citizenship have existed in effect (Solinger 1999). Whereas urban citizenship came with full provision of social welfare, rural citizenship essentially entailed self-responsibility in food supply, housing, employment, and income, and lacked most of the welfare benefits enjoyed by urban residents.

Historical roots

The urban–rural divide has its historical roots. By the early twentieth century, the old Western cliché of urban superiority became a new trend in Chinese society. Treaty Ports and the concentration of modern industries and commerce in urban areas brought job opportunities and material comfort as well as progressive ideas and advanced education. Migrants from the countryside, as outsiders, encountered varied forms of mistreatment in the city. One of the best-known examples was the low socioeconomic status in Shanghai held by migrants from northern Jiangsu in the republican period (Honig 1992). Disdained by local residents, they were given a common deprecating nickname and confined to undesirable jobs and overcrowded shack settlements. Discrimination against migrants also was not limited to urban areas, as shown in the case of the Hakka people throughout southeastern China. A migrant group distinguished primarily by dialect, they were often the objects of prejudice, and were described as uncivilized and poor.

Despite efforts to reduce the distinction between city and countryside after the Communist Party took power in 1949, the urban–rural distinction continued to widen. The pursuit of a development strategy that promoted heavy industries during the 1949–1978 period was a key root cause. This strategy aimed at rapid industrialization by extracting agricultural surplus for capital accumulation in industries and for supporting urban-based subsidies. Bound to collective farming, peasants were completely cut off from many urban privileges. When market reform commenced in 1979, rural China was among the first to benefit, through the dismantling of collective communes. Returning the responsibility for farming to individual households brought huge gains to peasants. As a consequence, around 1984, China became less dualistic and was probably the most equal across urban and rural areas, even more so than under state socialism (Naughton 2007). But such gains paled in the more recent phase of reform (since the mid-1990s) against rapid growth in the urban sector.

Today, income disparity between urban and rural areas has widened – the per capita income of the urban resident is more than three times that of the rural resident. At the beginning of market reform, this ratio stood at 2.6. Such a divide is now the single most important factor underscoring the growing level of inequality across China (Naughton 2007). As measured by a common economic indicator, Gini coefficient, total national inequality is consistently higher than either urban or rural inequality because the urban–rural gap is so large (Figure 5.1). Aside from income disparity, rural population has experienced shrinking public services. When collective communes dissolved as a result of market reform, the supply of public goods collapsed, particularly for the provision of basic health care. Under state socialism, peasants received such care under a system of "cooperative health services" (Naughton 2007: 244). Resources to pay for this system dried out with decollectivization. Now, rural residents pay a far higher share of their health care expenses out of pocket than urban residents.

Institutional roots and the household registration system

Mao's development strategies consistently favored industry at the expense of agriculture. The main enforcement mechanisms for this industrial, and urban, bias were a trinity of institutions, including the unified procurement and sale of agricultural products, the commune system, and household registration (Yang and

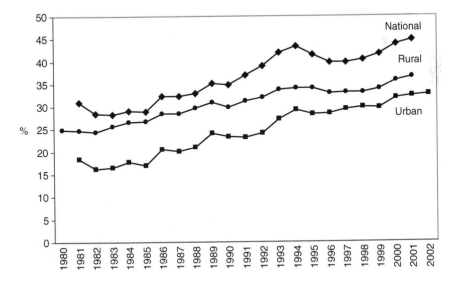

Figure 5.1 Changing inequalities, measured by Gini coefficient (percent).

Source: Adapted from Naughton 2007.

Cai 2000). Household registration, in particular, has long been associated with population management and the provision of social welfare.

Introduced and implemented in 1958, the household registration system (*hukou*) was designed to control population movement and in particular rural–urban migration. Individuals must register in one and only one place of regular residence. *Hukou* contains two related parts: place of registration and type of registration (urban v. rural). Those residents with local urban *hukou* are entitled to such urban amenities as public schools, welfare benefits, state sector jobs, and public housing. As a result of market reform, free or low-cost public housing is no longer available to urban residents. But municipal authorities continue to use household registration as a basis for providing urban services and maintaining infrastructure.

Until recently, housing provision exemplified how household registration institutionalized the rural–urban divide. Housing had long been a form of social welfare to urban residents. The dominant route, prior to 1999, was through a system of low-rent public housing. This urban welfare housing system, however, did not apply to local residents with rural *hukou* or peasants in the countryside, who did not have access to either municipal or work unit public housing (more in Chapter 10). Traditional family houses and private housing constructed on land allotted by production brigades were the norm for them, even in rural pockets within cities.

Hukou is largely based on employment for adults and, for children, on their mother's registration status. Local public security bureaus are in charge of enforcing the system. Population movement with official *hukou* change is strictly controlled, particularly to large cities. Given that *hukou* involves the place and type of registration, changing from rural to urban *hukou* is considerably more difficult than changing the place of registration. But there are at least four ways to adjust one's registration status. First, urban enterprises and institutions may recruit qualified employees and sponsor their registration changes upon prior official approval. Second, high-school graduates may enter university after successfully passing the national college entrance exams and their registration will be moved to the city under collective management of the university. After graduation, they may secure urban jobs that come with permanent *hukou* changes. This, in fact, is the most common way in which rural youth can achieve upward socioeconomic mobility.

The third method is through marriage to urban residents, although associated *hukou* changes may have a waiting period of up to 5 years. A more common trend recently is marriage migration – particular for non-local rural women to marry local peasants. In suburban and rural areas of cities, local authorities have allowed young men to request local rural *hukou* for their wives and children when they marry non-local women. Granting local *hukou* to non-local men married to local women, on the other hand, is more controlled. All such *hukou* conversions need to be approved first by village committees, then by township governments, and finally by county or district public security bureaus. Last, rapid urban expansion during the 1980s and 1990s often entailed acquisition of farm land outside urban cores. As a form of compensation, local peasants saw their registration status converted to urban and some even were assigned state sector jobs. But this practice has ceased as cash compensation has become the prevalent method.

Despite repeated calls, reform of the household registration system has been a slow process. Some cities have experimented with less drastic measures to loosen *hukou*'s hold. One such measure, introduced in 1992, was the issuance of an interim residence permit, called blue-stamp *hukou*, which could eventually lead to permanent status. In general, the primary candidates for a blue-stamp household registration were three groups of migrants: investors, property buyers, and professional or skilled workers. The higher the administrative status of a city, the higher the price for a blue-stamp household registration would be (Chan and Zhang 1999). For instance, in Shanghai in the late 1990s, cash purchase of a housing unit worth 100,000 yuan or more would entitle a non-local resident a blue-stamp household registration that could become permanent after 5 years. The size and price requirements vary by geographical location within the city. The variations are: 320,000–350,000 yuan for central areas, 180,000 yuan in three inner suburban districts, and 100,000–160,000 yuan in other suburban areas (Wu 2002). The practice of issuing blue-stamp *hukou,* however, has been discontinued nationwide.

More recent efforts to expand *hukou* reform include relaxing limitations on migration to small towns and cities, streamlining *hukou* registration in some provinces and large cities, and instituting many individual reforms aimed at addressing the abuse of migrants. In 1997, the State Council initiated an experimental program to allow rural migrants who had moved to designated small towns and cities to obtain local *hukou*. In 2001, the State Council expanded this program to include all small towns and cities. Since 2001, many provinces and large cities also have begun to allow migrants who satisfy certain criteria to obtain local *hukou* in urban areas. As with the State Council decision on small cities and towns, these measures generally require that applicants possess a "stable place of residence" and a "stable source of income." Many provincial and municipal regulations define these terms stringently, often based on educational or financial criteria (CECC 2005). Defenders of the system contend cities are unable to provide the services migrants demand in the absence of a nationwide and transferable social security network. The household registration system is likely to stay in place for the near future.

Population mobility and migration

Driving forces of migration

Population mobility was extremely low during the 1960s and 1970s, controlled through a system of food rationing and household registration. Particularly discouraged was rural–urban migration and moving to large cities from other urban areas. Such restricted mobility also was in line with China's urban policy to contain the growth of large cities. The only large movements of population were in the form of forced migration during the Cultural Revolution (1966–1976). City residents, particularly urban youth, were assigned to work in rural areas and frontier regions (see Chapter 4). This mobility was aptly called "up to the mountains and down to the villages" (*shangshan xiaxiang*) (Bernstein 1978).

The start of transition from a planned to a market economy in 1979 paved the way for a drastic shift in the country's population mobility. By the early 1980s, rural collective communes were dissolved as peasants gained rights to use individual plots of land. Agricultural productivity soared as a result. Peasants were no longer as tied to the land. At the time, about 80 percent of the population still lived in the countryside. Quite quickly, a large reservoir of rural surplus labor formed, since arable land had always been scarce in China on a per capita basis. Some farmers went to work in local or nearby township and village enterprises. Others looked to cities to augment their income from the land. Responding to this emerging trend, the central government issued a circular in 1984 that permitted migration of peasants and their families to market towns. This was the first break in the policy of strictly controlling rural–urban migration. As this happened, peasants did not confine their moves merely to market towns. They were attracted to the booming construction business and factory jobs in larger cities as well.

This new mobility also is a reflection of the rapid process of industrialization in China. After 1983, the spread of market reform brought industry to the forefront. Since then, it is the urban industrial economy that is largely responsible for China's unparalleled growth performance. Migrants have provided the manpower for much of the urban economic expansion. In 1978, the manufacturing sector's share in China's gross national product was close to 30 percent. By 2000, it was about 40 percent. By contrast, the agricultural sector's share went from 42 to 17 percent. During the same period, the share of agricultural employment declined from 70 to 50 percent, while that of manufacturing increased from 18 to 23 percent (Naughton 2007). The remaining employment is in the growing service sector.

Much like in other developing countries, migration is a product of persistent economic disparities between urban and rural areas in China. Despite an overall impressive record of progress in terms of economic growth since the start of market reform, income disparities between urban and rural areas, as well as regional imbalances, remain large. Higher living standards and income levels in cities prove to be a strong pull factor for migrants. These factors, combined with the partial relaxation of household registration mechanisms, have helped generate millions of peasants on the move. Unlike permanent migration to cities that require changes in household registration, most migrants move without such changes and hence are called, in the Chinese media, "floating population" or "labor migrants."

Patterns of migration

Migration takes place in two forms: through permanent migration (*qianyi*) with formal changes of household registration (*hukou*) and through temporary movement (officially called "floating population" or *liudong renkou*) without official changes of *hukou* from the origin to the destination. The latter group, which makes up the bulk of China's internal migration, is expected to return to their home place eventually. The notion of temporary migrants is unusual in China's contemporary context, as it does not necessarily denote a timeframe but an official designation (Chan 1996). For government authorities, the distinction between permanent and temporary migration is still important, since permanent migration entitles migrants to urban amenities enjoyed by local residents while temporary migrants have restricted access to these amenities. The focus in the rest of the chapter is on these temporary labor migrants, and not on permanent migrants.

Much of the migratory flow involves circular movements of rural labor and is characterized by movement from, and continuing linkages with, a rural home base. These linkages manifest themselves in frequent home visits, family ties, possession of farmland, remittance of money, or intention to return to rural life. In particular, many return home during harvest seasons, important family events (e.g., getting married, child birth) and the most important holiday, Chinese New Year. Migrant circulation is primarily stimulated by the security related to home ties and by official restrictions imposed at urban destinations. In particular, most Chinese living in rural areas do not want to give up their farmland because of the security it offers. A specific type of circulation, seasonal or temporary migration, may reasonably characterize most of China's rural–urban migrants who search for work to augment agricultural income and thus tend to invest very little financially and socially in cities.

The changes in the magnitude of migration reflect the ebbs and flows of the Chinese economy. The early 1980s witnessed the start of this mobility, which was largely confined to movements within the coastal region. It was estimated that no more than 2 million farmers were involved, based on their county of origin. During the mid-1980s (1984–1988), town and village enterprises began to absorb a large amount of rural labor from within their provinces. In the meantime, interprovincial migration increased steadily as well. The period between the late 1980s and mid-1990s witnessed perhaps the first height of labor mobility, at the end of which time about 75 million people were on the move and about 25–28 million of them were interprovincial migrants. The scope of migration also widened significantly, involving multidirectional flows. The period since 1997 showed a relative stabilization of migration, largely due to a macroeconomic slowdown as a result of the East Asian economic crisis. By the mid-2000s, the repercussions from the crisis had dissipated. As such, the volume of migration expanded to another height, at an estimated 150–200 million. Official estimates by the National Population and Family Planning Commission put the total number at 211 million in 2010. However, reliable estimation of migration volumes is extremely difficult for at least two reasons. First, definitions of urban population and urban places are complex and have changed frequently (see Chapter 4). Second, available data are collected primarily based on administrative hierarchy rather than urban and rural definitions (Fan 2008).

While many migrants move a short distance within a province, long-distance migration is on the rise. A comparison of the 1990 and 2000 censuses shows an increase in interprovincial migration from 32 to 54 percent of the total (Roberts 2007). Much interprovincial migration originates from the central and western regions and flows to the coastal region. While most interprovincial migrants still use coastal areas as their destinations, flows to western areas have been increasing slowly. Six provinces (all but one coastal), Guangdong, Zhejiang, Beijing,

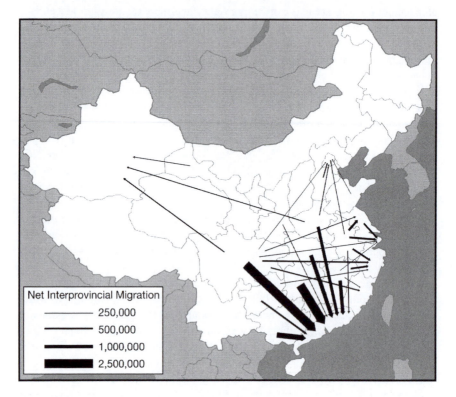

Figure 5.2 Net interprovincial flows of migration, 1995–2000.

Source: Fan 2005 (with permission).

Shanghai, Fujian, and Jiangsu, together attract about three-quarters of all flows (Figure 5.2). These largest migration flows display a high degree of consistency, at least during the 1990 and 2000 census. A contiguous zone spanning south central and southwestern China has been the major source of such flows, including Anhui, Jiangxi, Hunan, Guangxi, Guizhou, and Sichuan provinces. Sichuan is the single largest source of interprovincial migration (Fan 2008). Clearly, most of the prominent destination provinces are more developed economically while most origin provinces are relatively poor.

Characteristics of migrants

Similar to trends elsewhere in the world, China's labor migrants tend to be concentrated among the most economically active group (particularly between the ages of 15 and 34). The average age is about 27 years (Fan 2008). Overall, male migrants outnumber females (about 2 to 1), especially among those migrating across provincial boundaries and during earlier periods. Since the 1990s, however, women have become more mobile. The migration stream, as a result, is now more balanced in gender. Women migrate not just for economic reasons – many find jobs in the booming export sector in coastal cities – they also move to escape the drudgery of life in the countryside. They migrate married and single, with and without their husbands and their children. According to recent official estimates, about one-fourth of the women migrating with their husbands have brought their children with them (Roberts 2007). As such, family migration also is on the rise.

Migrants primarily move from rural origins to urban destinations. The mobility rate is highest for rural areas in the coastal region, followed by the central region and then by the western region. Rural migrants are likely to be better educated than the average rural population, representing a form of "brain drain" for rural China. They move primarily for economic reasons, rather than for social and cultural reasons. While female migrants tend to work in export-oriented, light-industry factories and in service occupations (e.g., domestic and hospital help), male migrants are more broadly employed across construction sites, factories, and businesses. The non-economic purposes of migration include pursuing education, visiting friends/relatives, joining family, and marriage. The latter category, a small but not inconsequential group, consists of largely peasant women who use marriage to achieve mobility in hope of advancing their well-being.

Despite their official temporary status, over time migrants have stayed longer and longer in the cities. The only evidence concerning settlement in the cities comes from successive surveys of floating population carried out by municipal authorities. For instance, in Shanghai about 70 percent of migrants stayed for less than 1 year between 1988 and 1993. But this proportion dropped to 51 percent by 1997, while those staying from 1 to 5 years rose from 24 to 39 percent, and more than 5 years from 4 to 10 percent (Roberts 2007). By 2005, according to the National One Percent Population Survey, about a third of migrants (34 percent) in Beijing and Shanghai had lived there for more than 5 years. The increasing duration of residence indicates that a substantial number of migrants are settling in cities. This clearly illustrates the working of push and pull factors behind China's migration flows. Farming is perceived to be a dead end for youth entering the labor force; migration at least offers opportunity, however remote the probability of success.

Migrants in the city

Barriers and access

Increasing mobility, however, has been a mixed blessing for migrants and a challenge to the state. On the one hand, the influx of low-cost labor is highly desired for the full functioning of the urban economy, and thus for China's ability to compete strongly in the global marketplace (see Chapter 6). Hence, temporary migration (without change of *hukou*) has become an alternative to permanent migration (with *hukou* change) in meeting urban labor force needs. On the other hand, the *hukou* system is one socialist institution that is slow to change in the course of transition, and remains a way for the state to exclude many migrants from acquiring urban citizenship and its attendant social rights (Solinger 1999).

The notion of temporary migrants denotes not a timeframe but an official designation – *hukou* defines whether a migrant is permanent or temporary. To permanent migrants, access to urban amenities and state sector jobs is fully guaranteed. But this access often is not available to temporary migrants. Without local *hukou*, most migrant children cannot be enrolled in urban public schools, although a small number can do so, at significantly higher expense than for local children. These and other similar restrictions inevitably increase the costs and hardships borne by migrants.

Recent *hukou* reforms have begun to address the urban–rural divide, but changes tend to occur in towns and small cities. The reforms enable rural migrants with stable jobs and residences to register as urban residents and to obtain social services, primarily their children's education. Almost all the large cities, however, continue to place significant limits on eligibility for urban registration. Such unwillingness runs true even under the administration of Hu Jintao (President) and Wen Jiabao (Premier), who have made fairness and reducing income disparities a hallmark of their administration. In a March 2010 report, Wen did not mention giving migrants equal treatment outside their provinces, or in China's largest cities (Reuters 11 March 2010). While recent reforms loosen the *hukou* system for the more privileged migrants, they do not address the primary problem still facing most migrants: the continued linkage of *hukou* registration to public services. Implementations also vary in scope across cities in response to a 2003 national circular calling for local authorities to abolish discriminatory measures against migrants.

Housing and settlement patterns

Settlement patterns are an important determinant of the future socioeconomic standing of migrants. Where and how they live is likely to affect their general level of satisfaction with urban living and the ease or difficulty of adapting to the new environment. Housing reforms implemented since the 1980s seem largely to overlook the needs of migrants even though they have broadened housing choices for urban residents (Table 5.1). A local urban *hukou* continues to be an important qualification for accessing several types of urban housing, particularly those that are more affordable. For instance, both the "economic and comfortable housing" and affordable rental units are reserved for local urban residents only. On the secondary housing market, where older housing units are traded, migrants can purchase housing. Commercial housing, the only real property sector open for migrant ownership, is not affordable for most migrants. In addition, a local urban *hukou* is required to qualify for bank mortgages for new commercial housing. As a result of these restrictions, the urban–rural divide in housing continues even after rural migrants move to cities.

Given this larger context, migrants display different housing behaviors from not only local residents but also migrants in other developing countries. Most migrants are not homeowners in the cities. They do not consider the security offered by ownership as a key factor in making housing decisions. Most rent old

Table 5.1 Migrant access to urban housing

Type of housing	Qualification	For migrants	
		Own	Rent
Commercial housing	Anyone, but only those with local urban *hukou* can qualify for bank mortgage loans	Yes	Yes
Economic and comfortable housing	Local urban residents with low or medium income can purchase at subsidized price		Yes
Public housing	Sitting local urban tenants can purchase and trade units on secondary housing market		Yes
Low-rent housing	For rental to local urban residents with the lowest income		
Resettlement housing	For local urban residents relocated from areas undergoing redevelopment		Yes
Private housing	Pre-1949 urban housing units passed on within family and housing in rural areas	Yes	Yes
Dormitory housing	Housing managed by local enterprises or institutions		Yes
Migrant housing complex	Housing managed by local agencies for migrants	Yes	

Source: Based on Wu 2006.

housing units from local residents or stay in dormitories provided by their employers (often seen in factories and on construction sites). Overall, migrant housing conditions are poor – overcrowded, temporary, with limited amenities (e.g., kitchen and bathroom), and located in precarious environment (Figure 5.3). The few lucky ones can afford to become homeowners (Wu 2002). China's migrants, however, do share some behaviors with their counterparts elsewhere. Low cost and proximity to the workplace are higher priorities than physical quality and space. Accommodations are used mainly as places to sleep and prepare for the next day's labor, given the long hours most work at what are almost inevitably physically exhausting and dirty jobs.

Large-scale squatter settlements are not a viable option for China's migrants, unlike in many other developing countries (particularly in Latin America), largely due to municipal authorities' intolerance. However, migrant settlements or communities have existed in some large cities. Migrants rent from local residents or live in market areas constructed by local governments or private businesses. Often called "urban villages" (*chengzhongcun*), these neighborhoods were formally rural but have become enveloped by urban uses expanding outward. Although located physically within the city, the local peasant residents of *chengzhongcun* have rural *hukou* status. By virtue of this status, they have stakes in collective land rights. Many expand their homes or build additional structures to rent out to migrants and earn supplemental cash. Migrants, on the other hand, see such housing as more affordable. In some "urban villages," migrants from the same general area of origin cluster and form a "daughter community" based on place identity (Box 5.1).

Figure 5.3 Migrant housing in a Shanghai factory.

Source: Weiping Wu.

Box 5.1 Wenzhou migrants in Beijing's Zhejiang Village

The ongoing, intense struggle over housing and the use of urban space by Wenzhou migrants in Beijing's Zhejiang Village illustrate how a migrant community has formed on the basis of a shared place identity. Successive waves of migrants from around the Wenzhou region in Zhejiang province have moved to an area called Dahongmen, just south of the Third Ring Road in Beijing. Many have been engaged in garment manufacturing (in small workshops) and trading. They first rented from local residents in this urban–rural transitional area. Then a group of migrants with more economic and social capital invested in the development of large, private housing compounds. They gained access to land and obtained limited infrastructure resources by buying off local village and township cadres and forming informal economic alliances with them. Built on extended kinship ties, clientelist networks with local cadres, and voluntary groups, a shadow migrant community and leadership structure emerged (Xiang 2000; Zhang 2001, 2002).

Within the community, allocation of housing, production and marketing space, and policing and social services proceeded through networks centered on the migrant bosses of housing compounds and market sites. This explosive growth outside party state structures worried local authorities and led them to order successive demolitions of many housing compounds. The migrants, desperate to stay in the lucrative urban market at the center of the national transportation network, continued to rebuild their community within months of each raid. This example shows that, rather than demanding specific welfare benefits from the state, migrant entrepreneurs aspire to urban citizen status mainly to gain a secure space of their own for business and living quarters. Housing space therefore figures as the central element in their bid for urban citizenship (Zhang 2002).

Migrants' growing demand for housing and their limited access to the mainstream urban housing distribution system contribute to the chaotic situation of the rental market (Wu 2002). As cities scramble to develop effective rental regulations, an increasing amount of deleterious building and informal rental activity continues, largely in the form of unauthorized construction and leasing of unsafe dwellings. This problem is particularly serious in urban–rural transitional areas where land is more readily available, the migrant population is more concentrated, and local residents have more incentive to rent out rooms due to the loss of agricultural income. Even when regulations about rental housing take shape in some cities, concerns for adequate housing conditions and rental rights tend to be secondary.

It is no exaggeration to say that, once in the city, migrants continue to be on the move. The majority have moved at least once, while many have done so multiple times (up to ten moves) within a span of 4–6 years (Wu 2006, 2010). Such mobility behavior, however, may not be the result of voluntary or predictable actions, as most migrants express little willingness to move again when asked. The frequency of moves in the first year is particularly high, with multiple moves for many. But there is a slow process of settling down for migrants, even though longer-term migrants still experience much higher mobility rates than local residents. The majority of moves are related to work, triggered by events such as job change, change in business location, and completion of work projects. For migrants in the construction sector, they live and move with work. For self-employed migrants, such as those operating food stalls and convenience stores, their moves are primarily determined by how profitable a location may be for business.

Migrants may be moving frequently, but not very far. Most moves are within the same general geographic area within the city. They tend to make short-distance residential moves to minimize unfamiliarity with the environment. Worse, few migrants make the transition from renters to owners even after years of living in the city (Wu 2010). Getting stuck in the private rental sector allows little room for improving their housing conditions. The main explanation would lie with local controls, which force migrants (even those with families in tow) into more of a renter's existence than they may otherwise prefer. Specifically, the system of granting only temporary residence permits to migrants discourages them from making substantial investment to alter their residential choices in the city. Many, however, remit most of their income back to the countryside to build better housing for their families.

Integration into the urban fabric

Inevitably there are variations among migrants. Some have gained access to limited benefits in the city by signing employment contracts with urban enterprises. Others with capital and skills have resorted to self-employment and prospered. But if residential mobility is any indication of socioeconomic mobility, it is clear that most migrants are drifting in the bottom layer of the urban society. Some resort to criminal activities because they have no social safety nets in cities, face fewer employment opportunities, or use unlawful means to revenge unfair treatment they have experienced. Because migrants are less entitled to local police protection, they are as much victims of crime as they are perpetrators (Solinger 1999).

Even though migrant workers are valued in urban enterprises and managers often prefer them for their work ethic and low costs, bias against them remains. They are paid less, work for longer hours, and have limited or no entitlements to health care, housing, and pension benefits. Local authorities worry about the heavy pressure on urban services and excessive open unemployment. Current economic slowdown further reduces labor market demand. These factors together are likely to pit the interests of urban workers against those of migrant workers. Migrants also are perceived as competitors of furloughed urban workers, especially as re-employment efforts intensify in cities affected by large-scale state sector reforms. Enterprises hiring labor migrants are required to obtain quotas from municipal labor bureaus, although some companies circumvent such rules in hiring migrants as a cost-cutting strategy. As a result, the majority of migrants are restricted to jobs undesirable to the local population, such as in construction, domestic services, factory and farm labor, and retail trade. Compared to urban residents, these tend to be low-paying occupations.

The more positive consideration for migrants is that it has become much easier to stay and work in urban areas for extended time. Most cities now have temporary registration for migrants and allow for much more leeway for their self-employment. Due to their significant presence in small-scale commercial activities and services, migrants are contributing to the development of the private as well as informal sectors in cities. Some such activities yield an adequate decent livelihood (e.g., small traders and food vendors) while others involve a daily struggle for a subsistence living (e.g., scavengers). For the latter, urban life is marginalized and precarious, with low and uncertain earnings, and worsened living conditions.

Within the increasing migrant influx to cities, more and more are families with children. The number of school-age migrant children is steadily rising and their education or lack of education is becoming a major urban problem. There is also evidence that school-age children are involved in the urban informal economy, in flower selling, shoe shining, newspaper sales, food services, begging, and garbage recycling. Being robbed of a normal childhood, these children are a particularly vulnerable population. The opportunity to go to school does not come easy for them. Even for those children attending school, their education experience is often interrupted by constant relocation of their families. In migrant schools, it is common to see older children in lower levels and decreasing numbers of students towards higher grades.

Given the restricted access to urban public schools, migrant schools are the inevitable alternative. These schools are mostly private and located in urban–rural transition areas with a high concentration of migrant population. Not officially registered in the city, the schools have limited budgets and shabby facilities (Figure 5.4). Most of them offer instruction only up to the sixth grade. Teachers often do not have formal training and experience high turnover rates. For school administrators, opening schools is an economic endeavor; few consider migrant children's interest as their motivation. As migrant schools are not formally recognized by municipal governments, many of them experience forced closure or relocation (Box 5.2).

Figure 5.4 A migrant school in Beijing.

Source: Weiping Wu.

***Box 5.2* Educating migrant children**

For many children of migrant families, life in the city is marked with poverty, instability, and hardship. The opportunity to go to school does not come easily for them. Migrant girls, in particular, are less likely to be enrolled in schools. Even for those children attending school, their education experience often is interrupted by constant relocation of their families. As more and more migrants stay longer in the city and bring their families with them, their need for child education increases. But their children either are not accepted by urban public schools or have to pay significantly higher fees to attend these schools. As a result, many migrant families leave their children at home in the countryside to attend school. This entails a long-term separation for family members.

Given the restricted access to urban public schools, migrant schools are the inevitable alternative. These schools are mostly private and located in urban peripheries with a high concentration of migrant population. The largest one in Beijing, for instance, enrolls over 1,300 students while the smallest has less than ten. Because these schools are not part of the urban school system, they are not entitled to any allocation of resources. Most follow curricula and use textbooks from the home origins of the school administrators.

There has been a serious concern among researchers and activists that, given China's urban demographic transition, future urban standards of living will depend on the productivity of the virtually illiterate and unskilled migrant population growing up in cities. With the increasing influx of migrant families, the frightening prospect is that a generation of uneducated, unemployed children may become a new urban underclass. Since education is a critical precondition for human development, it is imperative that authorities at central and local levels begin removing barriers to education for migrant children. This can be done by opening access to some urban public schools at lower fees, as the declining urban birth rate has resulted in a smaller local student population and in turn underutilized school capacity. It is also important for migrant schools to be recognized and regulated so that instruction quality and the learning environment can improve.

Conclusion

Under China's plan-to-market transition, the urban–rural divide has widened. This, combined with loosened control of population movement, has contributed to enormous tides of migration. But such rising mobility intersects with an institutional structure at urban destinations that separates migrants from local residents through a household registration system (*hukou*). As such, migrants have limited access to local public schools, welfare programs, state sector jobs, and the mainstream housing distribution system. As the second generations of migrants grow up in cities, their future as an urban underclass is a serious challenge for the Chinese state. Given the magnitude of migration and its impact on cities, it is important to explore ways in which migrant access to urban amenities can be broadened. To respond to the need of migrants and their quest for citizenship rights properly entails that the linkage between *hukou* and the provision of urban services be discontinued. With more tolerant migration policies, over time urban ties will surpass rural ties and many migrants may choose to settle permanently at urban destinations.

Bibliography

Bernstein, Thomas P. 1978. *Up to the Mountains and Down to the Villages: The Transfer of Youth from Urban to Rural China.* New Haven, CT: Yale University Press.

Chan, Kam Wing. 2007. "Misconceptions and Complexities in the Study of China's Cities: Definitions, Statistics, and Implications." *Eurasian Geography and Economics* 48(4): 383–412.

Chan, Kam Wing. 1996. "Post-Mao China: A Two-Class Urban Society in the Making." *International Journal of Urban and Regional Research* 20 (1): 134–150.

Chan, Kam Wing and Man, Wang. 2008. "Remapping China's Regional Inequalities, 1990–2006: A New Assessment of de Facto and de Jure Population Data." *Eurasian Geography and Economics* 49(1): 21–56.

Chan, Kam Wing and Zhang, Li. 1999. "The Hukou System and Rural–Urban Migration in China: Processes and Changes." *China Quarterly* 160(December): 818–855.

Congressional Executive Commission on China (CECC). 2005. "Local Governments Resist Reforms to Household Registration System." CECC Virtual Academy (retrieved on 16 March 2008 from http://www.cecc.gov/pages/virtualAcad/index.phpd?showsingle=32168).

Fan, C. Cindy. 2005. "Interprovincial Migration, Population Redistribution, and Regional Development in China: 1990 and 2000 Census Comparisons." *The Professional Geographer* 57(2): 295–311.

Fan, C. Cindy. 2008. *China on the Move: Migration, the State, and the Household.* London: Routledge.

Fan, C. Cindy and Mingjie, Sun. 2008. "Regional Inequality in China, 1978–2006." *Eurasian Geography and Economics* 49(1): 1–20.

Honig, Emily. 1992. *Creating Chinese Ethnicity: Subei People in Shanghai, 1850–1980.* New Haven, CT: Yale University Press.

Naughton, Barry. 2007. *The Chinese Economy: Transitions and Growth.* Cambridge, MA: MIT Press.

Roberts, Kenneth. 2007. "The Changing Profile of Chinese Labor Migration," in Zhongwei, Zhao and Fei, Guo (eds.) *Transition and Challenge: China's Population at the Beginning of the 21st Century.* New York: Oxford University Press, pp. 233–250.

Solinger, Dorothy J. 1999. "Citizenship Issues in China's Internal Migration: Comparisons with Germany and Japan." *Political Science Quarterly* 114(3): 455–470.

Wu, Weiping. 2002. "Migrant Housing in Urban China: Choices and Constraints." *Urban Affairs Review* 38(1): 90–119.

Wu, Weiping. 2006. "Migrant Intraurban Residential Mobility in Urban China." *Housing Studies* 21(5): 747–767.

Wu, Weiping. 2010. "Drifting and Getting Stuck: Migrants in Chinese Cities." *City: Analysis of Urban Trends, Culture, Theory, Policy, Action* 14(1): 10–20.

Wu, Weiping and Rosenbaum, Emily. 2008. "Migration and Housing: Comparing China with the United States," in Logan, John (ed.) *Urban China in Transition*. Oxford, UK: Blackwell Publishing, pp. 250–267.

Xiang, Biao. 2000. *A Community That Crosses Boundaries: The Living History of Beijing's Zhejiang Village* [*kuayue bianjie de sheqiu: Beijing zhejiangcun de shenghuo shi*]. Beijing: Sanlian Publishing House.

Yang, Dennis Tao and Cai, Fang. 2000. *The Political Economy of China's Rural–Urban Divide*. Working Paper No. 62. Stanford, CA: Center for Research on Economic Development and Policy Reform, Stanford University.

Zhang, Li. 2001. *Strangers in the City: Space, Power, and Identity in China's Floating Population*. Stanford, CA: Stanford University Press.

Zhang, Li. 2002. "Spatiality and Urban Citizenship in Late Socialist China." *Public Culture* 14(2): 311–334.

6 Cities in the global economy

Around 1979, after breaking ties with the former Soviet Union in the late 1950s and being closed economically and politically for over two decades, China opened its doors to the outside world. The domestic policy shift was reinforced by changes in the international arena: increasing global economic integration through trade and foreign investment. The spread of market reform also brought industry and the urban sector to the forefront. Since the mid-1980s, it is the urban industrial economy that is largely responsible for China's phenomenal record of economic growth. Cities have become engines of growth in China's rapid rise in the global marketplace. Urban Chinese now hold 70 percent of the country's wealth, command incomes that are three times the rural average, and generate much of the demand for new consumer goods coming on to the market.

This chapter outlines the gradual liberalization of trade and investment, the role of the Special Economic Zones (SEZs), the growth of export-oriented industries, and shifts towards more knowledge-based and consumer activities in large cities. For the latter, global firms are important drivers. Inter-city competition for foreign capital also is growing. The following questions should help guide the reading and discussion of the materials in this chapter:

- China adopted a gradual and incremental approach to opening up. What was the rationale behind such an approach? What were the important elements of this approach?
- What are the Special Economic Zones in China? How are they similar to and different from export-processing zones across Asia?
- What are the comparative advantages China has to engage in export-oriented development? What cities have dominated such development and why?
- As China moves beyond labor-intensive manufacturing, what types of other global investment activities are taking place in its large cities? How have such activities transformed the economies of these cities?

Toward an open economy

Process of gradual reform

Chinese economic reform has proceeded through a series of phased actions. In general, these actions were not the results of a grand strategy, but immediate responses to pressing problems. There are several guiding principles, however. The first is pragmatism. The criteria for success are determined by experiment rather than by ideology. The second is incrementalism. Instead of announcing and implementing a national program, typically, an idea is implemented locally or in a particular economic sector, and if successful it is gradually adopted throughout the nation (Naughton 2007). One of the first actions in the late 1970s and early 1980s was opening up to the world economy. The rationale was that China should take advantage of the global trend of offshore production to attract foreign investment to its capital-starved economy. Such investment would allow China to make full use of its large reserve of inexpensive rural surplus labor to produce labor-intensive goods for export and, ultimately, foreign exchange earnings. The government also recognized the importance of advanced foreign technology for stimulating growth. One channel through which technology transfer often happens is foreign direct investment (FDI).

The open-door policy had a strong spatial component. From the outset, it was clear that development could not happen in all places at once. Resources were limited. More importantly, certain policies needed to be experimented with in limited areas before being implemented nationwide given that China had been a closed economy for 30 years previously. Gradual reform allowed certain places to get rich first. These included the SEZs and coastal open cities (Figure 6.1). The

Figure 6.1 Special Economic Zones (SEZs) and 14 coastal open cities.

Source: Based on Wu 1999.

primary role of the SEZs was to experiment with and digest Western technology and management techniques so that inland enterprises could learn from their experience.

Supported by growing local enthusiasm for opening up, particularly from Guangdong province, radical reform policies were first introduced in the SEZs in 1979 and the early 1980s. In essence, these zones were designed as experimental sites as well as growth centers for China's new era of development. Their planning, to a large extent, followed the experience of export-processing zones in other Asian countries, particularly those in Taiwan and South Korea (Pepper 1988). A zone policy became attractive to the Chinese government for two principal economic reasons. First, in a purely economic sense, it was the second best method to a free trade regime. Although there is no such thing as completely free trade, a liberal trade regime can create an environment that facilitates flow of capital and goods, and encourages competition both domestically and internationally. Since China was unable to adopt a liberal trade regime nationwide at the time, a zone policy was the quickest way to promote exports, by creating an enclave to attract FDI into labor-intensive manufacturing industries. Second, export processing could provide a gateway to the international community for China. Indeed, the SEZs were the key attraction China used to draw FDI from countries where globalization of production and industrial relocation had already increased the capital available for overseas investment. With the introduction of FDI also came the opportunity for China to enter global export markets.

The SEZs, in particular the first and largest one (Shenzhen), represented in miniature the very essence of market reforms (Pepper 1988). Although geographically far from the center of China, Shenzhen was not politically peripheral. The central government, the reformists in particular, made a tremendous political investment to ensure its success. In fact, it became one of the political battlefields between the reform and conservative factions in the central government. Its success or failure, at least in the early 1980s, would determine the fate of the reform. Shenzhen benefited enormously from the attention by China's top leadership. Each of the major reformist leaders, including Hu Yaobang (1983 and 1984), Deng Xiaoping (1984 and 1992), and Zhao Ziyang (1988), visited Shenzhen, using it as a tool to push for further reforms at various junctures. In particular, Deng Xiaoping, the paramount leader of post-Mao China, used his visits to advocate new measures of economic reform and endorsed SEZs' development several times during his tenure. Deng's first trip was a prelude to the designation of 14 coastal open cities in 1984, with a special emphasis on promoting foreign investment. The 14 cities were Dalian, Qinhuangdao, Tianjin, Yantai, Qingdao, Lianyungang, Nantong, Shanghai, Ningbo, Wenzhou, Fuzhou, Guangzhou, Zhanjiang, and Beihai (Figure 6.1). Many of them, especially those which had been developed as Treaty Ports, focused primarily on re-establishing foreign investment patterns of the early twentieth century and historical linkages, such as the fostering of German and American investment in Shanghai, Taiwanese investment in Xiamen, and Hong Kong investment in Guangzhou. The following year saw the declaration of the Yangtze River, Pearl River, and southern Fujian Deltas as Open Economic Zones (see Chapter 4). In April 1988, the fifth SEZ was established in Hainan after its new designation as a province.

In the same year, the coastal development strategy, officially called the "outward-oriented development strategy," was launched in the coastal areas under the firm endorsement of then premier, Zhao Ziyang. This policy had a much larger scale and wider range, embracing 12 provinces and cities under the direct control of the central government. They were Liaoning, Beijing, Tianjin, Hebei, Shandong, Jiangsu, Shanghai, Zhejiang, Fujian, Guangdong, Guangxi, and Hainan (Wu 1999). The coastal development strategy called for the more prosperous coastal provinces to be transformed into major centers of foreign economic activities and be integrated with the world economy. The strong spatial orientation of the reform policy was, however, de-emphasized to some extent after the 1989 democratic movement as inland areas pressed for more attention from the central government. In the Eighth Five-Year Plan (for 1991–1995), the focus was placed more on the development of particular industries than regions.

Deng Xiaoping made another trip to Shenzhen in 1992, signaling the intensification of market reform and the opening of Shanghai. Because of its strategic role as an industrial center and revenue contributor to the central government, Shanghai was largely bypassed during the early round of reforms. The attention given to Shanghai in the early 1990s represented in part the attempt by the center to promote development in the further north part of the country to balance the rapid advances taking place in the southeastern region. Spearheading Shanghai's opening was the large-scale development of the Pudong New Area project, on farmland to the east of the city. Pudong was basically another SEZ, enjoying 18 super-special policies and preferential status (Yusuf and Wu 2002). Since then, the entire country has been moving toward a genuinely open economy. Gradually, special policies in SEZs and coastal open cities have been applied throughout China. The distinctive status enjoyed by SEZs and coastal open cities also has faded.

Special Economic Zones and "zone fever"

The first four SEZs – Shenzhen, Zhuhai, Shantou, and Xiamen – were designated in 1979. They are geographically insulated but economically open areas, where special and flexible economic policies are carried out primarily to promote foreign investment, technology transfer, and exports. "Special" implies that these policies may not extend to the rest of the country. "Economic" has two meanings. First, it distinguishes China's policy from export-processing zones in other Asian countries: SEZs would not merely be export zones, but would encompass a broader array of economic activities such as agricultural production and commerce. Second, they would not be "special administrative regions." The government at the time did not want to tie the success or failure of the zones too closely to non-economic questions (Wu 1999).

There were also certain political considerations. As a result, the zones were not selected on the basis of whether there was a strong industrial base, an adequate urban infrastructure, or a technologically innovative capacity. First, the zones needed to be easily separated from the vast inland areas, since drastically different policies were to be experimented with in the zones. All four SEZs were located along the coast, which made physical separation from the inland areas easier. Fences were built around them and checkpoints were stationed to inspect traffic. Administrative procedures were used to control population inflows to the zones as well. Non-SEZ residents had to apply through local police departments for entry permits, usually valid for a month for a legitimate visit. Second, the SEZs were not to be built into major industrial centers at first, so as to avoid significant losses if the experiment should fail. Third, the central government intended to use these zones as intermediary or "buffer" zones for future reunification. The locations of the SEZs were carefully chosen in proximity to Hong Kong, Macao, and Taiwan, perhaps with the hope that the integration with these external economies would eventually facilitate or lead to political reunification. Last, the central government recognized that the overseas Chinese community was a potential source of productive capital. Towns along the southeast coast in Guangdong and Fujian have been the ancestral homes to many overseas Chinese. The hope was that historical and cultural links would lure them back. The SEZs are close to the setting-off points for three of the most important dialect groups among overseas Chinese: the Cantonese (spoken in Shenzhen and Zhuhai), which predominates in Hong Kong; the Fujianese (spoken in Xiamen), which is used by 85 percent of Taiwan's population and much of Singapore's; and the Chaozhou (commonly known as Teochew) dialect from around Shantou.

Not long after the launching of the SEZs, they began to be emulated throughout China, creating a "zone fever." The concept was appealing: localities could cordon off a limited area to offer special incentives to foreign (and later domestic) investors. Invariably called Economic and Technology Development Zones or Special Development Zones, such zones were miniature SEZs, primarily engaged in export process and other manufacturing development. Some were quite successful in attracting large multinational corporations (MNCs), as in the case of Tianjin Economic-Technological Development Area (home to Motorola and Samsung operations in China) and Shanghai's Hongqiao and Caohejing Economic and Technology Development Zones (home to many large MNCs).

The next wave of zones was in the form of science parks, also called high-tech zones. They would allow research institutions and firms to cooperate and interact by placing them in close proximity to each other. Again, there would be special incentives for businesses. The national Torch Program, begun in 1988 primarily to jump-start high-tech industrial development, was instrumental in helping to launch science parks in nearly all provinces. Some target more conventional manufacturing firms, while others promote high-tech enterprises and knowledge-based services. Cities also can apply for national or provincial designations. At the end of 2007, there were a total of 54 national high-tech parks (NSD Bio Group 2009). There are two additional programs: four Asian Pacific Economic Cooperation (APEC) Science and Technology Industries Parks, which granted preferential status for APEC members interested in multilateral research and development (R&D) projects; and eight International Business Incubators, to promote the participation of foreign small and medium-sized technological enterprises in the Torch Program zones.

Overall, thousands of special zones, named in one way or another, are scattered around the country. Besides the nationally designated ones, there are provincial-, county- and city-sponsored development zones, with development plans and functions ranging from high-technology production to the promotion of tourism. As the Chinese economy grew, they became not only magnets for foreign investment, but the primary loci for domestic investment as well. In developing these zones, local officials have adopted the attitude of "if you build it, they will come." In reality, the competition for foreign investors is fierce not only among cities but also among different zones in the same city.

The most prominent recent example is Zizhu Science Park, located in the far southwest suburb of Shanghai. It is an R&D base spearheaded by the Minhang district government with subsequent municipal backing. With a combination of corporate and government incentives, two of the best universities in Shanghai – Shanghai Jiaotong University and East China Normal University – have relocated to the park. While the relocation has generated logistic difficulty, the universities have become the star attractions for a number of R&D firms, including Intel Asia and Pacific R&D Co., Microsoft China R&D Group, and Omron Sensor and Control R&D (Shanghai) Co. In Beijing, one of the most successful science parks is affiliated with Tsinghua University. The park not only attracted the R&D centers of major global firms such as Google and Procter & Gamble, it also hosted many small startups by overseas returnees.

Becoming "factory of the world"

Foreign investment

Dating back to at least the 1840s, China experienced an increasing presence of foreigners and their investment in industry, trade, and transport (see Chapter 2). Manufacturing activities began, as did the building of railways. A succession of "unequal treaties" signed with external powers – starting with the Treaty of Nanking of 1849, which ended the Opium War – led to the opening of Treaty Ports and enclaves along the coast and major waterways. Foreigners helped lay the groundwork for legal institutions that gave private Chinese capital a modicum of property rights, which are widely perceived as the basis of a modern market economy. After 1913, foreign investment increased in volume and diffused beyond commerce and finance into railways and industry, a shift that accelerated urbanization by adding layers of muscle to the economy of coastal cities. The bulk of foreign investment was in cotton textiles, cigarette production, oil pressing, pig iron production, and machinery (Yusuf and Wu 1997).

Innovations in transport and communication, transferred from overseas, transformed China's economic landscape. Motorized water transport and railways significantly increased the volume and patterns of production and trade. Improved transport brought down the costs of moving commodities, opened fresh trading routes, and induced the growth of new transportation hubs (including Tianjin in the north along the Bohai Bay, Hankou and Chongqing along the Yangtze River, and Shenyang and Harbin in the northeast). The most drastic changes occurred in and around major urban centers, where virtually every new form of transport could be seen, including railways, trucks, buses, steamships, and even aircraft. This stage had run its course by the middle of the twentieth century. The Communist victory in 1949 completely altered the economic as well as the political parameters of urban development. China's economy turned inward.

As China embarked on economic reform in 1979, a turnaround began. At the time, industries in Western developed countries, particularly manufacturing industries, had already begun to rationalize their production process through relocating to places where production costs were lower and/or access to potential markets was greater. Modern communications and transportation technologies further made it possible for production to expand geographically. Such globalization of production first happened in labor-intensive manufacturing industries. Subsequently, the need to supply MNCs in manufacturing with services in trading, design, marketing, transportation, communications, banking, and insurance led to foreign investment in the service sector. Particularly after 1992, foreign investment poured in. China has become one of the world's most important investment destinations. Such investment has three distinctive features. First, the bulk of it is in the form of FDI. Second, much FDI inflow is in the manufacturing sector. Third, the key sources of investment (close to 60 percent) are other East Asian economies, particularly ethnic Chinese firms from Hong Kong and Macao, Taiwan, and Singapore (Figure 6.2). Overall, FDI accounted for less than 2 percent of China's gross domestic product (GDP) before 1992 and averaged about 4 percent between 1992 and 2005 (Naughton 2007).

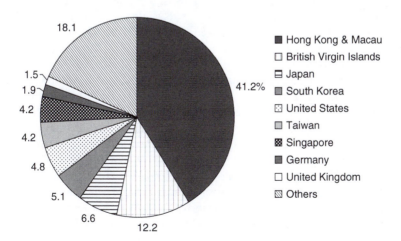

Figure 6.2 Main sources of foreign direct investment (FDI) in China, 1990–2009.

Source: Authors' own calculation based on data from http://www.uschina.org/ and http://www.china-profile.com/ (retrieved 19 February 2011).

China is appealing to MNCs for a variety of reasons. First, labor as a whole is attractive: low wage costs, low social costs, and a low level of militancy. A survey of 10,000 Japanese companies in 1990 revealed that one of the top motivations for companies investing in China was the low cost of labor (Zhan 1993). This is particularly true for labor-intensive manufacturing, such as garment, textile, footwear, toys, and household appliances. As a newly opened market with large potential, China also represents one of the last niches in the global investment competition. Some MNCs are even willing to invest a minimum amount in the beginning just to establish a foothold and perhaps brand recognition. Short-term profit maximization may not be their primary objective. Another attraction is the supportive investment policies that have served as a kind of guarantee of China's commitment to an open investment regime. With deepening market reforms, risks associated with investing in a country with socialist orientation appear to have subsided.

Small-scale capital from overseas Chinese, particularly from Hong Kong and Taiwan, has made up the bulk of FDI. Such ethnic capital is highly selective in choosing production sites where they can find strong kinship and social ties (Ma 2002). But as urban China is further exposed to globalization, the power of cultural affinity seems to decline. An increasing number of investors from Hong Kong and Taiwan have been investing in the Yangtze River Delta and further north since the 1990s, instead of the usual sites in the Pearl River Delta and the neighboring Fujian province. Accompanying this spatial shift is a gradual change in the content of investment from ethnic Chinese. During the early stage (1980s and early 1990s), most investors were small firms engaged in export processing. More recently, they are expanding to services, particularly since 1992, when new sectors were opened to foreign participation.

Overall, manufacturing is the largest beneficiary of FDI, consistently receiving about two-thirds of the total inflow. This is in contrast with patterns in other developing countries, where manufacturing accounts for about one-third of FDI stock (Naughton 2007). FDI also has become China's predominant source of technology transfer, albeit mostly in labor-intensive industries. It brings management know-how, marketing channels, and other tacit expertise. After the opening of the service sector to foreign participation in 1992, FDI is highly concentrated in real estate, specifically in property development. Hong Kong developers, for instance, have been instrumental in constructing grade-A office towers, high-end shopping centers, and serviced apartments tailored for expatriates. Walking along Huaihai Road, a major commercial fair in Shanghai, shoppers will encounter the Central Mansion (emulating Central, Hong Kong's premier shopping district), Hong Kong Plaza, Lippo Plaza (emulating Hong Kong's Lippo Center Shopping Mall), and Shuion Plaza (named after the name-sake Hong Kong developer). Compared to other developing countries, FDI in China's wholesale and retail trade, transport and telecommunications, and finance is clearly underperforming. These service sectors likely will see growth given China's entry into the World Trade Organization in 2001 (Naughton 2007). In fact, after China granted wholesale trading rights to foreign firms in 2003–2005, major cities have already seen the rapid expansion of international stores such as Walmart (USA), Carrefour (French), and Tesco (UK).

Although FDI concentrated in Guangdong province until recently, corporate headquarters often are located elsewhere. To them, several factors are critical, according to a survey of executives in about 2,500 foreign corporate headquarters: proximity to central government units, preferential policies designated to the location for businesses, superior business environment and culture, and superior urban infrastructure (Zhao 2003). As such, Beijing and Shanghai have been the more favored locations. Beijing, in particular, seems to greatly benefit from the dense "regulatory" information. The survey shows that Beijing alone housed 57 percent of all headquarters and 44 percent of the representative offices of foreign financial institutions. Shanghai's shares were 31 and 25 percent respectively, whereas Guangdong's cities accounted for a far less share. Here, the importance of Hong Kong should not be underestimated, as about one-third of the headquarters of foreign companies in China have senior-level administrative offices in Hong Kong (Zhao 2003).

Several cities in the Lower Yangtze Delta have showed spectacular success in attracting FDI (Figure 6.3). Suzhou and its satellite city Kunshan stand out, both

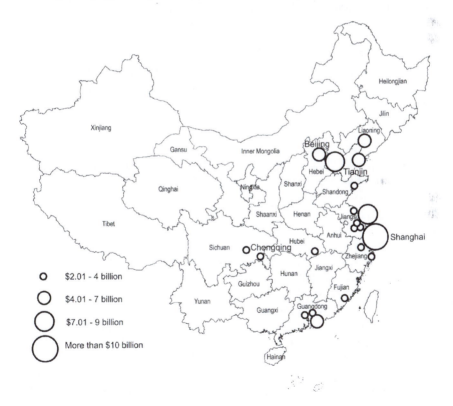

Figure 6.3 Main urban destinations of foreign direct investment (FDI) in China, 2008.

Source: National Bureau of Statistics of China 2009.

Note: Based on the amount of actually utilized FDI.

within an hour's drive of Shanghai. Suzhou, with a diverse industrial base, has used aggressive incentives to lure FDI from a wider range of countries in Asia and the West. FDI has risen dramatically since the early 1990s. In 2008 alone, more than US$8 billion of FDI was utilized in the city, making it the second largest destination in urban China and trailing only Shanghai. In particular, the China-Singapore Suzhou Industrial Park contains significant investment from Singapore government and firms. A satellite city of Suzhou, Kunshan's rise began around the same time, and it has been consistently ranked among the fastest-growing counties in both Jiangsu Province and China. Specifically, the growth of foreign-invested enterprises has made Kunshan one of the largest places for Taiwanese investment. Its proximity to Shanghai allowed Taiwanese investors to use Shanghai's advantages fully in headquarter functions and access to domestic and international markets. It also is among the first to implement a series of preferential policies to better serve foreign investors. Like other cities, at the very beginning Hong Kong was the major source of FDI. In the mid-1990s Taiwan replaced Hong Kong, becoming the largest source of investment in Kunshan. The other important sources are the USA, Japan, and Singapore (Wei 2010).

Export processing

FDI in China is closely associated with ethnic Chinese investment from Hong Kong and Taiwan. As such, their paths of industrialization have had a significant impact on the nature of foreign participation in China. Rapid industrialization in Hong Kong and Taiwan was stimulated by the so-called export-led growth, starting in the late 1960s. It was based on the production of labor-intensive commodities (electronics, textiles, and clothing), financed largely by FDI from industrialized countries. But by the late 1970s and early 1980s, their competitiveness in the export markets of consumer goods was seriously challenged by several other lower-cost developing countries in the region, including Thailand, Malaysia, and Indonesia. In particular, rising costs of labor indicated that their export-led economy had reached a point where further growth could not solely depend on domestic factor inputs. Coincidentally, China began its course of opening up at about the same time. Quickly, south China became the perfect site for their industrial relocation.

Such relocation was immediately taken on by largely labor-intensive industries and/or low-end processing operations in garments, textiles, footwear, electrical machinery, electronics, toys, plastics, and metal products. For instance, between 1989 and 1994, over 75 percent of all exports from Hong Kong to China in several commodity groups (clocks and watches, garments, toys, and textiles) were related to outward processing in China. By the early 1990s, most electronics and toy manufacturing in Hong Kong had been relocated to China. Many of Taiwan's footwear firms moved their production by the early 1990s. A similar restructuring of the electronics industry began around 1990, followed by many successive waves of relocation. Eventually the notebook computer industry was transplanted into China during 2002–2003 (Naughton 2007). These manufacturing industries also have been characterized by a large number of small and medium-sized firms, while other

Asian investors (such as Japanese and Singapore) have relatively large contract size. Likely, the territories' small sizes limited the amount of land available for industry and precluded the development of large-scale or land-intensive industries.

As labor-intensive industries moved to south China and later further north, the technology- and knowledge-intensive stages, such as designing, testing, and marketing, tended to stay back in Hong Kong and Taiwan. Manufacturing in Hong Kong and Taiwan started moving decisively away from labor-intensive production into high-value-added exportable products that could compete on quality. Overall, Hong Kong and Taiwan have deindustrialized substantially. Altogether, they have lost about a million manufacturing jobs, while Guangdong and Fujian provinces have gained about 5 million (Naughton 2007). Hong Kong, in particular, has moved into a variety of business services, such as finance, marketing, and management. Taiwan, on the other hand, is more successful in becoming a producer and exporter of high-tech products.

Export processing was clearly the motor of China's global expansion in the early stage of reform. Together, foreign-invested firms contributed close to 60 percent of exports (Naughton 2007). Riding on this wave, coastal provinces prospered much earlier than the rest of China. Other types of firms in these provinces, including town and village enterprises, were allowed to engage in export-processing contracts. From the mid-1990s on, China began to build on the achievement of coastal provinces and move towards an open economy nationwide.

A high degree of industrial monoculture characterizes export processing in China, where a single (or closely related) industry tends to employ the majority of workers. By far the most important component of domestic value added is wages paid to local production workers. Foreign firms engaged in export processing rarely locate technology- or knowledge-intensive activities in China. The most obvious impact export processing has on surrounding regions is the creation of a new labor force, particularly encouraging the participation of women. In the case of China, nearly all of the laborers working in export processing come from rural areas, as migrants (see Chapter 5). Young rural women, in particular, are the backbone of many factories in the Pearl River Delta producing garments, toys, small electronic appliances, and footwear (Box 6.1).

Box 6.1 **"Factory of the world": manufacturing Dongguan and a new form of urbanism**

In many ways, the Pearl River Delta region is like no other region in China. With less than 1 percent of the country's land area, it accounts for close to 10 percent of China's GDP and about a third of exports. Its stunning growth record since 1979 stems from export processing, making it truly the "factory of the world." Within the region, Dongguan is like no other place. Located in the heart of the delta and about an hour away from

Shenzhen SEZ, Dongguan was a village poorer than its neighbors back in 1979. But its fate changed overnight with China's opening. Given its geographic proximity and social connections with Hong Kong (it was reported that one in five young people in Dongguan managed to cross over to Hong Kong; see Lin 2006), it quickly became a haven for Hong Kong businesses to relocate their labor-intensive production. Local authorities quickly took action to lure investors. In addition to heavy investment in fixed assets and infrastructure, Dongguan also was the first prefecture-level city in China to install a digital telephone system (in 1987) to connect its towns and villages with 250 cities domestically and 150 countries internationally. Now, more than 18,000 Hong Kong firms and another 3,800 Taiwanese firms are located there, making garments, footwear, toys, electrical equipment, plastic and paper products, communications equipment, and others for the global market (Yang and Liao 2010).

Dongguan also has produced a unique form of urbanism, or so-called "peri-urbanism" (Lin 2006). The spontaneous and explosive growth of rural and export-processing industries has created a strong and diverse economy to sustain a large, dense population. This enables urbanism to occur outside urban centers. Foreign investment has led to the intensive intersection of industrial and urban activities with the agricultural economy and rural households. Local governments, at municipal, township, and village levels, have functioned as both decision-makers and investors directly involved in urban development. Such peri-urbanism also was enabled by the large army of migrant workers who have outnumbered the local population by a factor of three and sustained export production in Dongguan (and beyond). The movement of migrants to small towns and peri-urban areas has become much easier. On the other hand, the peri-urban region is probably the only place they can afford to live (see more on migrant settlement patterns in Chapter 5).

Moving toward knowledge-based and consumer economies

China's integration into the global marketplace now goes far beyond export processing. The country's vast population and rising prosperity present huge market potentials for foreign firms, products, and services. Its entry into the World Trade Organization (in 2001) further increased the openness of the Chinese economy. Even before that, foreign consumer goods and brands had already landed in large cities. For instance, as the first big American fast-food chain to open in China (in 1987 and about a 5-minute walk from the Tian'anmen Square in Beijing), KFC (Kentucky Fried Chicken) is now ubiquitous to urban consumers (Figure 6.4). So are McDonald's, Pizza Hut, and many others, in hundreds of cities, large and small. Foreign department stores and big-box retailers also have made their presence felt, helping transform the urban economic space (more in Chapter 8).

Figure 6.4 KFC in Shenzhen, 2008.

Source: Wikimedia Commons, retrieved 11 January 2012 from http://www.yum.com/brands/china. asp (donated by Chintunglee).

"KFC continues to be the number one quick-service restaurant brand and the largest and fastest growing restaurant chain in mainland China today, with nearly 3,500 restaurants in more than 700 cities. Yum! opens nearly one new KFC every day in mainland China (Q3 2011)." http://www.yum.com/ company/china.asp (retrieved on 12 December 2011).

With these new consumer activities, the service sector has seen huge growth in foreign participation.

Centers for trade and financial services

Historically, foreign trade concentrated in a set of Treaty Port cities, along the east coast and the Yangtze River. The arrival of Western powers in the nineteenth century greatly spurred the growth of cities. By 1911, about 90 cities along the entire coast, up the Yangtze River, and in north China, were opened up as Treaty Ports or open cities. Foreign presence widened China's trading horizons and was responsible for shifting the economic focus of the country to the coastal areas. Trade grew steadily in bean oil, leather, flour, eggs and egg products, ginned cotton, and hog bristles, replacing traditional exports of tea, silk, and porcelain. The domestic market and the export sector also supported a large cotton textile industry concentrated in Shanghai and Tianjin (Yusuf and Wu 1997). Between 1949 and 1979, in the same way foreign investment ceased, trade also stalled.

During the reform era, trade and export processing were inextricably linked in China. As such, provinces and cities first engaged in manufacturing for the global market also became centers of trade early on. Opening up provided an important impetus to the southern coastal provinces of Guangdong and Fujian. Together with Hainan province, their share of China's total trade skyrocketed from 16 percent in 1978 to 46 percent during the mid-1990s (Naughton 2007). Cities there, such as Guangzhou, Shenzhen, and Xiamen, eclipsed the traditional trading centers in the Lower Yangtze Delta (e.g., Shanghai, Hangzhou, and Suzhou). But, since Deng Xiaoping's 1992 southern tour, the Lower Yangtze region has begun its own export-led and trade-related growth. The importance of coastal cities to the north and northeast, however, has steadily declined (particularly as compared to their roles during China's early maritime trade). Overall, Guangdong province remains the single largest exporting province, accounting for about one-third of the country's trade.

Cities along the southeast coast and in the Lower Yangtze Delta also have become centers for the burgeoning financial sector in China. Shenzhen, in particular, benefited from special policies implemented in the SEZs. The Shenzhen foreign exchange center (swap market) was among the first established in the country and was allowed to offer the widest access for both foreign and domestic enterprises. In 1991, the Shenzhen Stock Exchange was set up and, except for the Shanghai Stock Exchange, was the only such stock market permitted in China since the Tianjin Exchange was shut down in 1952. The Chinese stock market expanded rapidly until 2001, although access was tightly controlled by the state through a quota system that favored large state enterprises (Zhang 2003). Overall, markets for stocks, other financial instruments, and foreign exchange are now functioning in China, along with adequate mechanisms for clearing transactions and settling payments. On the other hand, authorities have restricted retail banking activities, securities trading, and insurance, and only recently permitted foreign banks to make loans in domestic currency. First in 1996 foreign banks housed in Shanghai's Pudong area were allowed to deal in Chinese currency, Shenzhen followed suit in 1998 and major coastal cities by 2005, and in 2006 restrictions were lifted completely (He and Fu 2009). This gradual process of opening mirrored the spatial strategy used in the early reform stage to liberalize trade and FDI regimes.

Regional distribution of foreign banks and branches is highly concentrated in national and regional financial centers. Beijing and Shanghai dominate; next comes the Pearl River Delta, followed by several other coastal cities (He and Fu 2009). Beijing and Shanghai, in particular, have pursued the development of the financial sector and related business services. Both municipal governments have designated a part of the city as a financial district. In Beijing, the financial district is in the southeast, just outside the Second Ring Road and adjacent to a concentration of foreign embassies. In Shanghai, this is in the new Lujiazui Central Business District of Pudong (more in Chapter 8). The central government and municipal authorities have taken the first steps by inviting international financial entities of all kinds to establish offices. Commercial banks, investment and brokerage houses, insurance firms, other market makers, and telecommunications companies have responded. By the end of 2004, about 160 financial institutions and offices were housed in Shanghai's financial district, including offices of China's major state banks and such global banking heavyweights as Citibank, Chase, HSBC, Banque BNP Paribas, and Industrial Bank of Japan (Zeng and Si 2008).

There are both complementarity and competition between Beijing and Shanghai in vying for the position of China's premium financial center. Beijing's financial sector produced almost 13 percent of the municipal GDP in 2007, compared to Shanghai's 10 percent. Beijing is the headquarters city for the financial sector. It is home to nearly 700 institutions, including 234 banks, 59 security institutions, and 117 insurance companies. Most importantly, Beijing is the command center of the country's financial regulatory system, anchored by the head offices of the People's Bank of China, China Banking Regulatory Commission, China Securities Regulatory Commission, and China Insurance Regulatory Commission. On the other hand, Shanghai hosts the largest stock exchange in the country. Based on market capitalization, the Shanghai Stock Exchange is the fifth largest in the world. The sentiment among business watchers surveyed for the Global Financial Centers Index clearly favors Shanghai. Among East Asian centers with the greatest likelihood of becoming more significant, Shanghai comes in first and Singapore second. Beijing trailed a distant fourth, after Hong Kong. China's central government also has nominated Shanghai to become the country's leading financial center by 2020 (Yusuf and Nabeshima 2010).

Hosts to multinational research and development

Since the 1980s, strategies for enhancing research and innovation capabilities have come to occupy a more important position in China's development. A series of ambitious initiatives have been launched. Perhaps one of the most significant measures is the dismissal of the Soviet model of functionally specialized organizations with minimal horizontal linkages between research and production. As reforms proceed, the positions of key actors in the national innovation system – public research institutes, universities, and enterprises – undergo drastic change. The traditionally weak enterprise sector is expected to play a more significant role in technological innovation and to develop closer relationships with research institutes and universities.

An important new trend is the growing interest by MNCs in doing R&D in China. Recently, the number of multinational R&D centers has multiplied rapidly, including such manufacturing heavyweights as Intel, IBM, General Electric, Coca Cola, Motorola, and General Motors. Information technology firms followed suit, including Microsoft and Lucent (now known as Alcatel-Lucent). More recently (since 2000), the global pharmaceutical industry has discovered China as the most fitting partner in R&D among developing countries. Making a presence are such giants as Roche, Novartis, Eli Lilly, Bayer, Pfizer, and AstraZeneca. Affiliates of US MNCs, in particular, have begun spending significant amounts on R&D, making China the only developing country in which such affiliates do so (Naughton

2007). By 2005, there were reportedly as many as 750 foreign-invested R&D centers on the Chinese mainland ("Foreign Investors Select China as Research and Development Base," *Xinhua News*, 9 February 2006).

The rationale behind such R&D investment is multifold: establishing a foothold in the burgeoning Chinese consumer market, taking advantage of the sizeable lower-cost and skilled labor force, and servicing their subsidiaries' manufacturing functions. Increasingly, more MNCs are setting up corporate technology centers in China engaged in cutting-edge R&D and making it an important node of their global networks of innovation. Some R&D functions continue to be associated with servicing MNCs' offshore production and customers, as well as developing or customizing products for the Chinese market (Sun et al. 2006). For Chinese authorities, there is growing awareness that indigenous technology development programs lag behind. As such, they are more willing to provide protection of intellectual property rights, a sore point for many MNCs previously. There is also a set of technology policies offering additional subsidies and financial support packages to so-called high-tech firms, both domestic and foreign. These include tax breaks, subsidized credit, procurement preference (for domestic firms), and accommodation of startups supported by venture capital (Naughton 2007).

The majority of these R&D programs tend to be wholly foreign-owned enterprises, and headquartered in the USA or Japan. Concentrated along the east coast, they are located in such major cities as Beijing, Shanghai, Guangzhou, and their neighboring localities (e.g., Hangzhou, Suzhou, and Shenzhen). For instance, among the more than 700 foreign R&D firms established in China by 2005, close to one-third were in Beijing and about 23 percent in Shanghai (Wang and Du 2007). These places draw their advantages from extensive international exposure, deep human capital resources, better physical and business infrastructure, and more mature R&D networks (through concentration of universities and public research institutions). They have begun to create the institutional attributes that boost the productivity of R&D (Jefferson and Zhong 2004).

As an increasing number of MNCs establish R&D centers in these key centers, they compete with local firms for talent, but often with promises of better pay and work environment. This can put small domestic startups at a particular disadvantage in recruitment. In addition, horizontal mobility for knowledge workers has increased as a result of the growing demand, creating difficulty in retention for many firms. As a result, some medium-sized cities close by are rapidly catching up as new centers for knowledge-based firms and foreign R&D. A good example is Hangzhou, the capital city of Zhejiang province and a short 193 km (120 miles) south of Shanghai. Its top-ranked university (Zhejiang University) produces graduates of no lesser quality than those from universities in Shanghai. More importantly to firms, it remains feasible to retain quality knowledge workers (Wu 2007).

Globalization has brought with it another form of international commerce in the exchange of human capital, particularly of highly educated workers. Such an exchange is facilitated by the networking of global production. Most prominently, a transnational community of Western-educated (particularly in the USA) Chinese engineers and entrepreneurers has begun to operate across regions. Some believe that these "new Argonauts" (Saxenian 2006) have helped introduce the Silicon Valley model to China (Leng 2002; Zhou 2008). Numerous ethnic Chinese technology associations in the Silicon Valley region connect technology and talent between the USA and China. In response to such transnational talent flows, the Chinese state has concentrated on promoting specific cities as hubs to receive them. The first priority for Shanghai, for instance, is to attract overseas Chinese talent, and then skilled foreigners. In addition to reforming residency policy, cities like Shanghai have set up special returnee areas within major industrial parks with simplified bureaucratic procedures and high-quality infrastructure support (Leng 2002). These transnational talent circuits also have been instrumental in helping to develop China's indigenous R&D, as seen in the case of Zhongguancun (Box 6.2).

Box 6.2 Zhongguancun: China's "Silicon Valley" in the making

Zhongguancun, in northwest Beijing, is the heart of the largest concentration of China's universities (more than 60, with such elite institutions as Peking and Tsinghua University) and the Chinese Academy of Sciences. From the mid-1980s and well into the 1990s, major universities and public research institutes in the area began to establish spin-off firms, in part to commercialize their R&D results and in part to supplement budget shortfalls caused by shrinking central government spending on research. Some of China's leading high-tech companies emerged during this time. These included Lenovo (originally know as Stone and affiliated with the Chinese Academy of Sciences), Founder (affiliated with Peking University), Ziguang (affiliated with Tsinghua University), and Tongfang (affiliated with Tsinghua University). They also formed the backbone of China's first science park (Zhou 2008).

The development of Zhongguancun has followed the Silicon Valley or the "technology plus capital" model (Cao 2004: 650). The initial spin-off firms were quite successful commercially, turning research findings into marketable products in personal computing and peripherals. The area was even nicknamed the "Electronics Street." Its scope broadened after its official designation as an Experimental Zone for New Technology and Industrial Development in 1988 – to integrate technological development with industry and commerce. But it took another ten years to evolve into a science park. It now comprises a total area of 100 square kilometers, with more than 17,000 firms and institutions claiming to conduct R&D in information technology, integrated optical–mechanical–electronic technology, biotechnology, pharmaceuticals, new materials, and energy-saving and green technologies. Critics, however, believe that technological development is not the key driving force of the area's growth. It mainly serves as a distribution, processing, and trading center for foreign technology companies (Cao 2004). Others call it "an audacious experiment by a late industrialized country" (Zhou 2007). All agree, however, that Zhongguancun's path is rooted in a synthesis of indigenous development and multinational investment. In addition to spin-offs, another group of firms with a rising profile is overseas returnee-founded companies, which have emerged as major players in institutional innovation and technological progress. Because of their small size and heavy dependence on knowledge workers, they tend to form close relations with universities and public research institutes through both formal and informal arrangements.

Conclusion

Today, China is a key player in the global economy. The expansion of its participation in international trade and investment has been one of the most outstanding features of the country's economic development. Much of this remarkable transformation has taken place through urban-centered initiatives. Cities host a large concentration of export-based manufacturing and, more recently, have seen the rise of such knowledge-intensive functions as financial services and R&D. With increasing integration into global systems, foreign investment plays a key role in shaping the Chinese urban system. FDI helps keep the highest level of urbanization along the east coast. Growing global linkages, on the other hand, come with tradeoffs. With more openness and interdependence, urban China reverberated much more from the recent global economic downturn than it did in the 1997 East Asia economic crisis. Shrinking global demands for consumer goods led to squeezed margins for the manufacturing sector, and subsequently rising unemployment levels for both urban and migrant workers. The housing market also took a hit (though briefly, in comparison to Western countries). While China has largely recovered from the recent downturn, the future growth of its cities will inevitably be tied to the world economy. Continued urban prosperity will depend on the extent and speed with which domestic firms improve their technological competence and stay competitive in the global marketplace.

Bibliography

Cao, Cong. 2004. "Zhongguancun and China's High-tech Parks in Transition: 'Growing Pains' or 'Premature Senility'?" *Asian Survey* 44(5): 647–668.

He, Canfei and Fu, Rong. 2009. "An Empirical Study of the Location Choices of Foreign Banks in China [*waizi yinhang zai zhongguo de quwei xuan ze*]." *Acta Geographica Sinica (dili xuebao)* 64(6): 701–712.

Hsiung, Deh-I. 2002. "*An Evaluation of China's Science and Technology System and its Impact on the Research Community.*" A Special Report for the Environment, Science and Technology Section of U.S. Embassy, Beijing, China.

Jefferson, Gary H. and Zhong, Kaifeng. 2004. "An Investigation of Firm-Level R&D Capabilities in Asia," in Shahid, Yusuf, Altaf, M. Anjum, and Nabeshima, Kaoru (eds) *Global Production Networking and Technological Change in East Asia*. New York: Oxford University Press, for the World Bank, pp. 435–475.

Leng, Tse-Kang. 2002. "Economic Globalization and IT Talent Flows across the Taiwan Strait." *Asian Survey* 42(2): 230–250.

Lin, George C. S. 2006. "Peri-urbanism in Globalizing China: A Study of New Urbanism in Dongguan." *Eurasian Geography and Economics* 47(1): 28–53.

Ma, Laurence J.C. 2002. "Urban Transformation in China, 1949–2000: A Review and Research Agenda." *Environment and Planning A* 34(9): 1545–1569.

National Bureau of Statistics of China. 2009. *China Statistical Yearbook for Cities 2009*. Beijing: China Statistics Press.

Naughton, Barry. 2007. *The Chinese Economy: Transitions and Growth*. Cambridge, MA: MIT Press.

NSD Bio Group. 2009. "Research Report on Chinese High-Tech Industries." Prepared for U.S. China Economic and Security Review Commission (retrieved on 1 November 2010 from http://www.uscc.gov/researchpapers/2009/NSD%20BioGroup%20Final%20Report%20'Sunrise'%20Report%2002June2009.pdf).

Pepper, Suzanne. 1988. "China's Special Economic Zones: The Current Rescue Bid for a Faltering Experiment." *Bulletin of Concerned Asian Scholars* 20(3): 2–21.

Saxenian, Annalee. 2006. *The New Argonauts: Regional Advantage in a Global Economy*. Cambridge, MA: Harvard University Press.

Sun, Yifei, Du, Debin, and Huang, Li. 2006. "Foreign R&D in Developing Countries: Empirical Evidence from Shanghai, China." *China Review* 6(1): 67–91.

Wang, Chengyun and Du, Debin. 2007. "The Comparison of the R&D Investment Location Advantage between American and Japanese MNCs in China [*zaihua mei ri kuaguo gongsi R&D touzi quwei de bijiao*]." *Human Geography (renwen dili)* 22(2): 1–5.

Wei, Yehua Dennis. 2010. "Beyond New Regionalism, Beyond Global Production Networks: Remaking the Sunan Model, China." *Environment and Planning C* 28(1): 72–96.

Wu, Weiping. 1997. "Proximity and Complementarity in Hong Kong – Shenzhen Industrialization." *Asian Survey* 37(8): 771–793.

Wu, Weiping. 1999. *Pioneering Economic Reform in China's Special Economic Zones: The Promotion of Foreign Investment and Technology Transfer*. Aldershot, UK: Ashgate.

Wu, Weiping. 2007. "State Policies, Enterprise Dynamism, and Innovation System in Shanghai, China." *Growth and Change* 38(4): 544–566.

Yang, Chun and Liao, Haifeng. 2010. "Industrial Agglomeration of Hong Kong and Taiwanese Manufacturing Investment in China: A Town-Level Analysis in Dongguan." *Annals of Regional Science* 45(3): 487–517.

Yusuf, Shahid and Nabeshima, Kaoru. 2010. *Two Dragon Heads: Contrasting Development Paths for Beijing and Shanghai*. Washington, DC: The World Bank.

The book explores the contrasting development options available to Beijing and Shanghai, and proposes strategies for each city based on the current and acquired capabilities of each, the experiences of other world cities, the emerging demand in the national market, and likely trends in global trade. The authors weave economic growth, urban development, and technological innovation into their analysis.

Yusuf, Shahid and Wu, Weiping. 1997. *The Dynamics of Urban Growth in Three Chinese Cities*. New York: Oxford University Press, for the World Bank.

This book compares three cities – Shanghai, Tianjin, and Guangzhou – in the context of the changes that swept China's economy, history, and reform programs from the early 1980s through the mid-1990s. The authors consider the interplay among geography, size, and industrial structure that determines the industrial vigor of cities, concluding that each of these factors must be made to work for the city through effective policy-making.

Yusuf, Shahid and Wu, Weiping. 2002. "Pathways to a World City: Shanghai Rising in an Era of Globalization." *Urban Studies* 39(7): 1213–1240.

Zeng, Gang and Si, Yuefang. 2008. "Study of the Financial Services Cluster in Liujiazui District of China [*Shanghai lujiazui jinrong chanye jiqun fazhan yanjiu*]." *Areal Research and Development* [*diyu yanjiu yu kaifa*] 27(3): 39–43.

Zhan, Xiaoning James. 1993. "The Role of Foreign Direct Investment in Market-Oriented Reforms and Economic Development: The Case of China." *Transnational Corporations* 2(3): 121–148.

Zhang, Le-Yin. 2003. "Economic Development in Shanghai and the Role of the State." *Urban Studies* 40(8): 1549–1572.

Zhao, Simon X.B. 2003. "Spatial Restructuring of Financial Centers in Mainland China and Hong Kong: A Geography of Finance Perspective." *Urban Affairs Review* 38(4): 535–571.

Zhou, Yu. 2007. "China's High Tech Industry and the World Economy: Zhongguancun Park." *Japan Focus* (online journal, published in December). Retrieved on 8 March 2011 from http://www.japanfocus.org/-Yu-Zhou/2661.

Zhou, Yu. 2008. *Inside Story of China's High-Tech Industry: Making Silicon Valley in Beijing.* Lanham, MA: Rowman & Littlefield.

Part III
Urban development

7 Urban restructuring and economic transformation

The Communist victory in 1949 altered the economic parameters of urban development. The larger socialist plan of modernization emphasized heavy industries and regional self-sufficiency. Cities no longer were financial, trade, and business centers. Market transition and globalization, however, put cities on an entirely different path of growth. In addition to the production of low-cost consumer goods, business and consumer services non-existent before 1979 mushroomed. As urban economies restructure and diversify, the monopoly of state-owned enterprises (SOEs) has evaporated in virtually all sectors.

This chapter outlines the reorganization of the urban industrial sector and growth of the tertiary sector. In a similar fashion, SOEs have undergone sweeping transformation. Such economic restructuring has had a tremendous impact on the urban labor force and lives of urbanites. This chapter explores the interplay between growth and equality both within and across cities. The following questions should help guide the reading and discussion of the materials in this chapter:

- Changing employment patterns are a good indicator of urban industrial restructuring. How would you characterize such patterns in urban China since 1979?
- What are the key undertakings during each of the two phases of SOE reform?
- The non-state sector has grown steadily. What are the common types of business in this sector?
- What are the main sources of finance for urban economic restructuring?
- How have labor market and SOE reforms affected the urban labor force?
- Sustained urban economic growth requires more educated and skilled workers. How has China's higher education tried to fulfill such demands?
- Growth is uneven. What are the key dimensions of disparity within and across cities?

Revamping the urban economic landscape

Economic restructuring

China's large cities started their industrial life as producers of textiles and light manufacturing in the late nineteenth and early twentieth centuries. By then, the old

Western cliché of urban superiority had become a new trend in Chinese society. The concentration of modern industries and commerce in urban areas brought job opportunities and material comfort as well as progressive ideas and advanced education to cities along the coast and Yangtze River (see Chapter 2). But under the system of centralized planning after 1949, cities were assigned quite narrow roles. While they were important providers of industrial products and technology, they ceased to be financial, trade, and business centers. Financial and information functions were transferred to the central government.

Following Soviet-style command economy, China's economic strategies were characterized by a high rate of investment in capital goods, priority to heavy industry and military, investment in basic needs/education and health, and limited importance of foreign trade. At the macro level, household income was modest and savings small, and there were persistent shortages of goods. Consumption was neglected, as was growth in services. This shift to centralized planning began to transform the character of large coastal cities. One major tenet of Maoist socialism was that cities must be cast in the mold of producers, rather than consumers that lived off the surplus of the rural economy.

The practical implications of this were far-reaching and involved a large increase in the share of manufacturing relative to services. In particular, heavy industries were the ones targeted for expansion because of the overriding importance attached to metallurgical products, machinery, and petrochemicals. These industries served as the touchstone of economic strength and consequently received the largest allocations of capital through the state plan. There was an equally strong impetus towards local self-sufficiency by simultaneously developing the widest possible range of subsectors so that the maximum number of input–output relationships could be contained within cities (Yusuf and Wu 1997).

During the early stage of market reform after 1979, the necessity for balanced industrialization was recognized, which gave equal importance to consumer goods. There was an easing of the emphasis long given to heavy industries. Coastal cities responded by expanding light manufacturing. Moreover, China's export surge relied on the rapid growth of traditionally manufactured consumer goods (see Chapter 6). As the 1980s proceeded, the central government began giving economic priority to the coastal regions. The objective of equalizing industrial production throughout the nation was abandoned.

Starting in the 1990s, many cities embarked upon a program of restructuring and specialization. In contrast to their manufacturing prowess, these cities had a narrow range of producer services. State-owned banks provided financial services; foreign trade corporations and other government organs handled all forms of commerce; research institutes affiliated with industrial bureaus were the source of consulting services. Industrial enterprises struggled to be self-sufficient, doing in house what firms in market economies have been purchasing for decades. By then, the central and local governments came to recognize more fully the contribution of services to growth and employment. On June 16, 1992, central authorities issued a strongly worded directive to accelerate tertiary-sector development (Yusuf and Wu 1997).

The tertiary sector, since then, has grown steadily. By 2007, it accounted for 46.2 percent of value added across urban China while manufacturing was a close 50.5 percent. Compared to 1990, this was a 13.2 percent growth for services, and 9.9 percent decline for manufacturing (Association of Mayors of China 2009). Overall, urban employment in the service sector has consistently outnumbered manufacturing jobs since 1999 (Figure 7.1). Beijing, in particular, boasts a tertiary sector worthy of 73 percent of the city's total value added, nearly on a par with cities in the West. Investment in business services, logistic services, and creative industries has been particularly substantial. For those large cities with continuing eminence in manufacturing, the shift is towards industries with more knowledge and technology content (more in Chapter 6). But along the coast and close to resource bases, many cities still rely on labor-intensive manufacturing, particularly that related to export processing. There, the manufacturing sector accounts for close to 60 percent of total value added.

Economic restructuring in some cities also is driven by the ambition to become an "international city." Beijing's evolving city plans during the reform era presented an example of such conceptualization. Early on, the city retained its commitment to services primarily as the center of China's political and cultural systems. But while still maintaining a commitment to domestic functions, recent plans advocate aggressive development toward full participation in the global economy and the world system of cities. Beijing's functional orientation has been redefined from a political and cultural center to a center for politics, culture, and international affairs and finance. The 2005 revision to Beijing's urban master plan repeatedly calls for the development of a "modern international city." Similar transformations have taken place in cities throughout China – not only the largest one, but a wide range of other cities as well (Gaubatz 2005).

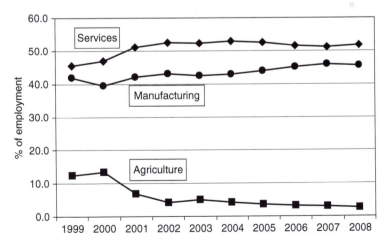

Figure 7.1 Share of employment in three sectors, prefecture-level cities, 1999–2008.

Source: National Bureau of Statistics of China 2000–2009.

State-owned enterprise reform

Now the factory of the world, urban China has gone through an "industrial revolution" since 1978 (Naughton 2007: 298). This is propelled by two kinds of forces: by the creative restructuring of large SOEs as well as collectively owned enterprises, and by the emergence of small and medium-sized new firms. The absence of significant early attempts to privatize large state firms distinguishes China from other transitional economies. Having served collectively as the "cash cow" for the command economy, SOEs began to show the strain under market reform. Faced with only soft budget constraints, loss-making SOEs continued to receive financing from state banks or industrial bureaus. When SOEs incurred losses, they would impose direct costs not only on the central government's budget but also on the finances of the local economy. The preferential access state enterprises enjoyed to financial resources also had serious "crowding-out" effects, starving private companies.

SOE reform has been through two major phases, repeating the incremental path of overall market transition. Before 1994, a contract responsibility system was implemented to expand enterprise autonomy (held by managers) and increase profit incentives. This began to reorient firms to respond to market competition. A combination of increased competition, improved incentives, and more effective monitoring of performance improved SOE performance over the 1980s. During this period, privatization played almost no role; but a variety of ownership began to develop in the 1980s – town and village enterprises and private firms. After 1994, however, SOE restructuring entailed painful closing of unprofitable enterprises and furloughing of workers in the millions. The number of industrial SOEs dropped from 120,000 in the mid-1990s to only 31,750 in 2004. Laid-off workers totaled 40 percent of the SOE workforce, and the urban collective workforce shrank by two-thirds (Naughton 2007). Between 1996 and 2000, SOEs and urban collectives shed some 47 million jobs (DFID 2004). Together, the share of these two types of enterprises in total urban employment fell from 100 percent in 1978 to 78 percent in 1994 and then to 23 percent in 2009 (Figure 7.2). The drop in SOE employment since the mid-1990s has been particularly drastic.

Corporatizing SOEs was another key undertaking after 1994. The 1994 Company Law provided a framework, allowing for new mixed-ownership forms and eventual privatization. One was through selling off some corporate shares and forming a share-holding corporation or joint ownership. Another was to organize corporate groupings to bring together firms into more formal networking arrangements, with overall financial control and strategic functions centralized in a parent holding company. An even more creative form was loose alliances that permit rationalization of production and joint financing of activities along with interfirm cooperation (Yusuf and Wu 1997). In addition, the central government adopted a policy of "grasping the large, and letting the small go": government control became concentrated in energy, natural resources, and a few sectors with substantial economies of scale (metallurgy, telecommunications, and military industry). In 2003, such control of SOEs was transferred to the newly established State Asset

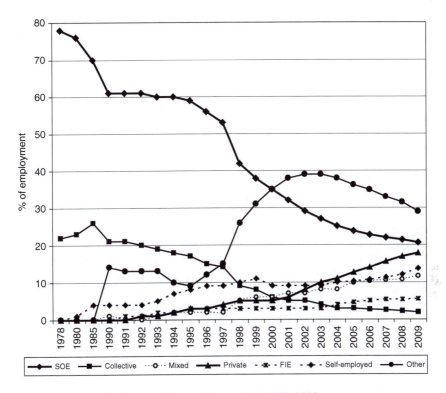

Figure 7.2 Urban employment by type of ownership, 1978–2009.

Source: Own calculations based on National Bureau of Statistics of China: China Statistical Yearbook 2010.

Note: Mixed ownership includes cooperative, joint ownership, limited-liability corporations, and share-holding corporations. SOE, state-owned enterprise; FIE, foreign-invested enterprise.

Supervision and Administration Commission. While managers of these remaining SOEs have achieved a great deal of independence, the Communist Party holds on to its personnel power. The party, as a result, continues to shape the incentives for SOE managers by controlling their career paths (Naughton 2007).

The non-state sector, since the mid-1990s, has grown steadily and gained more legitimacy. This sector is an amalgamation of mixed or blended ownership, private businesses, foreign-invested enterprises (FIEs), self-employment, and informal work. With increasing privatization of SOEs, blended firms have become common, particularly in the technology industry. The state interest often is in the minority position, and firms operate along commercial lines with a strong emphasis on profitability. Most of these and other private technology enterprises are in electronics, information technology, software, communications, and biotech industries. Another growing sector of employment is informal work, as seen in the rise of the "other" category in Figure 7.2. According to Park and Cai (2009), this

category includes unreported urban workers and unregistered informal employment, as well as undocumented work by migrants in urban areas. Since the number
of migrant workers tends to be undercounted, the size of informal employment may
be underestimated. Often, such employment is temporary, lacks a formal contract,
and does not provide social insurance benefits or other worker protection.

Financing industrial restructuring and growth

Urban economic restructuring is costly. SOE reform brings with it a drastic withdrawal of direct government funding in industry finance. With the exception of
the large SOEs under central control, privatization has been highly decentralized
and run by local governments. On the other hand, despite increasing local autonomy, Chinese cities still are subject to some strict fiscal rules. No municipality has
the authorization to establish its own taxes or other formal revenue sources. The
power to impose taxes and set rates is the center's prerogative.

While municipal governments have no formal authority to create new avenues
of financing, they have displayed ingenuity in finding new resources over which
they have more effective control. For instance, some cities created a "revolving fund" that was used for small-scale investments, predominantly in industry.
Ordinary fiscal grants for industrial investment were transformed into repayable
loans, and repayment of principal as well as interest payments were channeled into
the revolving fund. Loans were made at interest rates equal to or, up to a point,
below bank interest rates, typically for periods of around 1 year. Such funds operated in a legal gray area, facing opposition from officials of the banking system
(Yusuf and Wu 1997).

Bank credit has emerged as a major source, replacing the old system of direct
government finance. During the late 1980s and early 1990s, industrial financing
shifted from budgetary grants and retained funds (internal financing) towards a
dependence on bank credits. Increasingly resembling market economies, surpluses
generated in the household sector were loaned to the enterprise sector to finance
a large portion of investment. However, much of the loan financing was channeled through the state banking system, which was still subject to a high degree
of government control. The control was exercised in part through the earmark-
ing of funds for specific purposes. One such purpose was industrial restructuring.
About two-thirds of the total amount of bank fixed investment lending (for both
capital construction and renovation investment) was earmarked by the government, mainly to SOEs. Thus, although bank loans became a significant financial
resource, the system of allocating bank loans tended to replicate the old system of
administrative allocation of investment.

After the mid-1990s, however, state banks began to behave more like commercial banks and tightened their lending standards. Together with the closing of
many unprofitable SOEs, urban industry's overall financial position improved.
Still, most new finance continues to be from banks, rather than through capital
markets (such as the stock market and corporate bonds) – a unique characteristic of China's SOEs. Government and credit financing, however, has not always

come easy for non-state firms. Particularly in the early reform period, there were a limited number of financial tools available to support their growth. Often, bank lending was directed to foundering SOEs and lending to private firms was believed to be more risky. These firms resort to an assortment of investments, ranging from personal savings, loans from individuals, to small investment financing.

The importance of capital markets in industrial finance remains limited, even though markets for stocks and other financial instruments are now functioning. When China's first stock markets, in Shenzhen and Shanghai, opened in the early 1990s, nearly all of the listed firms were SOEs. Revenues from the initial public offerings (IPOs) went to the listing firms or their immediate supervisory entities. As such, the government never really sold the SOEs; instead, IPOs became a new source of funds for these firms. With SOEs dominating, the domestic stock markets did not turn out to be an important source of finance for non-state and private firms. The government also owned the largest proportions of total stock, and government policy has become the single critical driver of market fluctuations. Such limitations have affected the development of institutional investors – mutual funds, pension funds, and insurance companies have yet to become fully engaged. In addition, the market for corporate bonds is even more underdeveloped (Naughton 2007).

At the same time, foreign capital has become another source for industrial restructuring. One way is through merger and acquisition by foreign investors. Under three central government regulations issued since 2002, they can purchase both "state-owned shares" and "legal person shares" of SOEs. No limit is imposed on the amount of shareholdings that foreign investors may acquire, with some exceptions. For them, investing in SOEs is a quick way to obtain the required critical mass of customers to start a profitable business in China. On the other hand, since the early 1990s, an increasing number of Chinese firms are launching foreign IPOs, particularly large SOEs and technology firms. The Hong Kong Stock Exchange is the major arena for such listings, then the New York Stock Exchange and the NASDAQ. Other markets include the Singapore Exchange and the KOSDAQ, South Korea's NASDAQ equivalent. Given the overall strong performance of Chinese firms during the Asian financial crisis at the end of the 1990s, foreign markets are expressing more interest in Chinese IPOs. As such, foreign funds are an emerging source for satisfying the capital needs of some Chinese firms. More recently, municipal authorities have begun to acknowledge the importance of attracting foreign venture capital to finance domestic technology startups. State-backed venture capital firms also are being restructured in order to compete. For instance, Shanghai Venture Capital Corporation now enjoys considerable autonomy to operate according to market principles (Leng 2002).

Harnessing human resources

Urban industrial growth, much like that of the economy as a whole, also can be attributed to flows of capital and labor, the efficiency with which they are allocated across production activities, and the rate of technological change. While capital is important, the contribution of human resources should not be underestimated. China's urban industries have moved away from the practice of providing the workforce with tenured employment and guaranteed pensions along with other benefits. They face challenges of ensuring an adequate supply of entrepreneurship, skills, and labor.

Labor market reform

Labor markets did not exist under China's command economy. Instead, the majority of urban workers were in state work units and enjoyed a system of guaranteed lifetime employment – the so-called "iron rice bowl." This system lacked incentives for workers; hence labor productivity level was low and the problem of underemployment severe. Much industrial investment was not only capital-intensive (in heavy industries), but also relatively demanding technologically. SOEs were faced with having to employ more workers than necessary and used labor inefficiently. Not only did the "iron rice bowl" tie the hands of employers, it also sharply circumscribed the ability of workers to change jobs, and bounded them in a dependency relationship with their employers for almost all their living needs. As a result, at the macro level, there was little labor mobility and flexibility of resource use. Another consequence of lifetime tenure in SOEs was the compartmentalization of labor – workers were trained in a narrow specialization. The overall extreme rigidity of labor practice also grew out of the effort to control population movement (see Chapter 5).

Since 1986, labor market reform has proceeded in stages. The purpose was to improve labor productivity and flexibility by gradually displacing permanent employment and state allocation of jobs. First was the introduction of labor contracting, accompanied by somewhat greater scope for employee discharge and with "waiting for employment" or unemployment insurance. The greater use of fixed-term contracts could potentially increase labor mobility, but it did little to avert the historical legacy and expectations of permanent tenure. Rigid job definitions and work rules remained untouched by the contract system. However, after 1994, close to 50 million SOE workers were laid off as reform intensified in the urban industrial sector.

The mounting SOE layoffs were a driving force behind increasing job mobility. People now change jobs more regularly. Market forces also began to shape the ways in which workers were paid, with more incentives for those performing better and being more productive. After a drastic turn from the command economy in which education did not significantly increase income, today workers with better education are more likely not to face layoff and to get paid more. FIEs are particularly bidding up wages of highly educated and skilled workers. Guangzhou, as a pioneer of reforms in many aspects, is an appropriate place for observing the process of labor market reform (Box 7.1).

A foremost challenge is relocating workers displaced by the large-scale closure and divestiture of SOEs. This is particularly difficult for middle-aged workers close to retirement in urban core industries like textiles. Overall, furloughed

***Box 7.1* Breaking the "iron rice bowl" in Guangzhou**

Labor reform in Guangzhou started as early as 1977, ahead of other Chinese cities. From early on, municipal leaders recognized the need to enhance labor mobility, and to expand the authority of enterprises over employment decisions. A series of measures were introduced, particularly after 1986, to revamp the "iron rice bowl" system and to improve labor productivity. Two of the most successful changes in getting able people into better jobs were the circulation of information about job openings and requirement for applicants to take exams. A municipal labor service company was set up to facilitate recruitment and served as a center for training. By the early 1990s, contract employment was mandated for all employees. Accompanying labor market reform, the urban pension system also underwent restructuring. In 1983, a measure called for employers to provide pensions for contract workers. By 1989, the city developed regulations to assist furloughed workers by establishing a re-employment program. At the national level, this did not come about until 1994. Other regulations in Guangzhou called for disability benefits for non-local workers, and incentives for firms to retain, in addition to hire, laid-off workers. The implementation of these measures may be patchy across firms; even so, Guangzhou was clearly leading the charge of municipal reforms.

While local labor is plentiful, it is unlikely that the expansion of industry and construction could have boomed without voluntary and induced migration from poorer parts of the province as well as farther afield. Rural people from Guangxi, Sichuan, Hunan, and deeper in the interior have flocked to Guangdong province because of its prosperity. Many of them find their way to Guangzhou to work in factories, the many construction sites that have sprung up, and the myriad formal and informal activities that have proliferated. Factory managers have frequently sent supervisors to recruit young workers from distant villages once a few migrants from these places established a good reputation. Given the nature of factory work, fresh blood sustains productivity and fills gaps left by locals who tend to gravitate toward the less demanding jobs or alternatively begin pressing for higher wages.

To accommodate migrant workers, a system of temporary registration was first established to regulate entry into Guangzhou and to help enforce annually renewable employment contracts. Later, Guangzhou was ahead of others in experimenting with *hukou* reform, aimed at eliminating urban and rural distinctions. In 2004, it mandated that small cities and towns within the greater Guangzhou metropolitan area allow certain migrants of rural *hukou* to convert to local registration, if they have a fixed place of residence in small cities and stable source of income. Further, rural residents may obtain a *hukou* for Guangzhou city proper if they maintain a fixed place of residence there for a minimum of 5 years, have a stable source of income, and have participated in the city's pension program. However, the number of rural migrants qualified for these criteria remains small.

workers have been absorbed into low-skill service operations for which job prospects have multiplied and retraining is minimal. Specifically, these workers find employment in retail, repair and maintenance work, grounds keeping, household services, and cleaning services. Some of these jobs are through self-employment or in the informal sector. On the other hand, many older laid-off workers have taken the route of early retirement, often not by choice. They subsequently experience drastic reductions in income and standard of living. There is evidence that this segment of the population has emerged as the new urban poor (Naughton 2007). The situation of such workers, however, varies from city to city as local economic conditions vary substantially and local governments are responsible for their transitional support.

The rapidly aging population presents urban China, especially the large cities, with another challenge. One in every four Chinese will be aged above 65 by 2050, according to official projections. Large cities like Shanghai are about 20 years ahead of the national aging trend. They are witnessing a phenomenon similar to that occurring in countries with substantially higher income levels. A growing number of retirees demand a better pension system, as well as housing and medical benefits. The significant expansion in the share of old age cohorts in the population is partly the result of a low birth rate. This low rate stems from the one-child family-planning policy, enforced quite effectively in large cities. As such, natural growth of the urban labor force has averaged only about 3 percent annually in recent years. Nationwide, the growth of working-age population will drop off quickly and reach zero after 2015 (Naughton 2007).

The urban labor market for low-skilled jobs remains considerably segmented by *hukou* status. The projected decline in the urban workforce has been offset by an increasing volume of migrant workers. But they continue to be excluded from state and secure job opportunities (see Chapter 5). Municipal authorities, as well as urbanites, are still fearful that migrants would compete with urban residents. As such, a larger proportion of migrants work in the informal sector, setting up shops, conducting petty trade, collecting waste and recycling materials, providing household services, and engaging in construction work. Gender is another dimension along which the labor market segments. Female furloughed workers have fewer choices of employment than their male counterparts, and there is a rising gender wage gap (DFID 2004).

Education and skill formation

The Cultural Revolution (1966–1976) seriously interrupted China's educational system. Human and administrative skills necessary to run a market economy, as a result, were in short supply. However, socialist development brought with it a vast state-sponsored education and training establishment. Once market reforms were under way, the energy of this state system was unleashed. Non-state providers also have emerged, although they remain small in number and scale. Building up the higher-education sector has been a major priority, which is exclusively located in cities.

After the reinstatement of the National College Entrance Examination in 1978, higher education has undertaken a series of reform measures and experienced steady expansion. Through several rounds of policy and structural change, it has reverted from a system based on the Soviet model of specialized education to one that is closer to international norms. Universities gained more autonomy in enrollment expansion, curriculum development, faculty recruitment, and international exchanges. Since the mid-1990s, the oversight of many universities was delegated to local governments and there were many "mergers and acquisitions" among universities (Simon and Cao 2009). There are now two major divisions of higher education – regular higher-education institutions (RHEIs) and adult education institutions. By 2009, the number of RHEIs stabilized at 2,305. Drastic increase in student enrollments has taken place, leading to a student body of about 21.4 million in RHEIs in 2009 – 25 times more than that in 1978 (Table 7.1). China's achievement in transforming higher education from an elite to a mass type is unprecedented in the world. It took the USA 30 years (1911–1941), Japan 23 years (1947–1970), South Korea 14 years (1966–1980), and Brazil 26 years (1970–1996) to see the gross enrollment rate grow from 5 to 15 percent (Gao 2009; Simon and Cao 2009). It took China about 12 years to repeat the same feat (1990–2002). Even more impressively, China has been very successful in maintaining higher enrollment and graduation rates in science and engineering. Through all levels of tertiary education, engineering remains the dominant field of study. At the undergraduate level, the trend is particularly stable, accounting for around 37 percent of both incoming and graduating students each year between 1995 and 2007.

The expansion in university enrollment has led to a similar trend in secondary education, particularly since the turn of the twenty-first century (Table 7.1). China's senior secondary schools (also known as high schools) are bifurcated – the regular schools are feeders into higher education, whereas the other stream prepares students for vocational skills through specialized secondary schools, vocational schools, and craftsmen schools. The first stream, generally making up 50–60 percent of senior secondary education, has been sending an increasing share of its graduates to college (78 percent in 2009). Overall, there is a steady penetration of secondary education into the population, following the implementation of universal primary education in the 1980s and subsequently nine-year compulsory education.

However, the quality of higher education remains mixed. While longer-term projections suggest the need to produce a much larger number of highly educated professionals and highly trained technicians, there is an increasing level of unemployment and underemployment for college graduates (Gallagher et al. 2009). A 2005 survey shows that, overall, about three-quarters of them were able to find placement (including further educational opportunities) upon graduation. Those with graduate degrees tended to fare slightly better (over 80 percent), while those from two-year colleges had a lower placement rate (about 60 percent). This situation indicates a substantial mismatch between the quality of graduates and the demand of the job market. University curricula as a whole tend to be narrowly designed and

Table 7.1 General indicators of regular higher-education institutions (RHEIs) and senior high-school education, 1978–2009

	1978	1980	1985	1990	1995	2000	2005	2009
RHEIs								
Number of RHEIs	598	675	1,016	1,075	1,054	1,041	1,792	2,305
Number of RHEIs directly under Ministry of Education	—	—	—	36	35	72	73	73
Total undergraduate student enrollment (millions)	0.856	1.144	1.703	2.063	2.906	5.561	15.618	21.447
Total graduate student enrollment (millions)	0.011	0.022	0.087	0.093	0.145	0.301	0.979	1.405
Average size of RHEI (undergraduate + graduate)	1,450	1,726	1,762	2,005	2,896	5,631	9,261	9,914
Total full-time faculty (millions)	0.206	0.247	0.344	—	0.404	0.463	0.966	1.295
Faculty to undergraduate student ratio	4.15	4.63	4.95	—	9.83	16.30	16.85	17.27
Gross enrollment rate for age 18–22 (%)	—	—	—	3.4	7.2	12.5	21.0	24.2
Regular senior secondary schools								
Total student enrollment (millions)	15.531	9.698	7.411	7.173	7.132	12.013	24.091	24.343
As % of all senior secondary school enrollment	—	56.4	57.2	46.9	43.2	48.8	59.8	52.5
Teacher to student ratio	—	—	—	—	12.95	15.87	18.54	16.30
Gross enrollment rate for age 15–17 (%)	—	—	—	—	33.6	42.8	52.7	79.2
Admission rate to RHEIs (%)	—	—	—	27.3	49.9	73.2	76.3	77.6
Admission rate from regular junior to all senior high schools (%)	—	—	—	40.6	50.3	51.2	69.7	85.6

Source: National Bureau of Statistics of China: China Education Statistical Yearbook 2007 and 2009; National Bureau of Statistics of China: China Statistical Yearbook 2010; Gao 2009.

— Data not available.

delivered, rather than covering a broad range of knowledge and multidisciplinary approaches to problem solving. For example, a McKinsey Global Institute study reported that 33 percent of university graduates study engineering but they "focus more on theory and get little practical experience in projects of teamwork." A worrying conclusion of the study is that "fewer than 10 percent of Chinese job candidates, on average, would be suitable for work in a foreign company" (Farrell and Grant 2005). Representatives of multinational and domestic technology companies also revealed that they would feel comfortable hiring engineering graduates from only 10–15 universities across the country (Wadhwa et al. 2007).

As urban economies diversify, there is increasing demand for a wide range of technical and professional skills. Since the mid-1990s, technical and vocational education has become a more important component of the education system. It cuts across three levels: junior secondary, senior secondary, and tertiary, with training tailored for different degrees of proficiency and specialization. Recently, comprehensive training centers have been established or converted from employment training centers, providing vocational training, skill appraisal, vocational guidance, and employment-related services (Guo and Lamb 2010). Turning more market-oriented, these centers and other vocational institutes are exploring ways to combine learning and practice, connect with the enterprise sector, and emphasize competency skills.

Redistributing economic growth

Under market transition, China's economic growth has been spectacular. Its income prospects are bright, on the whole. The early part of reform (before the mid-1990s) unleashed productivity potentials through better incentives – a period generally known as reform without losers (Naughton 2007). But since then, the impact of economic changes has been less favorable for vulnerable sectors of the population. The relationship between inequality and growth is becoming a top political and social issue. In urban areas, social stratification has sharpened; across the country, regional imbalance continues to widen.

Urban poverty and inequality

When market transition first began, poverty levels were so high that inequality was not a major concern. Around 1983–1984, China was probably the most equal (especially in urban areas) that it had ever been. The household registration system (*hukou*) helped stabilize the urban population. Urbanites enjoyed social benefits subsidized by the government. But this has changed since. While the rural–urban divide remains a fundamental basis of inequality (see Chapter 5), urban poverty is increasingly becoming an important issue for policy and research. This is the result of rising unemployment and the monumental shift of the population to urban centers through migration.

Poverty is a broad concept: its characteristics include vulnerability, insecurity, and lack of access to services, in addition to income. Even so, poverty lines, when well constructed to measure expenditure or income, remain useful indicators. Research shows that, from 1986 to 2000, the incidence of urban income poverty doubled from 2 to 4 percent while consumption poverty (based on ability to buy things) doubled from 5 to 10 percent (Meng et al. 2005). This calculation, however, does not include migrant workers who by some estimate may experience poverty at double the rates. There are now three major groups of the urban poor: unemployed and furloughed workers, migrants, and chronic poor. The last group includes people with no ability to work, no savings, and no relatives to rely on. Often, they are elderly and disabled without family support.

Unemployment is a major contributor to poverty. The poverty rate for urban households with unemployed people is almost five times as high as those households without (DFID 2004). While massive unemployment associated with SOE reform had receded by the early 2000s, the recent global recession reignited the problem. Among a wide swath, three unemployed groups stand out. Migrant workers make up the first, who fuel the growth of many labor-intensive industries – hardest hit by shrinking export demand. By rough estimates, about 5–7 percent of migrant workers from Sichuan, Anhui, Henan, Hubei, and Hunan provinces returned home in 2008 for lack of work. Another estimate put the total number of unemployed migrant workers at 20 million nationwide. The second group is the increasing ranks of jobless urbanites, particularly from small enterprises. In 2009, for instance, their share reached 9.4 percent (scholars believe this was likely an underestimation). College graduates make up the third group. About one-quarter of the 6 million graduates in 2009 had difficulty finding employment (Association of Mayors of China 2009).

Government response to rising urban poverty is limited to the Minimum Living Standard Scheme, spurred by central directives issued in 1999 and 2000. While such directives lay down the broad framework, determination of the poverty line and financing method are up to municipal authorities. Such a scheme generally provides a subsistence allowance to urban residents below the defined poverty line to make up the household income gap. Again, migrants are not eligible. There is evidence that the supplement in many cities, particularly poor ones, is barely enough to cover food and clothing needs.

Regional dimensions

Regional disparity in economic performance and welfare has been long standing. Before 1949, more than 70 percent of the country's industries were concentrated within a narrow coastal belt in the east. This was largely due to the development of precapitalist industrialization in the early twentieth century – mainly in Treaty Port cities (see Chapter 2). Under the command economy, urban policy at the national level became more inland-oriented. From 1966 until the early 1970s, the Third Front strategy aimed at transferring producer goods industries from coastal centers to the deep interior and making provinces in central China more self-sufficient. Much of the industrial investment that took place during these years was in capital-intensive industries tucked away in small towns and medium-sized cities in Sichuan, Hunan, Hubei, and Gansu provinces. This stimulated the growth and development of interior centers, such as Wuhan, Zhengzhou, Xi'an, and Lanzhou (see Chapter 4). Despite such policy reorientations, economic disparity persisted across the three broad regions (east, central, and west). In fact, the pursuit of regional self-sufficiency may have severely distorted the allocation of resources and aggravated regional unevenness.

After 1979, the tune of socialist planning changed to one of economic efficiency and spatial deployment. Policies introduced since then favor a reorientation to coastal development, indicating China's new effort towards further technological and economic expansion. Consequently, when measured at the provincial level, the most pronounced disparities have arisen between coastal and interior regions. Some attempts to reverse the trend began around 1998, when the central government started redirecting fiscal resources toward equalization, including the Go West development strategy launched in 1999. A major driver of this transfer policy appears to be a concern for growing regional disparities that many consider have marred the country's impressive growth performance.

The degree of market transformation mirrors the varying rate of growth across the three broad regions. The central region lags behind the east and western behind central in the transition towards a market economy, measured by both higher proportion of SOEs in industrial production and higher proportion of government funding in total fixed investment. Coastal cities also have benefited the most from the shift to export-processing manufacturing and services. In addition to being magnets for FDI (more in Chapter 6), these cities host a large number of corporate headquarters for the top 500 domestic manufacturing firms. In particular, Beijing, Tianjin, Shanghai, Suzhou, Wuxi, and Hangzhou each is home to more than 20 such headquarters (Wu and Ning 2010). The manufacturing powerhouses, as such, are concentrated in the Lower Yangtze Delta, Bohai Bay region, and Pearl River Delta.

Economic restructuring and SOE reform have hit China's industrial heartland hard, particularly the northeastern region (Liaoning, Jilin, and Heilongjiang provinces). Cities there specialized in such heavy industries as energy production and raw-material industries. Suffering from obsolete factories and depleted resources, they benefited little from economic restructuring. The region's share of national industrial output, for instance, dropped from 16 percent in 1980 to about 9 percent in 2000 (Naughton 2007). Given their reliance on heavy industries, urban economies also were dominated by large SOEs. Shenyang's decline from an industrial powerhouse is a good case in point (Box 7.2).

Box 7.2 Shenyang: an industrial giant in decline

Capital city of Liaoning province, Shenyang is a quintessential industrial city. Its manufacturing roots date back to the early twentieth century when Japanese colonial forces industrialized the Manchuria region. Liaoning's reserves of coal and iron mines and the complex of machinery enterprises in Shenyang meant that resource costs were fairly low and production relatively cost-effective. Urbanization ensued, mainly as a result of Japanese investment. After 1949, Shenyang, and the northeast in general, continued to develop the manufacturing foundations left by the Japanese. Seen as the pillar of China's industrial development, the region received massive public investment under the command economy. Much of the industrial output came from heavy industries, which were dominated by SOEs. As such, Shenyang thrived as an exporter, to the rest of China, of heavy-machine tools, electrical machinery such as transformers and generators, and transportation machinery.

Shenyang's fate, however, changed drastically after 1979. The decline of its industrial might in recent decades is driven by the confluence of several factors. These include the loss of the northeast region's role as a supplier of capital goods (such as machinery and equipment), the city's location far from China's major urban markets, the presence of a very large state sector, and subsequent out-migration of skilled workers to more prosperous areas. It underwent a drastic process of industrial restructuring, with the manufacturing sector dropping from about three-quarters of GDP in 1978 to less than half in the early 2000s. Many loss-making SOEs were shed, merged, sold, or closed down. But SOE reform has taken a relatively slow pace because of the high concentration of large enterprises, particularly heavy and resources-based industries that were kept under state control during earlier periods of reform. Many large SOEs continue to face problems of outdated facilities, technology, and management. On the other hand, the city lacks the diverse, flexible, small-scale production clusters that are needed in globalized China.

The impact of SOE reform on urban workers is severe. Layoffs in the industrial sector have not been matched by increases in employment opportunities in other sectors. According to a 2001 World Bank Urban Labor Survey (cited in World Bank 2006), of nearly 4,000 workers in five large cities (Shenyang, Shanghai, Wuhan, Fuzhou, and Xian), Shenyang has the highest unemployment rate (14.5 percent) and smallest percentage of workers with employer-provided health insurance (53.5 percent) and of retirees with health insurance (42.8 percent, slightly better than Wuhan on this account). As such, assisting furloughed workers became a top local priority. By May 2001, there were over 600 grassroots-level organizations providing employment to about 90,800 laid-off workers. The municipal government also set up various markets employing over 170,000 workers. Additional efforts were undertaken to encourage workers to set up businesses, including tax reductions and exemptions, a temporary reduction in administrative fees, and credit support (Bidani et al. 2005).

A good indicator of overall development, the Human Development Index (HDI) continues to vary widely across different regions. HDI is calculated as the simple arithmetic mean of three basic dimensions of human life: life expectancy at birth, to represent the dimension of a long and healthy life; knowledge, measured by adult literacy rate and combined school enrollment ratios at primary, secondary, and tertiary levels; and real GDP per capita, to serve as a proxy for resources needed for a decent standard of living. The highest ranking of HDI in 2008, for Shanghai, was 44 percent higher than the lowest ranking, in Tibet. Overall the index for provinces in the eastern region was visibly higher than for the western region (Figure 7.3). The level of human development in Beijing and Shanghai was on a par with the Czech Republic, Portugal and the like, while the low level of Guizhou in the western region was similar to that of the Democratic Republic of Congo and Namibia (UNDP China 2010).

Regional rates of urban poverty also differ, from about 3 percent in the east to about 9 percent in the northwest of the country, compared to a national average of 4.7 percent, based on 1998/1999 data (DFID 2004). Beijing, Jiangsu, Zhejiang, and Guangdong were on the lower end (less than 2 percent), while Henan, Tibet, Shaanxi, and Ningxia were on the higher end (more than 8 percent). So, while average living standards in urban China have improved significantly as a result of the spectacular economic performance, this growth is accompanied by many other changes, including rising inequality.

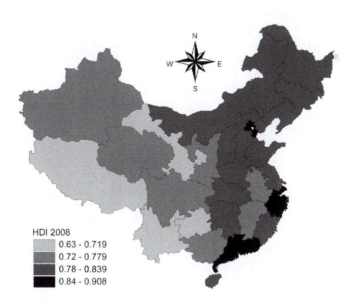

HDI 2008
- 0.63 - 0.719
- 0.72 - 0.779
- 0.78 - 0.839
- 0.84 - 0.908

Figure 7.3 Regional variation of Human Development Index (HDI), 2008.

Source: UNDP China 2010.

Conclusion

The revamping of the economic landscape of urban China is nothing short of drastic. Large cities are no longer singularly anchored in heavy industries and manufacturing activities. The tertiary sector has steadily grown in size, scope, and complexity. Such economic restructuring has entailed a sweeping – at times painful – reform of state enterprises. After laying off millions of workers, SOEs have gained more managerial autonomy and become more profitable in the face of rising competition from foreign firms and an emerging non-state sector. To power the burgeoning urban economies, human capital and skills are occupying an increasingly important space in municipal policy-making. Labor market and higher-education reforms are just some key aspects in this arena. At the same time as such welcoming transformations, urban China also is becoming more stratified both within cities and between regions – in income levels, in living standards, and in overall human development. Given the steady rise in urbanization, cities likely will become home to many of China's poor – a serious driving force for radical urban changes in the future.

Bibliography

Association of Mayors of China. 2009. *China Urban Development Report 2008*. Beijing: China City Press.

Bidani, Benu, Goh, Chor-ching, Blunch, Niels-Hugo, and O'Leary, Christopher J. 2005. "Evaluating Job Training in Two Chinese Cities." Upjohn Institute Working Paper No. 05–111. Kalamazoo, MI: W.E. Upjohn Institute for Employment Research.

Department for International Development (DFID). 2004. *China Urban Poverty Study*. Final Report. Hong Kong: GHK and International Institute for Environment and Development.

Farrell, Diana and Grant, Andrew J. 2005. "China's Looming Talent Shortage." McKinsey Global Institute (retrieved from http://www.mckinsey.com/mgi/publications/Chinatalent.asp on 27 October 2006).

Gallagher, Michael, Hasan, Abrar, Canning, Mary, Newby, Howard, Saner-Yiu, Lichia, and Whitman, Ian. 2009. *OECD Reviews of Tertiary Education: China*. Retrieved on 25 September 2009 from www.oecd.org.

Gao, Huibin. 2009. "History and Reality: Development and Problems in Higher Education in 30 Years of Reform and Opening-up [*lishi yu xianshi zhijian: gaige kaifang 30 nian gaodeng jiaoyu de fazhan yu wenti*]." *Higher Education Research and Evaluation* [*daxue yanjiu yu pinggu*] 2: 5–11.

Gaubatz, Piper. 2005. "Globalization and the Development of New Central Business Districts in Beijing, Shanghai, and Guangzhou," in Wu, Fulong and Ma, Laurence (eds) *Restructuring the Chinese City: Changing Society, Economy and Space*. New York: Routledge Press, pp. 98–121.

Giles, John. 2011. "China's Labor Market in the Wake of Economic Restructuring." Retrieved on 22 April 2011 from http://www.growthcommission.org/storage/cgdev/documents/LaborMarkets/Giles.pdf.

Guo, Zhenyi and Lamb, Stephen. 2010. *International Comparisons of China's Technical and Vocational Education and Training System*. London: Springer.

Leng, Tse-Kang. 2002. "Economic Globalization and IT Talent Flows across the Taiwan Strait: The Taipei/Shanghai/Silicon Valley Triangle." *Asian Survey* 42(2): 230–250.

Meng, Xin, Gregory, Robert, and Wang, Youjuan. 2005. "Poverty, Inequality, and Growth in Urban China, 1986–2000." *Journal of Comparative Economics* 33(4): 710–729.

Min, Weifang, Xiaohao, Ding, Dongmao, Wen, and Changjun, Yue. 2006. "An Empirical Study on the Employment of Graduates in 2005 [*2005 nian gaoxiao biyeshang jiuye zhuangkuang de diaocha fenxi*]." *Journal of Higher Education* [*gaodeng jiaoyu yanjiu*] 27(1): 31–38.

National Bureau of Statistics of China. Various Years. *China Statistical Yearbook for Cities.* Beijing: China Statistics Press.

National Bureau of Statistics of China. Various Years. *China Education Statistical Yearbooks.* Beijing: China Statistics Press.

Naughton, Barry. 2007. *The Chinese Economy: Transitions and Growth.* Cambridge, MA: MIT Press.

Park, Albert and Cai, Fang. 2009. "The Informalization of the Chinese Labor Market." Retrieved on 22 April 2011 from http://ihome.ust.hk/~albertpark/papers/informalization.pdf

Simon, Denis Fred and Cao, Cong. 2009. *China's Emerging Technological Edge: Assessing the Role of High-End Talent.* Cambridge, UK: Cambridge University Press.

United Nations Development Program China (UNDP China). 2010. *China Human Development Report. 2009/10: China and a Sustainable Future: Towards a Low Carbon Economy and Society.* Beijing: China Translation and Publishing.

Wadhwa, Vivek, Gereffi, Gary, Rissing, Ben, and Ong, Ryan. 2007. "Where the Engineers Are?" Issues in Science and Technology (Online). Spring (retrieved on 28 November 2009 from http://www.issues.org/23.3/wadhwa.html).

World Bank. 2006. *China Revitalizing the Northeast: Towards a Development Strategy.* Washington, DC: The World Bank.

Wu, Weiping. 2007. "State Policies, Enterprise Dynamism, and Innovation System in Shanghai, China." *Growth and Change* 38(4): 544–566.

Wu, Qianbo and Ning, Yuemin. 2010. "Headquarter Locations of Top 500 Enterprises of Chinese Manufacturing Industries [*zhongguo zhizaoye 500 qiang zongbu quwei tezheng fenxi*]." *Acta Geographica Sinica* [*dili xuebao*]: 65(2): 139–152.

Yusuf, Shahid and Nabeshima, Kaoru. 2010. *Two Dragon Heads: Contrasting Development Paths for Beijing and Shanghai.* Washington, DC: The World Bank.
     The book explores the contrasting development options available to Beijing and Shanghai, and proposes strategies for each city based on the current and acquired capabilities of each, the experiences of other world cities, the emerging demand in the national market, and likely trends in global trade. The authors weave economic growth, urban development, and technological innovation into their analysis.

Yusuf, Shahid and Wu, Weiping. 1997. *The Dynamics of Urban Growth in Three Chinese Cities.* New York: Oxford University Press, for the World Bank.
     This book compares three cities – Shanghai, Tianjin, and Guangzhou – in the context of the changes that swept China's economy, history, and reform programs from the early 1980s through the mid-1990s. The authors consider the interplay among geography, size, and industrial structure that determines the industrial vigor of cities, concluding that each of these factors must be made to work for the city through effective policy-making.

Yusuf, Shahid and Wu, Weiping. 2002. "Pathways to a World City: Shanghai Rising in an Era of Globalization." *Urban Studies* 39(7): 1213–1240.

8   Social–spatial transformation

Much like their counterparts in other former socialist countries, Chinese cities show-cased social egalitarianism and experienced few spatial inequalities before market reform. But that is more or less a bygone era. Under economic transition, market forces increasingly dictate the shaping and reshaping of urban landscape. There is evidence that spatial locations are now important within cities. Housing and land prices change by location, in a gradient similar to that in many Western cities.

Such social–spatial transformation is the central theme of this chapter. It outlines the alteration of urban neighborhoods, jobs–housing linkage, travel modes, and spatial configuration. The chapter also identifies factors and agents driving the simultaneous processes of inner-city redevelopment and suburbanization. The following questions should help guide the reading and discussion of the materials in this chapter:

- How would you characterize the urban form of the Chinese city under state socialism?
- The so-called cellular spatial structure has basically ceased to reproduce in the post-reform urban landscape. Why?
- What are the different types of residential neighborhood now existing in the Chinese city?
- What are the main changes in the urban economic space after 1979?
- Bicycles were the king of the roads until the 1990s. No more. What are the key modes of travel in the city today? In what ways are travel and urban form interrelated?
- How is inner-city redevelopment in China similar to and different from that in the West (particularly the USA)?
- Chinese cities are expanding spatially and undergoing suburbanization recently. But this process is distinct from suburbanization in the North American context. How so?

Changing urban forms

Socialist city revisited

Under the command economy, the Chinese city looked a lot like many other social-ist cities. Different from their capitalist counterparts, these cities allowed no private ownership of land and had no land market (see Chapter 10). The principles of socialist urban development called for standardized urban form. Generally, industries and residences were planned to be separated from each other by green space but sufficiently close by to minimize work-related travel. The egalitarian concern for equal access to amenities also dictated that shopping activities and public parks be located throughout the city, made possible by the absence of a price-driven land market.

Skyscrapers were absent from city centers. There was no profit motive for their construction. Instead, large public squares stood in the centers of most cities, flanked by public buildings and monuments. Statues of Chairman Mao were a common sight too. Much of the urban landscape consisted of sprawling, low-rise development. The downtown functions associated with the headquarters of corporate economy or retail centers were almost irrelevant. Instead, city centers contained political–cultural–administrative uses as well as important residential areas. Population density in city centers tended to be high. Newly built-up areas outside the urban core had lower density (Gaubatz 1995; Yeh and Wu 1995).

With an ideological anchor in social egalitarianism, Chinese cities were less spatially segregated than cities in many other countries. There was little significant variation in housing types and standards among different parts of the city – residential heterogeneity was very low (Brunn et al. 2008; Smith 2000). Municipal public housing was often constructed in groups to form a neighborhood unit to promote social interaction. Starting in the 1950s, workers' villages were developed in close proximity to industrial areas, mainly outside the urban cores of large cities. A neighborhood or residents' committee (*juweihui*) oversaw each neighborhood or workers' village for the purposes of security and political control (Lo 1980; more in Chapter 13). In the planning of residential neighborhoods, cities used fairly uniform standards for the provision of service facilities such as schools, daycare, and retail stores.

"Work unit compounds" dominated the urban landscape. Not only were state work units the key elements of the command economy (see Chapter 6), they also were an important provider of public housing (see Chapter 10). The result was a "cellular" structure: a unique neighborhood surrounded by gates and walls demarcating boundaries of work units. Inside the sprawling compounds of three- to five-story buildings, there was a high degree of social mixture and self-sufficiency. Because of this strong tie between work and residence, pre-reform cities were mainly organized around different land uses, rather than social grouping: factory workers lived near industrial areas, whereas teachers and intellectuals were clustered near schools and universities (Gaubatz 1995; Lo 1994; Yeh et al. 1995). This model of work unit compounds went much further in integrating working and living than the Soviet model. There, the principal urban unit, called a microrayon, was based on residential areas spatially separated from the workplace. But, in Chinese cities, there was differentiation among different work units. Large units generally built better housing and utility services, whereas employees of many small units relied on municipal public-housing programs. Households' access to neighborhood-based services depended on the work units' bureaucratic ranks.

Within work units, cadre status and job standing outweighed family size (Logan and Bian 1993).

Other urbanites continued to live in old housing stock and neighborhoods inherited from the pre-1949 period, mostly in downtown areas. The quality of living environment of such neighborhoods varied significantly. For instance, in Treaty Port cities (e.g., Shanghai, Guangzhou, Xiamen, Qingdao), they ranged from garden houses in former concession areas (where foreign traders had once lived) to shack dwellings in densely populated slum areas (see Chapter 3). Given the constraints on resources devoted to housing construction, redevelopment was limited. With the growing urban population, subdivision of old housing was common. In Beijing's inner city, for instance, many old courtyard houses built for one family now accommodated multiple families. As residents scrambled for living space, infilling in the form of unauthorized buildings within central courtyards and along alleyways transformed former courtyard houses into mazes of dense, haphazard construction. Residential densities shot up in these neighborhoods. For example, parts of Shanghai's central city had some of the highest population densities in the world, in the range of 50,000–60,000 people per square kilometer (Wu 2008).

Industry was a mainstay of socialist cities. Mao's development strategy placed industrialization as the major driving force for urban growth. Together with a system of administrative allocation of land (see Chapter 10), this led to a substantial amount of urban land devoted to industrial use. Generally, industrial land occupied about 25–30 percent of the total area, a higher share than that in Western cities (Yeh and Wu 1995). The location of industries largely influenced population distribution within the city. Without the agency of rent or price indicating accessibility or amenities, planners and architects made purely technical evaluations of land, based on the engineering costs of development, sanitary conditions of the site, and the like.

Overall, cities were compact. Limited spatial expansion on the outskirts took the form of satellite towns, to promote industrial development. Cities also had reasonably convenient and inexpensive public transportation in the form of extensive bus systems. Bicycles were another dominant mode of travel, followed by walking. There was a persistence of walking-scale urban life, as work unit compounds resulted in functionally mixed districts within the city. Residents could access work and services near their homes. Given the overall emphasis on industrial production, urban services were not a priority area of public investment. Living conditions were inadequate by modern standards, and transport facilities fell far short of levels reached by other East Asian countries. Private automobiles were absent.

Transformation under market reform

Since 1979, market forces increasingly drive urban processes in China, particularly in housing and land development (see Chapter 10). There is evidence that the importance of location, which was irrelevant in socialist cities without land markets, has led to the emergence of a land rent gradient similar to that of cities in capitalist systems (Ma 2003). There is a resurgence and continuation of the pre-1949 spatial division, particularly in former Treaty Port cities. In addition, many cities have seen the formation of a small number of wealthy housing areas in the suburbs. A new trend is toward residential areas mixing housing, services, and commercial activities, but not in the same way as the old work unit compounds. Deteriorating housing conditions in city cores and real estate development propaganda encourage the urban affluent to move to such areas. Some scholars (Gaubatz 1999a, 1999b; Wang and Murie 2000) propose there are now three rings of differentiated urban space: pre-1949 historic areas, a socialist planned work unit ring (1949–1985), and the new estates ring (built during the property boom years since 1985). But in the most rapidly developing cities, this concentric pattern is disappearing as older neighborhoods are torn down and replaced with new development.

Rapid expansion of the built-up areas and population deconcentration have transformed the once compact Chinese city into a more dispersed and often poly-centric form. New development often leapfrogs. With rising income and increas-ing private cars, no doubt there is demand for more spacious living in more far-flung corners of the city. To be sure, the linkage between work and residence has weakened. It all but disappeared by 1999 when state work units completely withdrew from housing provision (see Chapter 10). The cellular structure under state socialism has since ceased to reproduce in the urban landscape. Jobs have moved beyond the urban core as well. On the outskirts of the city proper, one can find high-tech development zones, and office and industrial parks. These areas also house a large number of wholesale markets for agricultural products because of their proximity to farms in the urban periphery.

Standardization of housing, a principle of urban development under state social-ism, is no longer the norm. Instead, there is an increasing level of differentiation in residential quality across the city. The case of Shanghai is perhaps most illustrative. Using a collection of 1,604 houses transacted in 2000, Fulong Wu (2002b) identi-fies significant variation of housing price in the city, similar to how urban space was divided into upper and lower ends before 1949. Semicolonial rule (1840s–1949) in no small way delineated where the upper ends were – in the French Concession and International Settlement, located both in and near the city core. Shantytowns were located along the boundaries of foreign settlements and in areas designated for Chinese residents (Lu 1995; see Chapter 3). Between 1949 and 1979, residen-tial differentiation markedly declined after years of building public housing and redeveloping shanty areas. Today, however, urban social space is becoming more heterogeneous, demarcated by education level and employment status.

There is also unprecedented residential mobility within the city. Urban resi-dents no longer are tied to their places of employment. Spatially, many residents now have the freedom to decide where they wish to live, leading to the resorting of population in the city. Housing commodification itself becomes a source of sociospatial differentiation (see Chapter 10). With the real estate market, house-holds are more likely to select their place of residence based on price and income levels. Of course, not all residents can fulfill their residential preferences. House-holds displaced through redevelopment, for instance, are very unlikely to return, as a result of rising property prices. Often unwillingly, they relocate to distant areas of the city and commute long distances to work.

Urban China now contains differentiated communities that can be called social areas. Western cities are commonly demarcated by socioeconomic status, family status, and ethnicity. But in the Chinese city, the main determinants of social space are more likely to be education, employment, income level, and *hukou* status. We begin to see zones of affluence and gated communities, mostly on the outskirts but also in select redeveloped downtown areas. Areas of concentrated deprivation also arise. First among these are pockets of dilapidated old housing in the inner city. Often they are awkwardly juxtaposed with modern, high-rise apartments and offices, since redevelopment guided by real estate markets is selective. Second are some of the former workers' villages close to industrial facilities, suffering from lack of upkeep and state enterprise downsizing (Feng and Zhou 2008; Wu 2004a; Yeh et al. 1995). Last are places in the urban fringes where migrants congregate. In many large cities, these areas are located just outside the city proper, as exemplified by the distribution of migrants in Shanghai (Figure 8.1).

Amid such ongoing social–spatial transformation, a widening range of neighborhoods now exist in urban China: traditional ones in the old core area developed before 1949; work unit compounds, largely associated with industrial development, built between 1949 and 1979; mixed-use suburban communities or satellite towns, developed from the late 1970s on; and rural–urban fringe or "urban villages" (*chengzhongcun*) formed more recently (Wu 2002a). Dotted among these are an increasing number of gated communities, catering to the more wealthy residents. Later parts of this chapter contain more detailed discussion of inner-city redevelopment, suburban and gated communities, and "urban villages."

Work and accessibility

Remaking of the urban economic space

Transformation is no less drastic in urban economic space. Cities have become engines of growth in China's rapid rise in the global marketplace (see Chapter 6). They have turned into the world's factory for producing low-cost consumer goods. Business and consumer services non-existent before 1979 have mushroomed. In the meantime, the downsizing of state-owned enterprises and rise of private enterprises (see Chapter 7) all have substantial imprints on urban spatial development. The major trend in the reconfiguration of urban economic space is the increasing separation and specialization of land use, including the relocation of industrial facilities from city cores, the creation of new spaces of production in the urban periphery, the building of the new central business district (CBD), and multiplying clusters of commercial and entertainment activities across the city.

Relocating factories from the central city to the periphery is a key method to solve problems associated with fragmented industrial land use (e.g., environmental pollution and inefficient infrastructure supply). This process, albeit slow and with mixed results, has freed up a significant amount of space for other land uses in the core.

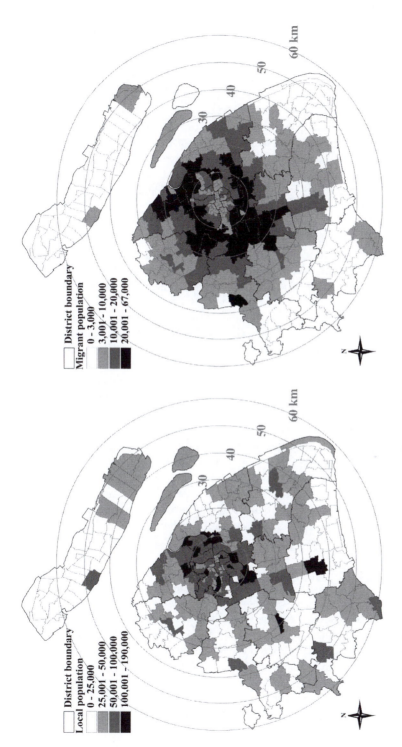

Figure 8.1 Distribution of local and migrant population in Shanghai, 2000.

Source: SFPCO 2002.

For instance, in Shanghai, between 1991 and 2004, land allocated to industrial use in the central city decreased by 42 percent, from 45 to 26 square kilometers (World Bank 2008). Increasingly, industries are located in the outskirts. Industrial relocation also may have contributed to residential decentralization, as population tends to increase in areas where major industrial development is in progress (Walcott and Pannell 2005). Industrial land use in the urban fringes remains fragmented, however, aggravated by the haphazard location of township and village enterprises.

To appeal to foreign investment and international businesses, many cities have built brand new industrial districts outside the inner city, variously called Economic and Technology Development Zones, Special Development Zones, High-Tech Development Zones, or science parks (more in Chapter 6). These zones often allow flexible planning control or virtually autonomous rights of land subdivision, and concession in land premium (Zhu 1994; Gaubatz 1999a; Wu and Yeh 1999), as the provision of cheap land is a basic instrument for local governments to induce foreign investment. Nearly all have been constructed on green-field sites, with modern office and factory facilities. However, planners have a rather passive position in dealing with foreign investors, who often negotiate directly with senior government officials.

Parallel to such new spaces of production is the birth of a new wave of CBDs in large cities. They house a variety of business activities and, most importantly, financial and business services that are the backbone of other major world cities. Often, such CBDs are brand new areas developed outside the traditional core (Box 8.1). In Shanghai, a new CBD can be found in the Pudong area across the Huangpu River from the pre-1949 CBD. In Beijing, several new districts have been built, vying for position as the pre-eminent CBD, including one area called Beijing CBD in the southeastern corner of the city near the foreign-embassy district. Harbin's new CBD is located to the north of the Songjiang River; in Kunming, a new downtown has been constructed east and south of the traditional city center. Much like "zone fever" (Chapter 6), there has been a rush to build CBDs, to appeal to global businesses and to cash in on the nation's real estate surge. The design of such districts and individual buildings often involves expensive international competitions and high-profile foreign architectural firms, the so-called "global intelligence corps" (Olds 1997). Experts from industrialized countries, such as the USA, France, the

---

***Box 8.1* Catching the CBD fever: Harbin's new downtown**

Much like the rest of the world, the Chinese city has caught the wave of mega-projects, all in the name of promoting local development and attracting global investment. Besides ambitious efforts to host world events (e.g., Olympic Games in Beijing and World Expo in Shanghai), building a new CBD has become a form of strategic marketing across urban China. Large cities on the east coast, such as Shanghai and Beijing, spearheaded the trend and created the likes of Wangfujing (central Beijing) and Lujiazui (in Shanghai's Pudong district). Cities below the first tier, and particularly those in the interior regions, have subsequently emulated this development model produced in the larger, more globally connected cities of the coast (Gaubatz 2008).

Few CBD projects are as ambitious as the one planned in Harbin, a north-eastern city in the historic Manchuria region. In early 2004, Harbin officials won approval to build a new city center called Songbei (literally: "to the north of Songjiang River"), a 740 km² area for residential high-rises, office towers, luxury villas, five-star hotels, shopping and entertainment complexes, trade zones, and industrial parks. To local developers, this was the most important thing that had happened to Harbin in a long time (Barboza 2005). The Songbei project, however, is as much a sign of the city's desperation as a symbol of its hope. Economic development in northern cities like Harbin is lagging far behind the booming coastal cities. The former manufacturing base is obsolete, droves of state workers have been laid off, and only a handful of multinational companies are willing to invest there.

For Harbin's new CBD, city officials lured investors like Stanley Ho, a Macao gambling tycoon, and China Poly Group, a state-owned military equipment maker with real estate and entertainment holdings. City officials and developers dubbed Songbei the "Pudong of the north" (Barboza 2005), after Shanghai's Pudong district, the financial center and economic power-house that rapidly developed during the 1990s. While too early to call the Songbei project a success (or failure), its development clearly is another example of how local governments collude with domestic and international business interests.

UK, Italy, and Japan, also participate in the planning process. The outcomes are often ultramodern glass towers of offices, hotels, and high-end apartments.

Another important force reshaping urban economic space is commercial development. Under state socialism, all cities on the coast, particular former Treaty Ports, had to demonstrate their contribution to the national economy as industrial producers. Their former commercial status associated with foreign participation and the use of local population in a service and consumer role were not acceptable to the Communist leaders (Yusuf and Wu 1997). As such, they ceased to be the financial, trade, and business centers. They ceased to orchestrate the regional commercial activities with a nested hierarchy of markets. On top of this, the priority of national development was on heavy industry and producer rather than consumer goods. Consequently, there was a dearth of commercial activities in cities and the urban core. Even the largest cities, such as Beijing and Shanghai, had only one or maybe two major commercial centers serving millions of people. Nearly all of the stores in such centers were state-owned, even the large department stores. With the exception of small shopping facilities in various planned neighborhoods, there was no other clustering of commercial, service, and entertainment establishments. The typical Chinese city, at that time, was basically monocentric or lacked any sort of center.

Market reform, however, has brought about a "consumer revolution" (Davis 2005: 692). With rising income and prosperity, the Chinese city has seen a giant wave of commercial development. Nearly all cities have gone on a building spree to add subcenters throughout old and newly developed urban districts. These centers are primarily commercial in character, with shops and offices displacing residential and industrial space. Beginning in the 1990s, global mass retailers (also called large box stores), such as Carrefour (French), Walmart (USA), and IKEA (Sweden), have broken into the Chinese market. More recently, high-end foreign department stores, including ISETAN and SoGo (both Japanese), Lane Crawford (Hong Kong), Marks & Spencer (UK), Parkson (Malaysia), and Printemps (France), have set up shop. Multiplex shopping and entertainment establishments resembling malls in the West have proliferated, albeit in more downtown locations and with a more compact layout. In the luxury goods sector, nearly all of the European brands have a presence (Figure 8.2). Today, the commercial landscape in the Chinese city is clearly polycentric and highly differentiated. Urban consumers from all walks of life can easily find what suits them in different subcenters, ranging from mass-market brands (both Chinese and foreign) to mid-market staples to brand-conscious luxury goods. Consumption is a big business.

Figure 8.2 Plaza 66 on West Nanjing Road, Shanghai.

Source: Weiping Wu.

Note: The structure in the middle (in 2011) resembles a hyper-enlarged Louis Vuitton trunk with the recognizable LV logo. On 21 July 2012, the company's largest store in China with an invitation-only private floor opened at this site with much fanfare.

In a parallel fashion, small businesses also are making their presence seen. At the neighborhood level, the commercial scene is a blend of modern shopping and small vendors (Figure 8.3). The latter cater to the everyday needs of nearby residents and passing commuters, selling Chinese-style fast food, fruits and snacks, small household tools, and toiletries. Then, there are the typical hair salons, massage parlors, and bicycle and motorcycle repair shops. These shops line up along the street-front space of older residential buildings downtown, or cluster around the edges of new residential compounds. Further out in the urban peripheries, migrants are a major force behind such small businesses. These areas do not have the same level of established services as the downtown. This may actually be an attraction for many migrants: opportunities to open small businesses are more abundant and competition with established local establishments is less fierce. Often, the clientele are migrants themselves, particularly for street vendors selling regional dishes. The street scene is lively and bustling.

Traveling the city

Urban transport policy, until the 1990s, encouraged walking and cycling. This policy was consistent with strict government controls on vehicle ownership, a denser pattern of urban development, and the cellular structure centered around work unit compounds. Many workers lived within a reasonable cycling distance (about 5 km) of their workplace. In the 1950s and 1960s buses carried a large proportion of urban trips. The use of bicycles began to rise in the late 1970s after economic reform provided people with a higher disposable income. In the 1980s, bicycle ownership increased explosively, especially in the cities. Of all bicycles registered in 1982, 40 percent were used in urban areas, with an average ownership

Figure 8.3 Urban commercial landscape: street scene.

Source: Joseph Hennen (with permission).

of one bicycle for every three people. This rose to one for every two persons in 1990 – almost every adult owned a bicycle. The majority of workers use them for their daily trips to work. By then, bicycles counted for more than half of urban trips, followed by walking, as indicated in a ten-city survey (Welleman et al. 1996). In cities of moderate density and ample streets, bicycles provided an inexpensive and convenient mode of transportation.

Today, Chinese cities are becoming less pedestrian and bicycle-oriented. The trend is clearly towards more motorized travel, as seen in the case of Shanghai. Non-motorized modes, walking and cycling, counted for just over half of travel in 2004; in 1982, they counted for 70 percent (Figures 8.4 and 8.5). There is evidence of similar changes in other large cities, such as Beijing, Chengdu, Hangzhou, and Ningbo (Figure 8.5). A multitude of factors are at work: the disassociation of work and residence, expansion of urban road networks, emergence of sprawl-style development, reduction in development density outside the urban core, rise of disposable income and car ownership, and construction of limited-access highways that do not permit bicycles. Road construction now receives the largest share of infrastructure spending, and per capita paved urban road space increased steadily between 1988 and 2008 from 3.1 to 12.2 square meters (National Bureau of Statistics of China 2009). Most cities have developed a ring-road structure to provide circulation, mobility, and access to all parts of the city. At the end of 2003, about

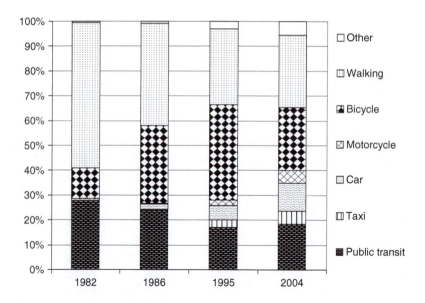

Figure 8.4 Changing modes of travel in Shanghai, 1982–2004.

Source: Association of Mayors of China 2009.

Note: The "car" category includes both private cars as well as employer and other vehicles.

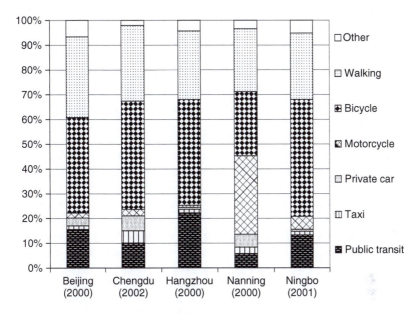

Figure 8.5 Modes of travel in other cities.

Source: Association of Mayors of China 2006.

11 percent of China's urban area was devoted to roads, compared to 20–30 percent for most industrialized nations (Cherry 2005).

More motorized travel has created unprecedented levels of traffic congestion and increase in travel time. Most transportation agencies rely heavily on supply-side approaches such as road capacity expansion and, in effect, encourage automobile transportation. The disassociation of work and residence and the increasing freedom of people to choose where to live also are creating more serious spatial jobs–housing imbalance that did not exist under state socialism. New residential developments are farther from employment opportunities; similarly, they are more distant from centers of shopping, entertainment, educational, and cultural facilities; so journeys, especially commuter trips, have become longer. For instance, average trip distances by all modes in Shanghai increased from 4.9 km in 1995 to 6.9 km in 2004. Many more trips, at longer distances, originate from outside the urban core (World Bank 2008). In Guangzhou, the average commuting distance was 4.6 km in 2001 and increased to 5.7 km in 2005 (Li 2010). But, in Beijing, commuters in formerly work unit housing are more likely to be working and residing in the same district than market-based housing commuters. They also are more likely to use non-motorized transport modes and spend less time commuting (Wang and Chai 2009).

Large cities along the east coast, in particular, have been experiencing rising motorization rates, at least twice as high as in inland provinces. Automobile ownership in Beijing, reaching 131 vehicles per 1,000 people in 2006 and the highest across China, is five times more than the national average. Trailing Beijing are

Shenzhen and Guangzhou. Shanghai, on the other hand, has a vehicle ownership of less than one-third that of Beijing (around 65 vehicles per 1,000 people), arguably reflecting local policies that restrict the number of license plates and certain vehicle types, and promote public transport options. In general, all cities have seen increases in motorization, in both absolute numbers and the ratio of private motorized vehicles over total fleet (Schipper and Ng 2004; Darido et al. 2009). Higher disposable incomes certainly are a major factor, as is individual desire for mobility. Perhaps more importantly, the Chinese government has promoted car manufacturing as a key industry and engine for economic growth. Every major international automaker has production facilities in China. Proposals to restrain car ownership and usage have encountered stiff resistance at nearly all levels of government.

The worsening urban transport problems are becoming an environmental liability (see Chapter 11). Increasing levels of motorization, mixed use of road, inadequate road maintenance, and shortage of travel management all contribute to frequent gridlock and excessive travel time throughout the day on most main corridors leading to central business areas and main industrial parks. This results in loss of work and leisure time, elevated fuel consumption and emissions, and high accident rates. Compounding the situation is a lack of consideration of environmental impacts for many urban transport projects.

Chinese cities, however, may have the opportunity to integrate transport and land use planning in a sustainable manner. Cities in China are generally denser than their Western counterparts, and this is associated with a higher usage of public transport and lower energy consumption per capita. Differences in gas prices, income, and vehicle efficiency explain only some of these variations. What is significant is the urban structure: cities with high concentrations of jobs and good public transport systems have much lower energy use than cities where jobs are scattered. Even at the present level, urban roads in China can be used more efficiently through careful planning and sophisticated traffic management systems. Feasible actions may include: developing a logical road hierarchy, improving street and junction layouts, segregating major traffic modes, automating and linking traffic signal systems, and reducing conflicting traffic movements (World Bank 1999). Shanghai provides an effective case of such actions. Implementing nearly all of these with the assistance of international lending agencies, the major downtown sections of the city have significantly reduced traffic congestion. On top of these, a network of more than ten subway lines and light rail provides an efficient alternative for travel.

Because of the increasing congestion on roadways and the decreasing effectiveness of mixed-flow bus service, Chinese cities have been looking to improved exclusive right-of-way transit solutions (e.g., heavy rail, light rail, and bus rapid transit). Many large cities have developed or are developing plans to invest heavily in bus rapid transit, light rail transit, subway, or some combination of those technologies. The central government allows subway systems to be built only in cities with a population greater than 3 million. Consequently, all 15 such cities have preliminary plans to proceed (Cherry 2005). By 2011, ten cities had operating subways. However, of a total of 46 active lines, 39 were located in just four cities: Beijing, Shanghai, Guangzhou, and Shenzhen. Other cities are concentrating their mass transit investment in lower-cost light rail systems.

Spatial reconfiguration and expansion

Inner-city redevelopment

After 30 years of disinvestment (1949–1979), urban core areas were rundown. There were chronic problems of residential overcrowding, dilapidated housing conditions, and high population density. Starting in 1979, especially after the deepening of reform in 1992, urban redevelopment began to speed up. Central areas, previously residential, were increasingly under pressure for redevelopment, often for commercial and office uses and in the form of high-rise buildings. There were also relocation of factories and redevelopment of formerly industrial land at premier central locations. Prompted by the desire to make city centers a haven for finance and business uses, municipal governments have actively facilitated these shifts. These changes have transformed the skyline of the Chinese city, with significant growth in the vertical dimension (Gaubatz 1999a, 1999b).

Urban redevelopment involves extensive demolition of "vernacular urban residential fabric" (Campanella 2008: 145). For instance, in Nanjing, redevelopment has left intact only a handful of its once numerous historic neighborhoods that originated from the Ming, Qing, and Republican eras. In Beijing, the central city was characterized by narrow alleyways, also known as *hutong*, whose origin dated back to the Yuan dynasty. Together with courtyard houses, they filled the residential space within the old city walls. While overcrowding and dilapidation overshadowed their charm, they were among the last remnants of traditional Chinese cities. But preservationists fought a losing battle against local officials and real estate developers. Between 1990 and 2002, at least 40 percent of Beijing's old city was demolished (Campanella 2008). Preparation for the 2008 Olympic Games served as another rationale for expedited redevelopment.

Redevelopment also has become a vehicle to promote heritage tourism and, in some cases, residential exclusion. For historic neighborhoods and areas spared from demolition, preservation really has less to do with restoring the buildings than it has to do with making them new tourist attractions (Figure 8.6). Beijing's *hutong* tours now are a must-see for foreigners, who get a narrow glimpse of the once splendid traditional neighborhoods. Wealthy Chinese and foreign expatriates have bought out and restored dilapidated courtyard houses, turning them into luxury residences, restaurants, or clubs. Beijing's municipal government also has tagged along, by issuing an official circular and using tax breaks and incentives to entice more such restorations (Campanella 2008). Going even further is Shanghai's *Xintiandi* (New Heaven and Earth) project. Completed by Hong Kong's Shui On Group and designed by an American architect, *Xintiandi* incorporated key elements of *shikumen* architecture (row houses built in the city's concession areas during the semicolonial period of 1840–1949). With its seemingly old

Figure 8.6 Heritage tourism in urban China.

Source: Michael F. Crowley (with permission).

look, high-end shopping and restaurant offerings, and location around the historic site of the Chinese Communist Party's founding meeting, *Xintiandi* is now a major tourist destination and a success story in urban redevelopment in the eyes of Chinese municipal officials. Its commercial appeal also has helped push up real estate prices in the surrounding area.

A great deal of displacement has occurred with redevelopment. Over the years compensation practice has changed from in-kind to monetized forms. Early on, residents displaced from redeveloped neighborhoods were compensated in accordance with the State Council's Ordinance on the Management of Urban Housing Demolition and Relocation (Shin 2007). Developers were to provide rehousing or relocation housing elsewhere, guaranteeing the continuation of residents' existing tenure in a relocation dwelling. Such practice led to high project costs and low profitability. In fact, it became more and more difficult for remaining areas of shack dwellers to attract real estate development if the standard of compensation was maintained. A major revision took place in the late 1990s. Compensation became monetized, based on two factors – the number of registered household members and formal dwelling space. Informal or self-built space was in principle not eligible for compensation.

There are also shifts in government policy toward redevelopment. In the early 1990s, the emphasis was on modest urban renewal to improve the quality of the housing stock and living environments of residents. But increasingly, vision for urban renewal changed to embrace a wider agenda for full-scale redevelopment. Given the lack of public resources for such redevelopment, local governments have resorted to market forces and private-sector resources. The process begins to bear some resemblance to gentrification in many Western cities where capital moves in and out of the built environment to yield profits. Developers thus have become key drivers of social–spatial reconfiguration of the Chinese city, through selected development of upper-market properties in prime sites and low-end uses in other areas (Tian and Wong 2007). New commercial housing is being built according to different standards, prices, and locations.

Inner-city redevelopment, moreover, is an important mechanism of population decentralization. Displaced residents often face acute affordability problems and cannot afford new homes in their redeveloped neighborhoods. They purchase either subsidized resettlement housing or market-based commercial units on the urban fringes, having to move outside the inner city. Resettlement housing tends to be of lower quality and in more remote locations. Hence, inner-city redevelopment also creates the demand for new suburban residential development. This effect is particularly immediate for second-tier cities, such as Chengdu, Wuhan, and Nanjing, where demand for new housing mainly comes from local buyers (Hsing 2010).

Urban redevelopment in China, in general, is different from the process of gentrification and displacement in the West (particularly in the USA). China's central cities are in no danger of irrelevance and decline. If anything, they are prime real estate. They are being redeveloped because of the enormous demand for downtown living and commercial activities. Such redevelopment is rarely the product of government subsidies; rather, private developers and quasi-public entities are the driving force. Given the large population and high density in inner cities, the scale of redevelopment also is much larger, particularly when in preparation for such mega events as the Olympic Games and World Expos. But the social impact is nonetheless similar: residents relocated far from city centers, work and social connections stretched in distance, and vernacular residential fabric replaced by irreverent structures. Rising social discontent, stemming from losing residents' right as a result of forced eviction, has led some cities to begin incorporating public participation in the process of redevelopment decision-making (see Chapter 13).

Suburbanization

Sprawl, in the form of suburban development, is one of the hallmarks of the Chinese city today. Major cities, both on the coast and inland, have experienced lateral, spatial expansion. Increasingly, new economic functions are concentrated on the outskirts, as well as substantial housing construction in new suburban areas and satellite towns. Compounding this process of spatial expansion is the large influx of rural–urban migrants, who have settled primarily outside the urban core. Hence, in the urban fringes there is an increasing juxtaposition of high-tech zones, new commercial housing projects, resettlement housing for central-city residents, migrant communities, and rural villages (Wu 2002b). Contemporary Chinese suburbanization, however, should not be confused with the processes of suburbanization taking place in North America during the latter half of the twentieth century. Because Chinese municipalities administer vast areas, suburbanization and sprawl take place within cities, rather than drawing population and resources outside municipal jurisdictions. Moreover, much Chinese suburbanization comprises pockets of high-rise dense development, rather than large tracts of single-family homes.

The process of suburbanization began in the 1990s. Studies of Beijing, Shanghai, Shenyang, and Dalian show that, between 1982 and 1990, inner-city population first experienced decline, while population in inner suburbs rose rapidly (Zhou and Meng 1997; Zhou and Ma 2000). This was a significant departure from controlled urban development under state socialism. The only exception then was the satellite-town program, launched after the 1950s and primarily for the purpose of industrial development. Another function of satellite towns in some large cities was to reduce excessive concentration of industries in city centers. They were usually built in outer suburbs, more as isolated development, and were less successful in attracting residents given the inferior living conditions and services. In other words, they ought to be seen as a form of suburban development instead of suburbanization (Zhou and Ma 2000). Today's suburbanization, on the other hand, is going at full throttle and underscored by a multitude of driving forces: employment expansion, residential dispersion, rural–urban migration, and a pro-growth development agenda that favors the transformation of central city areas from low-rent residential use to "high-rent" commercial land use.

Employment expansion into the suburbs has happened on several fronts. Industrial restructuring (as discussed in Chapter 7), for instance, has been accompanied by land use policies inducing a shift of industry away from the urban core and permitting mixed commercial and residential uses on prime urban land. This also stems from a better understanding on the side of municipal agencies of the problems of industrial pollution. Relocation of manufacturing and warehousing facilities leads to an industrial concentration on the outskirts. Many cities also have converted outlaying land to build industrial or high-technology development zones, hoping to attract new investment (as seen in the case of Guangzhou, in Figure 8.7). Such functional reorganization in urban land use is further facilitated by new transportation infrastructure, often in the form of ring roads, subway or light rail, and bus rapid transit.

Residential dispersion not only is a byproduct of employment expansion but also a response to the increasingly marketized land and housing sectors (see Chapter 10). In fact, the rising land market has paved the way for suburbanization. Land on the urban fringes is much cheaper: a study of Shenyang found that land prices in the outskirts were one-tenth of those in the city core (cited in Campanella 2008). Developers are quick to take advantage of the low prices there to construct more affordable housing units. As such, many areas in the inner suburbs have experienced rapid transition from rural to urban uses (often referred to as rural–urban transitional areas or *chengxiang jiehebu*). These areas were left undeveloped prior to market reform, a legacy of underurbanization during the era of state socialism. In addition to mid- and high-rise apartment buildings, suburbs house an increasing number of gated communities for the rising middle class (Box 8.2). Many have

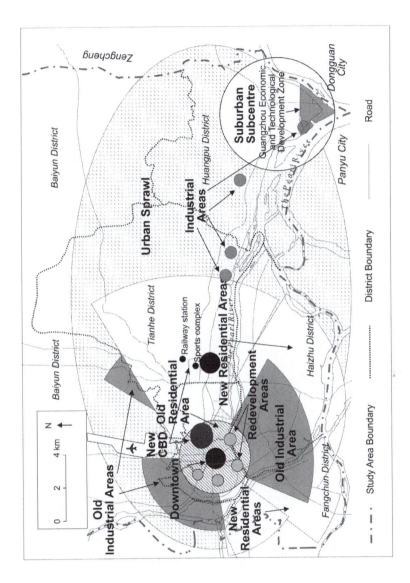

Figure 8.7 Urban expansion in Guangzhou.

Source: Wu and Yeh 1999 (with permission).

***Box 8.2* Gated communities on the rise**

In the West, gated communities are built for their prestige and exclusivity, a result of social polarization and segregation. Often, they also stem from the discourse of fear and private provision of services. Gating, on the other hand, has been a long-standing practice in China. Many residential complexes, including traditional housing built before 1949, work unit compounds and workers' villages built under state socialism, and new commercial housing built during the reform era, are walled and gated in one way or another (Huang 2005). Gating for these residences conveys a sense of boundary as well as neighborhood, and rarely is a result of social segregation.

Exclusive residential estates resembling gated communities in the West, however, have emerged in urban China recently. While most gated communities are located in the suburbs, there are a small number of such estates in central districts. These estates often contain unique architectural design, exclusive amenities and a luxury lifestyle, catering to the rising segment of population with higher socioeconomic status. Some particularly try to emulate foreign or Western building and living styles, as revealed by such names as "Continental European Style," "Orange County Project," and "Thompson Golf Villas." They form "transplanted cityscapes" emerging in urban China (Wu 2005: 295). By some estimates, luxury gated communities now number in the hundreds: more than 250 in Beijing, more than 150 in Shanghai, and more than 50 in Chengdu, the capital city of Sichuan province (Giroir 2007).

An extreme form of luxury estate also combines exclusive living with leisure activities, such as gated golf communities. While small in number, they consume an excessive amount of land and water in large cities already strapped in such resources. As a result, they have become the subject of intense public debate, even in the official media. But isolated local government efforts to rein in such developments have not made much difference, and the number of gated golf communities continues to grow (Giroir 2007).

Western names and building styles – Rose Garden Villas, Park Avenue, Orange County (all three are in Beijing), South California (in Guangdong), and Cambridge Impression (in Wuxi). Other such communities bank on the cachet of traditional courtyard homes. Cathay View, in Beijing, features homes in the range of US$800,000 with many of the architectural details of Ming and Qing courtyard houses. Developers see this as offering an authentic cultural experience; residents see this as demonstrating their own status (Campanella 2008).

A pro-growth development agenda also is driving urban expansion, as local governments face the need to generate more revenue under fiscal decentralization (see Chapter 9). The adoption of a two-tier structure of urban management in large cities has led to a highly aggressive role for district governments (see Chapter 13). They have gained substantial power to regulate development, including project approval and registration, and issuing of planning and building permits and land-leasing certificates. They also have a number of regulatory functions such as approval of residential and housing development plans, site occupation licens-ing, and levying penalties for illegal construction. With the authority of managing local revenue and land, many district governments turn into business partners with real estate and other companies. Available land for development, as a result, has become an important source of revenue for suburban district governments.

Another driver of spatial expansion is rural–urban migration. At the turn of the 1980s, when the migrant influx first began, the central city was the chosen residential location of most new arrivals. But with downtown redevelopment and higher-end real estate development, central-city housing is more costly and less attractive to migrants. Inner suburbs have become a more important receiving area for migrants. The spatial distribution of migrants, as a result, has undergone a gradual shift. This mirrors a trend in many cities elsewhere in developing coun-tries undergoing continuing urbanization (Wu 2008). The shift also coincides with the decentralization trend seen in local population. As such, inner suburban areas immediately flanking the central-city boundary are now residential cent-ers for both migrants and, to some extent, locals (Figure 8.1). "Urban villages" have emerged there as well, as urban expansion gradually engulfs formerly rural areas (see Chapter 5). Although municipal agencies attempt to impose control over "urban villages," chaotic land use, crowding, and irregular rental activities persist.

Despite suburbanization, the urban core remains the preferred place to live and work. To some extent, suburbs are subordinate to the central city in their func-tions. Hence, this is not a scenario seen in the North American context where sub-urbanization has come at the expense of central-city decay (Zhou and Ma 2000). Beyond the compact core, however, development tends to be non-contiguous and leapfrog. Given the long period of growth ahead, such development patterns will translate into large losses in terms of land consumption. A World Bank study (2008) shows that the intensity of land use in several large cities is actually quite low by international standards, even though it has increased over time. In particu-lar, land allocated to industrial use is between two and three times that in compa-rable global cities elsewhere with functioning land markets.

Conclusion

The Chinese city today looks and develops in drastically different fashion from the once egalitarian, low-profile, and walking-scale socialist city. With work unit compounds no longer functioning as the organizing units, the linkage between work and residence has all but disappeared. Commutes are longer and travel more motorized. Location is now more important, urban districts more specialized, and social–spatial differentiation more profound. Inner-city redevelopment has not only altered the skyline but also permanently displaced some of the vernacular residential fabric. Simultaneously, employment expansion and residential dispersion have powered a new-found process of suburbanization. Given China's unprecedented rate of economic growth and pace of urbanization, spatial expansion of cities will likely continue steadily. With an already intense population-to-land ratio, following the path of suburbanization and urban sprawl, as seen in some industrialized countries, really is not an option. Increased land use efficiency can come from more infill development, higher density, and more contiguous expansion.

Bibliography

Association of Mayors of China. 2006 and 2009. *China Urban Development Report 2005 and 2008*. Beijing: China City Press.

Barboza, David. 2005. "In China, A Drive to Uproot the Rust Belt." *New York Times* Friday, November 25.

Brunn, Stanley D., Hays-Mitchell, Maureen, and Zeigler, Donald J. (eds) 2008. *Cities of the World: World Regional Urban Development*, fourth edition. Lanham, MD: Rowman & Littlefield.

Campanella, Thomas J. 2008. *The Concrete Dragon: China's Urban Revolution and What it Means for the World*. New York: Princeton Architectural Press.
   The book surveys the driving forces behind the great Chinese building boom, traces the historical precedents and global flows of ideas and information that are fusing to create a bold new Chinese cityscape, and considers the social and environmental impacts of China's urban future. It is an accessible book for students and scholars with limited or little prior expertise in urban China, with interesting and well-researched vignettes throughout.

Cherry, C. 2005. "China's Urban Transportation System: Issues and Policies Facing Cities." Working Paper 20054. UC Berkeley Center for Future Urban Transportation. Berkeley, CA: University of California.

Darido, Georges, Torres-Montoya, Mariana, and Mehndiratta, Shomik. 2009. "Urban Transport and CO_2 Emissions: Some Evidence from Chinese Cities." Working Paper 55773. Washington, DC: The World Bank.

Davis, Deborah. 2005. "Urban Consumer Culture." *The China Quarterly*, 183(September): 677–694.

Feng, Jian and Zhou, Yixing. 2008. "Restructuring of Socio-Spatial Differentiation in Beijing in the Transition Period [*zhuangxingqi Beijing shehui kongjian fenyi chonggou*]." *Acta Geographica Sinica* [*dili xuebao*] 63(8): 829–844.

Gaubatz, Piper. 1995. "Urban Transformation in Post-Mao China: Impacts of the Reform Era on China's Urban Form," in Davis, Deborah, Kraus, Richard, Naughton, Barry, and Perry, Elizabeth (eds) *Urban Spaces in Contemporary China: The Potential for Autonomy and Community in Contemporary China*. Washington, DC: Woodrow Wilson Center Press and Cambridge University Press, pp. 28–60.

Gaubatz, Piper. 1999a. "China's Urban Transformation: Patterns and Processes of Morphological Change in Beijing, Shanghai and Guangzhou." *Urban Studies* 36(9): 1495–1521.

Gaubatz, Piper. 1999b. "Understanding Chinese Urban Form: Contexts for Interpreting Continuity and Change." *Built Environment* 24(4): 251–270.

Gaubatz, Piper. 2008. "Commercial Redevelopment and Regional Inequality in Urban China: Xining's Wangfujing?" *Eurasian Geography and Economics* 49(2): 180–199.

Giroir, Guillaume. 2007. "Spaces of Leisure: Gated Golf Communities in China," in Wu, Fulong (ed.) *China's Emerging Cities: The Making of New Urbanism*. London: Routledge, pp. 235–255.

Hsing, You-tien. 2010. *The Great Urban Transformation: Politics of Land and Property in China*. Oxford: Oxford University Press.

Hu, Xiuhong and Kaplan, David H. 2001. "The Emergence of Affluence in Beijing: Residential Social Stratification in China's Capital City." *Urban Geography* 22(1): 54–77.

Huang, Youqin. 2005. "Collectivism, Political Control and Gating in Chinese Cities." Unpublished paper.

Li, Si-Ming. 2010. "Evolving Residential and Employment Locations and Patterns of Commute under Hyper Growth: The Case of Guangzhou, China." *Urban Studies* 47(8): 1643–1661.

Lo, C.P. 1980. "Shaping Socialist Chinese Cities: A Model of Form and Land Use," in Leung, Chi-Keung and Ginsburg, Norton (eds) *China: Urbanization and National Development*. Research Paper No. 196. Chicago: University of Chicago, Department of Geography, pp. 130–155.

Lo, C. P. 1994. "Economic Reforms and Socialist City Structure: A Case Study of Guangzhou, China." *Urban Geography* 15: 128–149.

Logan, John R. and Bian, Yanjie. 1993. "Inequalities in Access to Community Resources in a Chinese City." *Social Forces* 72(2): 555–576.

Lu, Hanchao. 1995. "Creating Urban Outcasts: Shantytowns in Shanghai, 1920–1950." *Journal of Urban History* 21(5): 563–596.

Ma, Lawrence J. C. 2003. "Some Reflections on China's Urbanization and Urban Spatial Restructuring." Paper presented at the *Urban China Research Network Workshop on Urban Studies and Demography in China*, April 30, Minneapolis, MN.

National Bureau of Statistics of China. 2009. *Statistical Summary of Sixty Years of New China 1949–2008*. Beijing: China Statistics Press.

Olds, Kris. 1997. "Globalizing Shanghai: The 'Global Intelligence Corps' and the Building of Pudong." *Cities: The International Journal for Urban Policy and Planning* 14(2): 109–123.

Schipper, Lee and Ng, Wei-Shiuen. 2004. "Rapid Motorization in China: Environmental and Social Challenges." Paper commissioned for the ADB-JBIC-World Bank East Asia and Pacific Infrastructure Flagship Study (retrieved on 9 January 2012 from http://www.transportroundtable.com.au/courses/201008_hsst/references/schipper-bank-China_Motorization.pdf).

Shanghai Fifth Population Census Office (SFPCO). 2002. *Tabulation on Shanghai's Fifth Population Census [shanghaishi diwuci renkou pucha zilao]*. Shanghai, China: Shanghai People's Press.

Shin, Hyun Bang. 2007. "Residential Redevelopment and Social Impacts in Beijing," in Wu, Fulong (ed.) *China's Emerging Cities: The Making of New Urbanism*. London: Routledge, pp. 163–184.

Smith, Christopher J. 2000. *China in the Post-Utopian Age*. Boulder, CO: Westview Press. Written by a geographer, this book is a comprehensive textbook on China beyond the scope of spatial sciences. It covers additional subjects such as cultural change, population and health, agrarian reform, inequality and poverty, gender issues, and state–society relationships. The bulk of the book has an urban focus, with limited discussion on the morphological changes in cities. The historical context is confined to after the founding of the People's Republic of China in 1949.

Tian, Ying Ying and Wong, Cecilia. 2007. "Large Urban Redevelopment Projects and Sociospatial Stratification in Shanghai," in Wu, Fulong (ed.) *China's Emerging Cities: The Making of New Urbanism*. London: Routledge, pp. 210–231.

United Nations Center for Human Settlements (UN-HABITAT), with China Science Center for International Eurasian Academy of Sciences and China Association of Mayors. 2010. *State of China's Cities 2010/2011: Better City, Better Life*. Beijing: Foreign Languages Press.

Walcott, Susan M. and Pannell, Cliff W. 2006. "Metropolitan Spatial Dynamics: Shanghai." *Habitat International* 30(2): 199–211.

Wang, Donggen and Chai, Yanwei. 2009. "The Jobs–Housing Relationship and Commuting in Beijing, China: The Legacy of *Danwei*." *Journal of Transport Geography* 17(1): 30–38.

Wang, Yaping and Murie, Alan. 2000. "Social and Spatial Implications of Housing Reform in China." *International Journal of Urban and Regional Research* 24(2): 397–417.

Welleman, Anton G., Louisse, Cees J., and Ligtermoet, Dirk M. 1996. "Theme Paper 5: Bicycles in the Cities," in Stares, Stephen and Zhi, Liu (eds) *China's Urban Transport Development Strategy: Proceedings of a Symposium in Beijing*, November 8–10, 1995. World Bank Discussion Paper no. 352. Washington, DC: World Bank.

World Bank. 1999. *World Development Report 1999/2000: Entering the 21st Century*. New York: Oxford University Press.

World Bank. 2008. *The Fragmented City: Issues in Urban Land Use in China*. Washington, DC.

Wu, Fulong. 2002a. "Real Estate Development and the Transformation of Urban Space in Chinese Transitional Economy: With Special Reference to Shanghai," in Logan, John R. (ed.) *The New Chinese City: Globalization and Market Reform*. New York: Blackwell, pp. 151–166.

Wu, Fulong. 2002b. "Sociospatial Differentiation in Urban China: Evidence from Shanghai's Real Estate Markets." *Environment and Planning A* 34: 1591–1615.

Wu, Fulong. 2004a. "Urban Poverty and Marginalization under Market Transition: The Case of Chinese Cities." *International Journal of Urban and Regional Research* 28(2): 401–423.

Wu, Fulong. 2004b. "Intra-urban Residential Relocation in Shanghai: Modes and Stratification." *Environment and Planning A* 36(1): 7–25.

Wu, Fulong. 2005. "Transplanting Cityscapes: Townhouse and Gated Community in Globalization and Housing Commodification," in Wu, Fulong (ed.) *Globalization and the Chinese City*. Oxford, UK: RoutledgeCurzon, pp. 295–306.

Wu, Fulong and Li, Zhigang. 2005. "Sociospatial Differentiation in Subdistricts of Shanghai." *Urban Geography* 26(2): 137–166.

Wu, Fulong and Yeh, Anthony G.O. 1999. "Urban Spatial Structure in a Transitional Economy: The Case of Guangzhou." *Journal of the American Planning Association* 65(4): 377–394.

Wu, Weiping. 2008. "Migrant Settlement and Spatial Distribution in Metropolitan Shanghai." *Professional Geographer* 60(1): 101–120.

Yeh, Anthony G.O. and Wu, Fulong. 1995. "Internal Structure of Chinese Cities in the Midst of Economic Reform." *Urban Geography* 16(6): 521–554.

Yeh, Anthony G.O., Xu, Xueqiang, and Hu, Huaying. 1995. "The Social Space of Guangzhou City, China." *Urban Geography* 16(7): 595–621.

Yusuf, Shahid and Wu, Weiping. 1997. *The Dynamics of Urban Growth in Three Chinese Cities*. New York: Oxford University Press for the World Bank.

Zhou, Yixing and Ma, Laurence J.C. 2000. "Economic Restructuring and Suburbanization in China." *Urban Geography* 21(3): 205–236.

Zhou, Yixing and Meng, Yanchun. 1997. "Suburbanization in Shenyang: A Comparison of Suburbanization in China and the West [*Shenyang de jiaoquhua: jianlun zhongxifang jiaoquhua de bijiao*]." *Acta Geographica Sinica* [*dili xuebao*] 52(4): 289–299.

Zhu, Jieming. 1994. "Changing Land Policy and its Impact on Local Growth: The Experience of the Shenzhen Special Economic Zone, China, in the 1980s." *Urban Studies* 31(10): 1611–1623.

9   Urban infrastructure

Under China's command economy, cities suffered from disinvestment in non-productive sectors and saw their infrastructure and housing stocks deteriorate. The drastic turnaround during the reform period, consequently, is astounding. Aggressive infrastructure investment has enabled the rapid modernization of cities. As the Chinese economy moves ever closer to a market system, the role of state-owned enterprises and institutions in infrastructure development has diminished. Today, municipal authorities bear most of the responsibilities, using a widening range of financing mechanisms.

This chapter assesses the overall level of infrastructure services in urban China and outlines patterns and mechanisms for providing urban infrastructure. As economic growth becomes more concentrated in the coastal region, there has been a sharp rise in inter-regional disparities. This chapter also discusses how cities of different regional locations perform in infrastructure services and financing capacity. The following questions should help guide the reading and discussion of the materials in this chapter:

- Why was urban infrastructure a low priority for public investment under state socialism (1949–1979)?
- In what areas of infrastructure have Chinese cities made the most overall progress during the reform era (after 1979)? What areas present major challenges for urban China?
- How has fiscal decentralization affected public finance and the financing of urban infrastructure in particular?
- What are the key sources of finance for municipalities to build new infrastructure?
- What are extra-budgetary funds? Why have they become a common mechanism for financing urban infrastructure?
- Why has regional disparity in infrastructure services and financing capacity persisted during the reform era?
- What types of infrastructure services have attracted substantial investment from private and foreign sources?

The state of urban infrastructure

Legacies of state socialism

Traditionally, provision of infrastructure is divided among the central, provincial, and municipal governments. Infrastructure owned, constructed, and operated by the central government includes railways, pipelines, large ports and airports, large power plants, and now inter-city highways, some of which also can be operated by an authority at either provincial or local level. The central government has the authority to set investment goals, offer special incentives to investors, approve projects with foreign investment, and limit the scope of operation of certain infrastructure facilities. For instance, influential leaders of Shanghai were able to obtain approval from the central government for a joint venture with General Electric to build a 400-megawatt oil-fired plant that would provide the city with electrical power. In southern China, one of the most modern airports in the country was built in the city of Zhuhai, but the central government made it clear that the airport would be licensed for domestic flights only so as not to provide too much competition with the region's other airports (see Box 4.3). A similar principle would apply to port construction and operations whereby the central government designates a number of selected ports for handling foreign vessels.

The transport sector in particular has been subject to a heavy dose of central control. Central authorities formulate 5-year development plans and set priorities and targets. This is often done through consultation with the ministries involved, including the Ministry of Communications in charge of national highways, ports, and major waterways, and the Civil Aviation Administration. The central ministries also exercise control of the tariff structure of the national transport system and continue to play a crucial role in offering guidance and technical assistance to provincial and municipal governments.

Municipal governments are responsible for the provision of a full range of urban infrastructure, including local roads, public transport, water and sewers, utilities, telecommunications, and environmental protection services. Municipalities also can operate their own power plants. Municipal Construction Commissions (or similar agencies) are the main local institution overseeing urban construction and infrastructure matters. They also oversee the activities of a large number of public utility corporations and service bureaus.

Around the world, urban infrastructure generally includes public utilities (power, piped gas, telecommunications, water supply, sanitation and sewerage, and solid-waste management), municipal works (roads and drainage), and transportation sectors (public transit, ports, and airports) (World Bank 1994). In China, as a result of budgetary and administrative categorization, urban infrastructure (often called "urban maintenance and construction") includes public utilities (water supply and drainage, residential gas and heating supply, and public transportation), municipal works (roads, bridges, tunnels, and sewerage), parks, sanitation and waste management, and flood control (Wu 1999). Power, telecommunications, and other transportation sectors (ports, airports, and railway) are financed separately and not counted as a part of urban maintenance and construction. Before 1999, although a very small portion of urban maintenance and construction expenditure was for the maintenance of municipal public housing, the budget for housing construction normally was separated from that for urban maintenance and construction.

Prior to reform, the central government controlled investment in urban infrastructure, mainly through central budgetary allocation, supplementary project funding, and the public utility surcharge levied on enterprises and commercial users. There were no earmarked taxes or funds for urban construction, and user charges were minimal. As a result, there was no guaranteed steady flow of funds to urban construction from year to year. Indeed, central funding for the urban sector dried up from the 1950s to 1970s as Beijing focused on building inland defense facilities. Under the rigid central–local fiscal relations, municipalities also did not have the local finance systems necessary to support infrastructure projects (Dowall 1993; Chan 1998).

The infrastructure sector was a low priority when funds were allocated through municipal budgets. Because of the bias towards industrial production, capital outlays for urban construction were very limited, often not exceeding one-tenth of total municipal expenditure. Compared to agencies in charge of industrial production, those in charge of urban construction had little clout in the municipal decision-making circle. Take the example of urban bus systems. Since public transportation was considered a form of welfare, all bus services were non-profitable and relied on government subsidies. Municipal authorities would set aside a small, fixed portion of the annual budget for maintaining and operating public transportation, apportioning funds by a population account based on the system of household registration (Wu 1999).

Take another example of the city of Shanghai. Starved of capital throughout the Maoist era, the city's infrastructure was debilitated when economic reform commenced in 1979. Utility services, such as water, heating, and sewers, lagged far behind requirements. Per capita paved road space was well below the national average (Yusuf and Wu 1997). Designed for an age of pedestrians and rickshaws, the graceful tree-lined boulevards were jammed with cars, buses, bicycles, and pedestrians, causing extended delays at almost any hour of the day. Much of the housing stock that was built prior to 1949 lacked basic plumbing systems, so covered wood buckets were a common sight in bathrooms.

The attitude toward urban infrastructure, however, changed after 1979. The provision of urban infrastructure was critical in attracting foreign investment to the original four Special Economic Zones, coastal open cities, and beyond (Box 9.1). A major selling tool many cities used was the availability of land already equipped with infrastructure. The importance of infrastructure-led development is becoming even more apparent after 1988's urban land reform (more in Chapter 10). Urban infrastructure is considered to be a good form of public investment because it can increase revenues from land leases. In addition, the central government has significantly reduced its direct involvement in local development projects. Urban development is now shaped by local governments.

Box 9.1 **Upgrading Guangzhou's transport network to meet demands of growth**

The experience of Guangzhou and its hinterland in the early 1990s offers a good example of how urban transportation was modernized to meet the demands from rapid economic growth. As the capital city of Guangdong province, Guangzhou was in the heart of the Pearl River Delta that spearheaded China's market reforms. The city had the highest ratio of motor vehicles in relation to population in the country. But it was at the low end in terms of per capita road coverage and total road length. The surface transport network was not well developed, which inhibited door-to-door delivery advantageous for light industries. There was heavy congestion in the faster-growing areas as road capacity had not expanded in line with the drastic growth of traffic. The mixed use of roads by motor vehicles, tractors, bicycles, and pedestrians, without separation, also exacerbated the situation. Congestion and road conditions greatly reduced speeds and consequently increased vehicle operating costs. Moreover, much of the technology used for road transport was of 1950s' vintage. As a riverine city, water transport could have played a vital role. Unfortunately it had been allowed to decline since the 1960s and only accounted for 10 percent and 28 percent of passenger and freight traffic respectively (Wu 1995).

The city responded to the increasing severity of traffic congestion by embarking on the construction of ring roads. It tackled the shortage of public transport by building an 18.4 km subway, completed around 1998. The city invested about 35 billion yuan by the year 2000 into road and bridge projects, including the subway. An expressway around the city and four new expressways in the downtown area were planned. But an unfortunate pitfall in planning for these projects was the low priority placed on environmental considerations, particularly in terms of limiting greenfield development and possibly damaging impacts on the urban center. Two more initiatives also took place in the hinterland of Guangzhou. One was the intensive upgrade of the road system, and another the heavy emphasis on and investment in highway construction, using both domestic and foreign sources. The Shenzhen–Guangzhou highway, developed by Hong Kong's Gordon Wu, was a key accomplishment. Highways began to play an increasingly large role, as fast-growing industries needed the speed and flexibility of road transport. Moreover, Guangzhou was ahead of other municipalities in financing through the transfer of some ownership rights of bridges, tunnels, and waterworks to private companies through leases similar to the transfer of land use rights. These companies began to charge user fees in order to maintain and improve such infrastructure. This helped ease the financial burden of the municipal government and provided new sources of funding for public utility construction (Yusuf and Wu 1997).

Generally speaking, investment in urban infrastructure had for a long time taken up just a fraction of investment in fixed assets and GDP. The United Nations (UN) once suggested that investment for urban infrastructure should reach 9–15 percent of investment in fixed assets, or 3–5 percent of GDP. Since 1979, although China drastically raised the size of such investment in an effort to make up for "historical arrears," the ratio still fell far short of what the UN has recommended (World Bank 2000). It is only in the new millennium that urban infrastructure investment has reached the UN recommended level.

Overall progress in infrastructure services

Infrastructure plays an indispensable role in urbanization and development. There is a strong positive relationship between measures of urban infrastructure, especially water and sanitation, and GDP per capita (World Bank 1999). But for low-income developing countries (GDP per capita below US$1,000), the quality of water and sanitation infrastructure does not necessarily correspond with income level. Cities in some countries appear to be able to improve the delivery of services even under income constraints; better services are possible at low levels of income.

Rapid urbanization in the reform period has resulted in a very high demand for basic urban infrastructure in China. While the public sector has largely withdrawn from housing provision (see Chapter 10), its role in infrastructure development has remained critical for cities. On an aggregate level, China has made significant progress in urban infrastructure since 1979. By the mid-1990s, the level of services in urban China had already approached that in lower middle-income economies (Chan 1998), while the country's per capita GDP did not reach the same status until 2002. By 2009, most urban residents had access to faucet water, cooking gas, and public transportation (Table 9.1). Water supply coverage reached more than 90 percent in urban areas, which would put China in the group of upper middle-income countries. The most usual source for faucet water is surface water

Table 9.1 Average levels of infrastructure services in urban China, 1981–2009

	Water coverage rate (%)	Gas coverage rate (%)	Public transportation vehicle unit (per 10,000 persons)	Per capita road area (square meters)	Wastewater treatment rate (%)	Per capita public green space (square meters)
1981	53.7	11.6	—	1.81	—	1.50
1985	45.1	13	—	1.72	—	1.57
1990	48.0	19.1	2.2	3.13	—	1.78
1995	58.7	34.3	3.6	4.36	19.69	2.49
2000	63.9	45.4	5.3	6.13	34.25	3.69
2005	91.1	82.1	8.6	10.92	51.95	7.89
2009	96.1	91.4	11.1	12.79	75.25	10.66

Source: National Bureau of Statistics of China 2009, 2010.

drawn from rivers, lakes, and reservoirs (particularly in the south), which is often processed to a quality good enough for domestic and industrial use, but not for drinking.

One area of significant progress is the treatment of wastewater, the rate of which more than doubled between 1996 and 2006, from 23.6 to 55.7 percent (Ministry of Construction 1998, 2008). Prior to the 1990s, municipalities generally had no incentive to invest in sewage and wastewater treatment because they received little income for handling wastewater discharges. Low sewerage coverage, inadequate treatment facilities, and low water discharge fees resulted in contaminated groundwater and polluted surface water. In response to the growing demand for water, the central government made the development of water and wastewater infrastructure a national priority in the 1990s. Substantial investment went into the water sector, about US$9.5 billion in water supply and US$6 billion in treatment just between 1990 and 1998 (Bellier and Zhou 2003). As a result, urban wastewater treatment capacity grew 17 percent annually in terms of volume in the same period.

Yet the continuity of service and quality of faucet water remain a major concern. While demand is tremendous and growing, water resources are so limited and/or polluted that they cannot meet the demand, especially in northern cities. Because of a drier climate and less surface water, the predominant source in the north is groundwater, and it has been seriously over-extracted (see Chapter 11). In addition, as a result of increasing wastewater discharge and insufficient wastewater treatment, water pollution is getting serious. Although enormous efforts have gone into developing wastewater treatment infrastructure, the expanded capacity has not been enough to stop the contamination of water resources. For instance, about 24 percent of water in Beijing is not fit for human consumption. Almost 90 percent of the river water in Shanghai is considered to be polluted (Cornel and Zhang 2003).

Urban transport is becoming another major concern for cities. On the one hand, road construction now receives the largest share of municipal infrastructure spending, and per capita paved urban road space increased steadily since the mid-1990s and nearly tripled by 2009 (see Table 9.1). Most cities have developed a ring-road structure to provide circulation, mobility, and access to different quarters. But at the end of 2003, only 11 percent of China's urban area was devoted to roads, compared to 20–30 percent for most industrialized nations (Cherry 2005). On the other hand, rising income and automobile ownership have created unprecedented levels of traffic congestion and air pollution. Most transportation agencies rely heavily on supply-side approaches such as road capacity expansion and, in effect, encourage automobile transportation.

Clearly, Chinese cities face a choice: fall into some of the traps faced by other cities around the world, or seize the opportunity to integrate transport and land use planning. Even at the present level, urban roads can be used more efficiently through careful planning and sophisticated traffic management systems. The elements of such solutions are: developing a logical road hierarchy, improving street and junction layouts, segregating major traffic modes, automating and linking traffic signal systems, and reducing conflicting traffic movements. Shanghai provides an example of how a city can tackle some of the urban transport problems. With only about 10 percent of the urban area devoted to roads, the central city has inherited an antiquated transportation network from the early twentieth century designed for pedestrians and bicycles. Implementing nearly all of the foregoing methods with the assistance of international lending agencies, the major downtown sections of the city have avoided the kind of traffic congestion now frequently seen in Beijing.

Last but not least, municipalities are facing difficulties in the maintenance of urban infrastructure. Because infrastructure investment tends to be long-term, continuous improvements are particularly important to avoid an accumulation of problems in the future. Much of the infrastructure investment goes to new construction and capacity expansion. The maintenance of existing facilities receives less funding (10–20 percent of total expenditure). As a result, the responsibility for maintenance has become a heavy burden for many municipal governments.

To be globally competitive, cities also need to provide a wide range of new, additional infrastructure services, for example cable networks for advanced telecommunications, wired and energy-saving structures, and high-speed trains. Such infrastructure is particularly important to knowledge-based sectors, especially that critical to the direct operation of firms. New server hotels – large facilities constructed specifically to house servers of multiple firms – are becoming a significant factor in local electric demand. Most Chinese cities, with the exception of a few large ones, have only begun to offer these services.

Financing urban infrastructure

Public finance and fiscal decentralization

Public finance matters for the provision of urban infrastructure. China's record of investment in highways, ports, power plants, and a variety of urban services has been nothing short of astonishing. But its fiscal system remains in transition – the product of decentralization. Much like the macroeconomic reforms, fiscal decentralization has been gradual and incremental, responding to immediate problems with short-term fixes. There continues to be a mismatch of expenditures and revenues between levels of government. As a result, local governments often have to cope with funding shortfalls through a variety of off-budget mechanisms.

Prior to the economic reform, China's fiscal system was characterized by centralized revenue collection and fiscal transfers. The paramount importance of the central government derived from the control of fiscal revenue and expenditure of local governments. Given the three general levels of governments (central, provincial, and municipal), the lower levels were entirely subordinate to governments of higher levels in fiscal matters. All taxes and profits were collected by local governments, remitted to the central government, and then transferred back to the provinces and municipalities according to their expenditure needs approved by Beijing. Under this fiscal system, municipal revenues were shared with the central or the provincial government for redistribution. For many years revenue retention rates for municipal authorities were very low and insufficient to allow significant expansion of infrastructure and adequate maintenance.

The central government had direct control over local governments (both provincial and municipal governments) in three main areas: allocation of materials and resources, production planning for key industries, and budgetary control of revenues and expenditures. Large infrastructure projects, such as subway systems in Beijing, Shanghai, and Guangzhou, needed to be incorporated into provincial plans or plans of the central government. All infrastructure projects with national importance were reviewed by the then Ministry of Construction (now renamed as Ministry of Housing and Urban–Rural Development), while those with regional importance were reviewed and approved by provincial authorities. Most other urban infrastructure projects were approved by municipal authorities.

Central–local fiscal relations have been altered significantly by decentralization efforts since 1980. First, the central government introduced a new fiscal regime that visualized each provincial entity as a "separate kitchen" for fiscal purposes. This, together with subsequent fiscal reform by provinces, allowed many municipalities to retain higher rates of revenue and to allocate funding more freely. In 1980 a new system of fiscal contract was introduced, which designated separate types of taxes or revenue. Under this arrangement participating provinces and municipalities were allotted a share of revenue. They retained all income collected in excess of this share. In exchange for being given a bigger slice of revenue, they also were required to accept responsibility for most items of expenditure.

The concept of revenue sharing introduced in 1980 was somewhat murky. Although the central government remained in control, tax assignment and revenue-sharing arrangements often had to be negotiated with local governments, typically those at the provincial level. The normal practice was that the central government would designate revenue from certain taxes as central fixed revenue, and a portion of revenue from other taxes as local fixed revenue, with the remainder going into a pool of shared revenue (Tseng et al. 1994). The task of defining the share formula proved difficult, and several schemes were tried. Some large provinces and municipalities, to their advantage, negotiated lump-sum revenue remission schemes.

The central government would offer financial assistance to local governments through three types of grants: (1) quota grants under fiscal contracts – unconditional transfers to provinces to finance their budget deficits; (2) special-purpose grants – used to finance specific tasks in local governments' budgets; and (3) final accounts or settlement grants – compensatory transfers arising from revenue-sharing contracts (Tseng et al. 1994). This allowed the central government to redistribute a part of fiscal resources from surplus to deficit provinces and municipalities.

The 1994 tax reform introduced yet another set of new measures to further streamline central–local fiscal relations. Three areas of concern were addressed: providing adequate revenues for government, particularly the central government; building a more transparent tax structure; and improving central–local revenue-sharing arrangements (Wong 1997). Taxes were reassigned between the central and local governments, with a shift from a negotiated system of general revenue sharing to a mix of tax assignments and tax sharing. For the first time, local governments were assigned some taxes with significant revenue generation capacity as local taxes. Related to urban construction, an urban land use tax, a real estate tax, and an urban maintenance and construction tax are among the local taxes.

Investing in urban infrastructure

Behind the impressive progress in building China's urban infrastructure, its financing is fundamentally different from that in most other countries. In industrialized countries, borrowing is widely used as a key method because of the capital-intensive nature of much urban infrastructure, especially in terms of upfront costs (Chan 1998; Bird 2004). Most such borrowing is directly from a functioning capital market and relies on a system of municipal bond ratings. Large cities, in particular, tend to have better access to bond markets than small cities. Excluding borrowing, local taxes (particularly property taxes) are the most important source of infrastructure financing, counting on average a 40 percent share (Chan 1998). What follows are grants and subsidies, and other sources, including user charges. Although the situation in developing countries varies substantially, local property taxes dominate the revenue structure and loan financing tends to be a small source.

Municipal governments in China, on the other hand, have neither sufficient tax resources to finance infrastructure nor the authority to borrow on the international market (Wong and Bird 2004). Under reform, the decentralized fiscal system has mostly worked to the advantage of municipal governments, allowing them better incentives to mobilize local resources. At the municipal level, new mechanisms of financing arising from such fiscal freedom have contributed to significant expansion of infrastructure investment. However, the 1994 tax reform also has resulted in significant distortions. There continues to be a mismatch of expenditures and revenues between levels of government. The central government assigns heavy responsibilities for the provision of nearly all public services to local governments while not adequately supporting them through either revenue assignments or an intergovernmental transfer system. Before 1985, for instance, local government revenues were higher than expenditures, and local government remitted revenues to the central government. But by 1998, 35 percent of local government expenditures were financed by subsidies from the central government (Lin 2001).

There are four major sources of funds available for urban infrastructure development (Table 9.2). The first is budgetary allocation from central and local governments. The second is local fiscal revenues: urban maintenance and construction tax (collected by local governments as a surcharge on the combined value of value-added tax, product tax, and business tax – 7 percent in cities, 5 percent in towns, and 1 percent elsewhere), and public utility surcharge (collected by local governments at a rate of 5–8 percent from the turnover of water, electricity, natural gas supplies, public transportation, and local telephone service) (Chan 1998; World Bank 2000). Set by the central government, these rates are low relative to the financing needs of many cities. Exactions also fluctuate with output levels of

Table 9.2 Urban maintenance and construction revenues, 1990–2009 (billion yuan)

	1990	1993	1996	2002	2005	2009	2005 (%) Total	2005 (%) Sans borrowing	2009 (%)
Budgetary allocation									
Central budgetary allocation	1.09	2.70	1.04	7.60	6.22	10.66	1.15	1.71	1.58
Local budgetary allocation	1.98	5.95	8.63	39.27	79.59	166.04	14.68	21.96	24.67
Local taxes									
Maintenance and construction tax	6.51	9.80	15.78	31.60	55.13	77.19	10.17	15.21	11.47
Public utility surcharge	2.26	3.30	5.56	4.99	5.55	9.80	1.02	1.53	1.46
Fees and user charges									
Water resource fee	0.28	0.48	0.61	1.24	2.50	2.48	0.46	0.69	0.37
Infrastructure connection fee	—	—	—	8.66	14.29	32.56	2.64	3.94	4.84
User charges[a]	—	—	—	8.94	14.55	25.89	2.68	4.01	3.85
Land transfer fee	—	—	—	28.30	59.45	263.60	10.96	16.40	39.17
Borrowing[b]									
Domestic loans	0.88	4.46	9.57	87.39	166.99	—	30.80	—	—
Foreign capital[c]	0.25	1.38	5.59	6.11	9.27	—	1.71	—	—
Bonds	0	0	0	0.29	3.43	—	0.63	—	—
Stock financing	0	0	0	0.68	0.10	—	0.02	—	—
Self-raised funds	2.58	4.59	11.95	60.08	94.60	27.62	17.45	26.10	4.10
Other sources[d]	5.21	25.47	26.05	30.47	30.59	57.08	5.64	8.44	8.48
Total	21.04	58.13	84.78	315.62	542.25	672.94	100.00	100.00	100.00

Sources: Wu 1999 and Ministry of Construction 2004–2009.

Notes:
a User charges include primarily toll on roads and bridges, wastewater treatment fee, and garbage treatment fee.
b Data on borrowing no longer available after 2005.
c Foreign capital includes both direct investment and loans.
d A major component of other sources prior to 2002 is infrastructure connection fees.
— Data not available.

enterprises and do not apply to state institutions. As a result, local fiscal revenues have counted for a declining share of infrastructure financing.

The third and increasingly important mechanism of financing is the collection of fees and user chargers, including infrastructure connection fees (similar to impact fees in the USA), although most such non-fiscal revenues were recorded in the "other sources" category in the 1990s. Such income now counts for between 15 and 25 percent of urban maintenance and construction revenues. There is a downside to this because of the wide range of fee scales and fee items across cities. Some municipal authorities include a multitude of infrastructure services in the fee collection and often ask for exorbitant amounts of money. This is shown in the case of some 28 different fees imposed on various aspects of real estate development in Shanghai (Bird 2004). The increasing importance of land transfer fees as a source of local finance also is remarkable (see Chapter 10 for more details).

The importance of the fourth source, borrowing from both domestic and foreign sources, including foreign investment, is rising – from only about 5 percent in 1990 to more than 30 percent in 2005. Domestic borrowing includes not only cash and deposits but also treasury and other financial bonds, and equities. Bank credit also is emerging as an important source of borrowing. As such, China increasingly resembles market economies in which surpluses generated in the household sector are loaned to finance a portion of public investment. But banks sometimes are reluctant because infrastructure investment is large, with long terms and lower return rates, and regional in location (Zhang and Wu 2005). Borrowing on the international market is in principle not permitted at the local level, with exceptions approved only by the central government. Fiscal reforms also have selectively allowed certain provincial-level entities the right to issue construction bonds domestically. Aimed at capturing the high level of household savings, this practice has been growing but remains a small source of infrastructure financing. In addition, foreign capital (including government and commercial loans, direct investment, and international bonds) has begun to play some role in financing urban infrastructure.

The emergence of these new mechanisms (primarily the third and fourth sources) has resulted in the increasing importance of extra-budgetary funds, which are still in the public sector but are not subject to central or provincial budgetary control. This is an indication of a higher degree of municipal fiscal independence. The formal budget is only a part of the fiscal story and not necessarily the most important part. A broad definition of extra-budgetary funds is that they constitute all resources managed directly or indirectly by administrative branches of the government outside the normal budgetary process. Extra-budgetary funds generally include fees and/or funds that are not taxes, such as water resource fees, infrastructure connection fees, user charges, and land transfer fees (also called land use fees, see more details in Chapter 10). While budgeted funds are under strict supervision, extra-budgetary funds are easier to manipulate and soften the budgetary constraint for municipal governments. They provide considerable local autonomy and financial resources to achieve local policy goals (Wong and Bird 2004).

An additional source, self-raised funds, makes up 10–20 percent of infrastructure revenues and is an even grayer area in local financing. Though not collected as taxes or budgetary items, extra-budgetary funds nonetheless are specifically authorized by some government body. Sometimes called extra-extrabudgetary funds, off-budgetary funds, or extra-system revenues, self-raised funds are not specifically authorized as a fee or fund (Wong and Bird 2004). Such revenue is irregular and often non-recurring, raised by central ministries, local governments, enterprises, or public institutions. It is an indication that fiscal burdens to finance public services may be passed on to the enterprise sector. For instance, in Dongguan, Guangdong province, local authorities have created an energy and communications company to raise money from state, collective, and private sources for the construction of roads and power plants. The company pays interest on these funds and repays the capital by collecting user fees and tolls (Harral 1992).

Overall, China's experience in infrastructure financing is a significant departure from the international norm – the capacity of municipalities to borrow externally is lacking, local property taxes are nearly absent, and extra-budgetary funds allow for softer budgetary constraint. The local problems, to a large extent, stem from the changing central–local fiscal relations. As such, the operation of local governments in public financing often is chaotic, as they cope with funding shortfalls through a variety of off-budget mechanisms.

Challenges

Infrastructure and regional disparity

A long-standing issue in China is regional disparity in economic performance and income. Infrastructure services are no exception. There are noticeable differences in nearly all available indicators across cities of different geographic region and size (Wu 2010). Cities in the coastal (or eastern) region uniformly enjoy higher levels of service in all sectors, with a markedly higher percentage of the population having access to piped water and wastewater treatment. In many central and western provinces, local infrastructure, such as public transportation, roads, streets, water supply, and waste treatment, is in poorer condition. As city size increases, levels of service tend to increase. The largest cities (or super-large cities; see Chapter 4), most of which are located in the coastal region, particularly outperform others in wastewater treatment and bus services. Their superior position likely reflects their affluence and better ability to raise funds for building infrastructure. In general and around the globe, large cities tend to have much larger tax bases (per capita) than small ones because of their higher levels of economic activities and higher density of development (Bird 2004).

Fiscal decentralization has affected regional patterns of infrastructure financing. In particular, there are substantial distortions associated with the new tax sharing system created through the 1994 tax reform. By sharing value-added tax revenues with local governments at a flat rate, the system introduced a highly disequalizing feature to revenue sharing, ensuring that revenue-rich provinces keep more. Specifically, coastal provinces gained revenue shares relative to inland provinces. From 1993 to 1998, for instance, the ratio of provincial per capita fiscal expenditures in Shanghai (in eastern region) grew from 2.8 to 4.5 times the national average and in Beijing from 2.0 to 3.0. In contrast, the ratio fell in Gansu (in the western region) from 0.76 to 0.61, and in Hunan (in the central region) from 0.60 to 0.52 (Wong and Bird 2004). As wealth becomes more concentrated in the coastal region since the 1990s, there has been a sharp rise in inter-regional variation in fiscal spending, accompanied by a gradual deterioration in public services provided in the inland provinces (Wong and Bird 2004).

Overall, provinces in the eastern region have fared better. From 1996 to 2006, for instance, Beijing, Shanghai, Jiangsu, and Zhejiang enjoyed per capita public investment in infrastructure (as measured by urban maintenance and construction revenues) around twice the national averages (Figure 9.1). But there is substantial variation within the coastal region. Provinces on the southeast coast (such as Guangxi, Fujian, and Hainan) saw their revenue levels decline relative to national averages. Some provinces in the western region, particularly Chongqing (as the newly designated provincial-level city), experienced more rapid growth in urban infrastructure investment. The western region, in addition, relies much more on budgetary allocation and borrowing to finance urban infrastructure and has much less ability to raise funds from extra-budgetary sources (Figure 9.2).

The regional patterns in urban infrastructure financing have shifted in tandem with the progress of economic reforms. In 1996, provinces with the highest levels of per capita infrastructure investment, as compared to the national average, tended to be on the southeast coast (including Guangdong, Fujian, Zhejiang, and Shanghai). This was where market reforms took hold first. As reforms spread north and west, the Yangtze Delta area has grown rapidly and the Bohai area also is rising. Cities in Zhejiang and Jiangsu provinces and Shanghai and Beijing now invest more in infrastructure than other cities. Shanghai, for instance, has outpaced all Chinese cities in its development of transport networks and houses the world's only commercial maglev (magnetic levitation) train operation. Maglev trains, based on a technology originally developed in Germany, can travel up to 430 kilometers per hour (Figure 9.3). On the other hand, some of the poorest provinces, primarily in the central and western regions (such as Anhui, Henan, Hubei, and Gansu), continue losing ground in their capacity to finance urban infrastructure (Figure 9.1).

Private and foreign participation in urban infrastructure

Private involvement is increasing, and strongly encouraged, in infrastructure provision. But such participation is no equivalent of the wholesale privatization programs implemented in other former socialist countries. In the mid-1990s, the central government began to introduce foreign investment into the infrastructure sector (e.g., thermal power, hydropower, highway, water supply). Two important policy papers were issued: the *Circular on Attracting Foreign Investment through BOT Approach* (BOT refers to build, own, and transfer) and the *Circular on Major Issues of Approval Administration of the Franchise Pilot Projects with Foreign Investment* (Zhong et al. 2008). These two policy papers formed the first legal ground for private sector involvement and foreign investment in China's urban

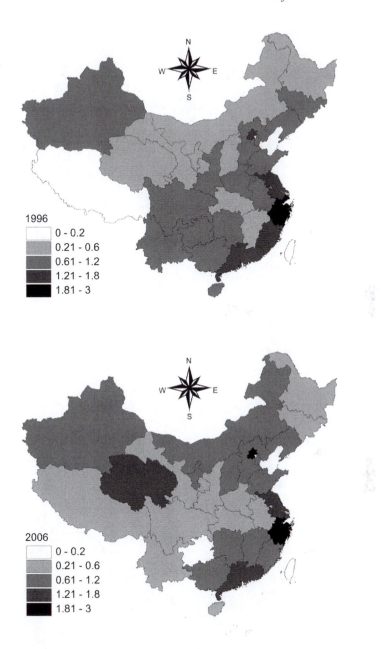

Figure 9.1 Index of per capita public investment in urban infrastructure by province, 1996 and 2006 (national average = 1).

Source: Wu 2010.

Note: Public investment in infrastructure is often called urban maintenance and construction revenues in China.

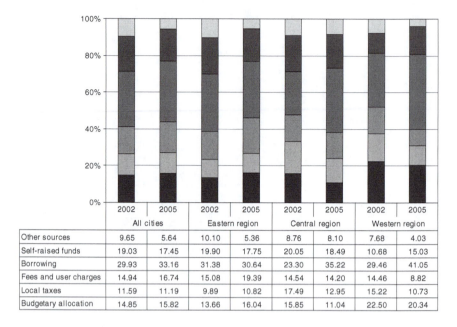

	All cities		Eastern region		Central region		Western region	
	2002	2005	2002	2005	2002	2005	2002	2005
Other sources	9.65	5.64	10.10	5.36	8.76	8.10	7.68	4.03
Self-raised funds	19.03	17.45	19.90	17.75	20.05	18.49	10.68	15.03
Borrowing	29.93	33.16	31.38	30.64	23.30	35.22	29.46	41.05
Fees and user charges	14.94	16.74	15.08	19.39	14.54	14.20	14.46	8.82
Local taxes	11.59	11.19	9.89	10.82	17.49	12.95	15.22	10.73
Budgetary allocation	14.85	15.82	13.66	16.04	15.85	11.04	22.50	20.34

Figure 9.2 Public investment in urban infrastructure by region, 2002 and 2005.

Source: Wu 2010.

Note: Public investment in infrastructure is often called urban maintenance and construction revenues in China.

Figure 9.3 Shanghai's maglev train serving Pudong Airport.

Source: Wikimedia Commons, retrieved 3 January 2012 from http://commons.wikimedia.org/wiki/
File:A_maglev_train_coming_out,_Pudong_International_Airport,_Shanghai.jpg (donated by Alex Needham).

infrastructure. Between 2001 and 2003, both the State Planning Commission and Ministry of Construction issued a number of other policies to guide private investment in public utilities and projects. Both domestic and foreign investment would be permitted into nearly all forms of infrastructure, particularly in water supply and wastewater treatment, through sole investment, cooperative enterprises, joint ventures, share purchase, or franchising (Song 2003).

Private participation includes a variety of forms, as shown by experience worldwide. There are at least four models. First, there are contracts for private enterprises to manage public services or facilities but with no ownership or investment stake. Second, there are leases: a private contractor pays a leasing fee and provides a service at its own risk. Then there are concession and BOT and BOO (build, own, and operate) schemes: a private entity provides a public service at its own risk, with the extreme of private ownership under BOO. Last is divestiture, in which the ownership of existing assets and responsibility of future expansion are transferred to the private sector (Bellier and Zhou 2003; Lee 2003).

When China started to open infrastructure to overseas investment in the late 1980s, foreign companies responded enthusiastically. China has since attracted money for truly commercial projects, such as joint ventures to build and operate roads and bridges. Costs are expected to be recovered through toll collection. Public–private partnerships (or joint ventures) are another common form of foreign participation. Both sides contribute funds or services (frequently by providing property or land on the Chinese side), and the public sector often is represented by a company directly or indirectly owned by the government (Bellier and Zhou 2003; Bird 2004). Since 1995, however, more concession-based projects have emerged, particularly in the form of BOT (Box 9.2).

Box 9.2 **Foreign participation in urban infrastructure: Chengdu Water**

The first water BOT project in China took place in Chengdu, the capital city of Sichuan province. Chengdu Water involved the construction and operation of a water treatment plant (with 400,000 m^3 daily capacity) and a network of water pipes. The Asian Development Bank (ADB) provided US$26.5 million in direct loans and assisted in project structuring and mobilization of additional debt from commercial sources worth a total of US$21.5 million. Procured through competitive bidding, Vivendi Utilities (France) and the Japanese group Marubeni became the co-sponsors, with US$31.9 million of sponsor equity. The total project cost, US$107.6 million, also included US$26.5 million of a direct loan by the European Investment Bank. After 18 years of plant operation, the ownership of the plant would be returned to the municipal government (ADB 2005). Considering its late entry into the Chinese water market, Vivendi's record was impressive. As of 2002, Vivendi was providing water services to around 8.5 million customers in China and implementing water projects in Shanghai, Chengdu, and Tianjin (Lee 2003).

In addition to open bidding, the successful launch of the project can be attributed to the separation of operational and regulatory functions by avoiding the joint-venture model, which tends to create conflicts of interest for local governments (Bellier and Zhou 2003). The project was the first instance in the nation to rely on municipal credit risk without the central government's guarantee. In this case, the Chengdu municipal government guaranteed to purchase bulk quantities of treated water and greatly reduced investors' risks associated with local competition and uncertainty in future demand. As it became clear that implementation problems for BOT projects were not adequately addressed or resolved at the national level, the municipal government also issued a local policy paper, titled the *Interim Provision on Administrating Concession Right of Chengdu* (Zhong et al. 2008). Officially signed in 1999, plant operation of the Chengdu Water project started in 2002. The foreign investors began to collect profits in 2003. The success of this project has since triggered a wave of private-sector participation in and around Sichuan province (including Chongqing and Yunnan).

The water sector perhaps has the highest degree of foreign and private participation. There are a number of powerful transnational corporations engaged in China since the 1990s. These include Suez, Vivendi, Bouygues (SAUR), Thames Water, and Anglian Water. Most such firms enter into partnerships with local municipal water authorities, as a strategy of best insurance to avoid legal, regulatory, and political risks. So far, Suez and Vivendi Group have firmly established their positions in water supply and sewage treatment services in diverse regions. In addition, there are a small number of Chinese private enterprises in the water sector, such as the Youlian Consortium and Beijing Sound Group.

Attempts to establish wholly foreign-owned water projects have been made in Shanghai. The slow but gradual shift of the government's policy toward water privatization in the city occurred during the 1990s. There had been chronic problems in water supply and sewage treatment services: management was inefficient and skills and facilities were out of date. There was a lack of finance, and raw-water sources were polluted. The Dachang plant, cosponsored by Bovis (France) and Thames Water Overseas (UK) in 1996 (through a 20-year BOT) marked the first step toward private participation, followed by Vivendi's unprecedented equity share contract with the Shanghai Pudong Water Supply Corporation in 2002 (Lee 2003). These projects show that well-structured, wholly privately owned projects with a strong international sponsor could be an alternative to joint ventures for investment in the water sector.

Another form of privately funded infrastructure services is premium water networks in some cities. Beginning in the mid-1990s, select residential communities began to bypass municipal faucet water supply systems through the construction of small-scale secondary pipe networks for purified drinking water. The idea was to replace networks of water vendors of bottled water with special pipe networks.

Two trial systems were designed and built in Shanghai in 1996–1997, followed by other cities in the south (Boland 2007). Some pipe networks were designed to serve a single building, while others served a cluster of high-rise apartments. One of the largest early systems supplied filtered drinking water to 30,000 residents in a residential development on the outskirts of Shenzhen. Such a practice was becoming a selling feature for new, high-end housing developments throughout the country. More recently, Guangzhou launched premium water supply initiatives as part of citywide water improvement plans to overhaul the city's entire water supply infrastructure.

Conclusion

How China accommodates its increasingly urban population is critical, not only directly for the well-being of an increasing number of its people, but also more indirectly for its sustained economic development. Over the last three decades, China has made great strides in providing basic services and perhaps performed better than most other developing countries with similar income levels. However, rapid growth and modernization are putting significant strains on the transport networks and water sources for nearly all cities. In addition, there is an increasing disparity among cities of different geographic locations in their ability to finance and provide urban infrastructure. Such unevenness will have a long-term effect on urban economic growth and well-being.

Bibliography

Asian Development Bank (ADB). 2005. *Chengdu Water Supply Project: Effective Public–Private Partnership at the Municipal Level.* Manila, the Philippines: Private Sector Department, Asian Development Bank.

Bellier, Michel and Zhou, Yue Maggie. 2003. *Private Participation in Infrastructure in China: Issues and Recommendations for the Road, Water, and Power Sectors.* World Bank Working Paper No. 2. Washington, DC: The World Bank.

Bird, Richard M. 2004. "Getting It Right: Financing Urban Development in China." International Tax Program Paper 0413. Toronto, Canada: Institute for International Business, University of Toronto.

Boland, Alana. 2007. "The Trickle-Down Effect: Ideology and the Development of Premium Water Networks in China's Cities." *International Journal of Urban and Regional Research* 31(1): 21–40.

Chan, Kam Wing. 1998. "Infrastructure Services and Financing in Chinese Cities." *Pacific Rim Law and Policy Journal* 7(3): 503–528.

Cherry, Chris. 2005. "China's Urban Transportation System: Issues and Policies Facing Cities." Working paper 20054. UC Berkeley Center for Future Urban Transportation. Berkeley, CA: University of California.

Cornel, P. and Zhang, W. 2003. "An Analysis of Water Situation and Water Consumption in Several Regions in China," in Wilderer, P.A., Zhu, J., and Schwarzenbeck, N. (eds) *Water in China.* London: IWA Publishing, pp. 79–87.

Harral, Clell G. (ed.) 1992. *Transport Development in Southern China.* World Bank Discussion Paper No. 151. Washington, DC: The World Bank.

Lee, Seungho. 2003. "Expansion of the Private Sector in the Shanghai Water Sector." Occasional Paper No. 53, School of Oriental and African Studies/King's College London, University of London.

Lin, Shuanglin. 2001. "Public Infrastructure Development in China." *Comparative Economic Studies* 43(2): 83–109.

Ministry of Construction. 1999–2009. *China Urban Construction Statistical Report*. Beijing: Ministry of Construction.

National Bureau of Statistics of China. 2009. *Statistical Summary for Sixty Years of New China 1949–2008*. Beijing: China Statistics Press.

National Bureau of Statistics of China. 2010. *Statistical Yearbook of China 2010*. Beijing: China Statistics Press.

Song, Jinzhou. 2003. "Privatization of Basic Service in Shanghai Metropolitan Region." Paper presented at the Workshop on Global City Regions as Changing Sites of Governance. Free University of Berlin, Berlin, Germany, August 8–10.

Tseng, Wanda, Khor, Hoe Ee, Kochhar, Kalpana, Mihaljek, Dubravko, and Burton, David. 1994. *Economic Reform in China: A New Phase*. Occasional Paper No. 114. Washington, DC: The International Monetary Fund.

Wong, Christine P.W. 1997. "Overview of Issues in Local Public Finance in the PRC," in Wong, Christine P.W. (ed.) *Financing Local Government in the People's Republic of China*. Hong Kong: Oxford University Press for the Asian Development Bank, pp. 27–60.

Wong, Christina and Bird, Richard M. 2004. "China's Fiscal System: A Work in Progress." Paper presented to the Conference on China's Economic Transition, Pittsburgh, November.

World Bank. 1994. *World Development Report 1994: Infrastructure for Development*. New York: Oxford University Press, for the World Bank.

World Bank. 1999. *World Development Report 1999/2000: Entering the 21st Century*. New York: Oxford University Press, for the World Bank.

World Bank. 2000. "Workshop on China's Urbanization Strategy: Opportunities, Issues, and Policy Options." Sponsored by the State Development Planning Commission, the Ministry of Construction, and the World Bank Group.

Wu, Weiping. 1995. "Financing and Management of Urban Infrastructure in China." Prepared for the Center for Urban and Community Studies, University of Toronto and Urban Management Division, the World Bank, November.

Wu, Weiping. 1999. "Reforming China's Institutional Environment for Urban Infrastructure Provision." *Urban Studies* 36(13): 2263–2282.

Wu, Weiping. 2010. "Urban Infrastructure Financing and Economic Performance in China." *Urban Geography* 31(5): 648–667.

Yusuf, Shahid and Wu, Weiping. 1997. *The Dynamics of Urban Growth in Three Chinese Cities*. New York: Oxford University Press for the World Bank.

Zhang, Yaoqing and Wu, Qingling. 2005. *Urban Infrastructure Provision and Management*. Beijing: Economic Science Press.

Zhong, Lijin, Mol, Arthur P. J., and Fu, Tao. 2008. "Public–Private Partnerships in China's Urban Water Sector." *Environmental Management* 41(6): 863–877.

10 Urban land and housing

Under economic transition, market forces are increasingly the dominant drivers of urban processes in China, particularly in housing and land development. The adoption of a leasehold system for urban land, instead of wholesale privatization, has distinguished China from other transitional economies. Reforms in the housing sector, on the other hand, have led to rising levels of private home ownership. The outcomes of these efforts are in stark contrast to how land and housing were allocated under state socialism (1949–1979).

This chapter outlines the changes in the urban land management system and delineates the various market and state actors engaged in land development. Reforms in the housing sector have been incremental: this chapter traces the multiple stages. It also includes a discussion of the challenges in ensuring a transparent land-leasing process and regulating the rising property markets. The following questions should help guide the reading and discussion of the materials in this chapter:

- How was urban land allocated and managed under state socialism?
- Housing was a form of social welfare before market reform. Who were the major providers of urban housing?
- How would you characterize the land leasehold system now in effect in cities? How are land use rights allocated to state units and to commercial users?
- How have land reforms affected the efficiency and patterns of land use?
- The role of the public sector has diminished with housing reforms. What are the key mechanisms of housing provision today?
- Many observers believe there are pervasive housing bubbles in the urban sector. What are the driving forces?

Legacies of state socialism

Urban land as state property

As in most other former socialist countries, state ownership of land was the cornerstone of China's command economy prior to 1979. Land was perceived as a means of production. Urban land in cities and towns, as well as industrial and mining sites, was owned by the state at large. Municipal governments, in reality, owned (and continue to own) urban land as they were entrusted by the state to allo-cate land. Central ministries, through their subordinate institutions or enterprises, also owned some land. Land in rural and suburban areas, on the other hand, was collectively owned by communes and their residents. Within the administrative boundaries of any large city, therefore, there would be inevitably a mixture of land designed as urban (state-owned, in the urban proper) and rural (collectively owned, largely in the urban edges).

Local government's role was (and remains) critical for the acquisition of land for urban development, especially when the conversion of farmland was involved. For instance, for a state-owned work unit in the city to gain access to the use of collectively owned land in the urban fringe, the local government would acquire the land, convert its designation from agricultural to non-agricultural use (i.e., from rural to urban), and allocate to the work unit. The local government also would make arrangements to resettle the displaced farmers. While a collective commune could use and benefit from rural land, it did not have the authority to dispose of or transfer the land. Only local governments could do so, as the state had the full right of land ownership. In 1978, use rights of agricultural land began to be contracted to individual households in the collectives. But ownership of rural land has remained collective even after the commune system was dismantled in 1983 (Lin and Ho 2005; Lin 2009).

In cities, the City Planning Bureau, Construction Commission, and Land Admin-istration Bureau made major urban development and land use decisions, such as uses, location, density, timing, and size. All construction projects must apply for a land use planning permit and a building permit from the Planning Bureau. Even projects under the central government must have such land use permission. The Planning Bureau would examine and approve the location and boundary, stipulate conditions for subsequent design, and issue licenses for construction (these steps continue to apply to new development requests today). Then the work unit would apply to the Land Administration Bureau for the allocation of land, free of charge and without any time limit.

Such free land use resulted in severe inefficiency in allocation. Since land was treated as a means of production owned by the state, it would be irrational to treat land as a commodity and levy fees on users. The result would be "tak[ing] money out of one pocket of the state and put[ting] it into another" (Lin 2009: 75). As such, there were no incentives for users, mostly state-owned work units, to economize on land. Similar to other former socialist cities elsewhere (e.g., Soviet Russia), land use in built-up areas tended to remain unchanged over time and there was little land redevelopment (Bertaud and Renaud 1997).

Planners and architects made purely technical evaluations of land, based on the engineering costs of development, sanitary conditions of the site, and the like. The socialist development strategy placed industrial growth above social welfare. As a result, large industrial areas were slated for land-intensive industries in prime areas of the city. When slippages in planning occurred, sizeable lots of land stayed vacant because planned development did not materialize. Given that many work units were involved in providing housing for their employees, it also was common to see mixed land use patterns with industrial and residential activities juxtaposed. Such patterns were reflected in the cellular neighborhood structure built around work units (more in Chapter 8).

Housing as a form of social welfare

Housing had long been a form of social welfare for urban residents. The dominant route, prior to 1999, was through a system of low-rent or free public housing produced and distributed by either work units or municipal governments. A work unit (*danwei*) was a specific kind of workplace in the context of state socialism, as an extension of the state apparatus and with the function of social organization and control (Wu 1996; more in Chapter 13). There were two general types of state work units: enterprises (*qiye*) and institutions (*shiye*). The overwhelming majority of the urban population worked for state work units and gained access to housing through their employers (in 1991, non-state employment counted for less than 7 percent of the urban workforce). Municipal governments became the providers for other urban residents, including those working in some small state work units and collective work units, and those unable to work. Once an urban resident obtained a public housing unit, s/he automatically gained access to household utilities (water, electricity, and sewerage) free of charge. This mechanism of household infrastructure provision remained intact throughout the 1980s and the first half of the 1990s.

This urban welfare housing system, however, did not apply to local residents with rural *hukou* or farmers in the countryside. Traditional family houses, or private housing, constructed on land allotted by the collectives (now called villages) were the norm for them. Even with reforms, these residents still do not have access to the housing provident funds established in cities and associated low-interest mortgage loans, as they have not been treated officially as urban population since 1964 (see discussion on the household registration system in Chapter 5). This separate system of housing provision stems from rural collective land ownership and has remained more or less intact until today.

State work units occupied a critical link in the provision of public housing. An estimate circa 1980 put the share of work unit housing in total housing stock at over 60 percent. This was a much higher proportion than in other former socialist countries such as Hungary and Poland (less than 20 percent: Wu 1996). These work units were responsible not only for allocating housing to their employees, but also for bargaining with their supervisory government agencies for investment to develop and manage such housing. After 1999, state work units are no longer involved in housing provision and distribution, but many of them still offer housing subsidies to their employees. The most important form of subsidy is their match of employee contribution to housing provident funds.

There were major problems with this system of public housing. The direct involvement of work units in housing led to low efficiency in the building industries and redundant construction. In addition, housing investment was unsustainable. Very low rents were a feature of the socialist redistributive system in which a "low wage, but high welfare" policy was pursued (Wang and Murie 1999; Wang 2003). For work units, the more housing they constructed, the heavier the financial burden. For municipal public housing, funding came from the municipal fiscal budget, and units were built and distributed by local housing bureaus. There was little return on the investment, which made the expansion and maintenance of housing very difficult. On the whole, the government minimized investment in non-productive sectors such as housing and service utilities in order to maximize output from industries (or productive sectors). The government allocated on average only 6.2 percent of total construction funds to urban housing between 1949 and 1978. Housing investment amounted to only 1.5 percent of gross national product (GNP) annually (Wang 2003).

As a result, acute housing shortages were widespread. In 1978, urban per capita living space was less than 4 square meters (about 43 square feet) nationally. In fact, between 1952 and 1978, per capita living space in urban China declined, from 4.5 to 3.6 square meters (Wang 2003). Crowding was especially acute in large cities, such as Shanghai and Beijing. Housing shortages, however, did not affect urban residents evenly. Because housing was part of the social wage, its distribution corresponded with the official rankings and employment affiliations of urban residents. Senior government officials had more generous housing than lower-level officials, official professionals more generous than non-official professionals, and employees of large state enterprises more generous than those in small state enterprises and non-state sectors.

Transition into land leasehold

Urban land reform

Unlike most other transitional economies, China so far has not initiated a privatization program for urban land. It has chosen to preserve state ownership but permitted market transactions of user rights (as distinct and separate from ownership rights) during urban land management reforms since the late 1980s. There were several motivating factors. To accommodate and facilitate economic growth, cities needed new ways of land management to promote a more efficient use and allocation of land resources. With the implementation of the open-door policy, foreign and overseas Chinese investors became important players in the real estate sector, and the old land allocation system would not be able to accommodate their needs. The non-state sectors, including collective, private, and joint ventures, also grew rapidly in size and were more exposed to market forces. In addition, many cities were starved of funds for infrastructure building; revenues generated through land leases could be a viable financing option. This could allow for greater local fiscal freedom (see Chapter 9). One Chinese researcher (Guan Qingyou) estimated that the percentage of land transfer revenues in local fiscal revenues increased from 0.24 percent in 1989 to 74.14 percent in 2010 nationwide (cited in Chen 2011).

A series of regulations were enacted to facilitate the launching of the land lease system. The possibility of transferring use rights of public-owned land, in effect the separation of use rights from ownership, was first written into the Constitution through an amendment in 1988. The State Council issued the *Temporary Regulations of Leasing and Transfer Right of Urban Land Use* in the same year. The basic message was that the right of land use could be transferred, leased, rented, and mortgaged. The rationale was that the leasehold system would not deter market-led reforms and freehold ownership (as practiced in most Western countries) may not be suitable for China. In fact, leasehold could provide municipalities with more latitude in land use planning as the ultimate ownership rights still lay in their hands. Several other countries, including Singapore, Sweden, and the Netherlands, have similar leasehold systems.

A milestone in China's urban land reform was the first transfer of land use rights in Shenzhen in September 1987, when a local company obtained such a right to the use of a lot for 50 years. The deal was contracted after negotiation, but the city has since tried two other forms of land transaction – bidding and auction. By 1990, about two-thirds of all usable land stock was leased out, although less than 10 percent of this was commercially transferred (Zhu 1994). The rest was assigned free of charge or with heavy subsidies. The rest of the country soon followed Shenzhen's experience.

Land use rights are assigned or leased in two parallel fashions: state allocation (to state units) and conveyance (to commercial users). State work units pay an allocation fee and have no time limits, whereas commercial users pay for a higher conveyance fee that is market-determined (as prices). Between 2003 and 2007, according to official estimates, an average of 250,000 cases of leasing land use rights occurred each year nationally. The majority of these cases – more than 65 percent – were through conveyance (Figure 10.1). In general, land conveyance ranges from 40

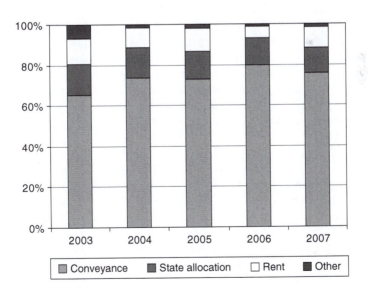

Figure 10.1 Land use rights assignment cases, 2003–2007.

Source: National Bureau of Statistics of China 2004–2008.

years for commercial land, 50 years for industrial land, and 70 years for residential land. Municipal governments also can reacquire land use rights from state units, with compensation, and then lease the land to commercial users at a higher price. Upon paying the higher conveyance fee, commercial users can then transfer the land use rights in a secondary market, rent to others, or use such rights as collateral (Lin 2009).

The setting of land prices remained a somewhat opaque process for much of the reform period. The State Land Administration published the *Procedures for Land Classification in Cities and Towns* in 1989 to direct land value classification. Each land parcel was supposed to be classified by aggregating a large set of factors, including transport conditions, infrastructure services, environmental conditions, commercial development, and population density (World Bank 1993). But due to a lack of formal, standardized land transaction procedures, informal land markets were operating behind closed doors. In reality, urban land may have three different kinds of price for commercial users: prices revealed by black-market property transactions, prices more clearly sanctioned through land lease contracts, and those that could be described as quasi-market prices (e.g., compensation for redevelopment projects and land acquisition from rural areas). The active informal or black markets persisted through illegal land occupation (without proper approval), illegal authorization of land use, and illegal transfer of land use rights (Lin and Ho 2005).

In a drastic attempt to correct these distortions, the central Ministry of Land and Resources issued a new regulation in April 2002. It mandated all land conveyance for commercial users be undertaken through transparent and competitive mechanisms: public tender, auction, or listing (so-called *zhao pai gua*). The goal was to reduce the dominance of negotiation in land leases. But many municipal governments ignored the central order. Between 2003 and 2007, between 70 and 80 percent of land conveyance cases were still done through negotiation – the least transparent, least competitive, and most easily manipulated format (Figure 10.2). This points to a recurring theme in China's urban transformation: while the central government formulates policies and oversees local practices, local governments and developers often resist top-down regulations. As a result, the urban land market is under a series of regulations but remains highly unregulated (Wu et al. 2007).

Consequences of land marketization

Marketization of land use rights has unleashed new forces in China's urban development, perhaps with impacts no less drastic than in other former socialist countries that have undertaken wholesale privatization. The consequences range from municipal governments securing rents from land leases and development projects, government agencies increasing involvement in land development, to urban expansion encroaching into agricultural land. A nascent property right awareness has emerged, at the same time, as urban residents and farmers mobilize to assert their stakes (Hsing 2010).

Since land is the most valuable commodity under the control of municipal governments, generating revenues from leasing land use rights and charging land use

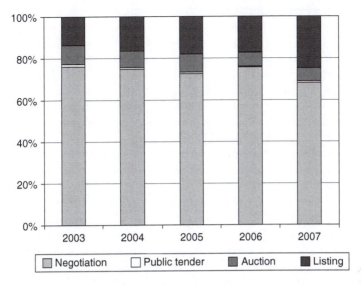

Figure 10.2 Cases of land conveyance to commercial users, 2003–2007.

Source: National Bureau of Statistics of China 2004–2008.

fees has become a popular practice. Many municipal governments are dependent on land conveyance for a considerable fraction of their revenues and have at the same time been increasingly involved in land and real estate development. For some, land use fees are considered self-raised funds because they are not specifically authorized as fees and the revenue varies from year to year. But for others, revenues from land leases are treated as extra-budgetary (see Chapter 9). Either way, income from land conveyance is retained at the local level and shielded from central government oversight. In their current practice, land leasing and transfer are driven by short-term interests of municipal governments. The use of state assets by the government and its agencies to raise off-budget revenues, in addition, is a major source of inefficiencies and distorted incentives (Wong and Bird 2004).

The establishment of the land lease system has strengthened the status of the local government as the most powerful manager of state land. With this status, local governments have adopted pro-growth policies, to bolster their performance records and to increase revenue incomes. State-owned development companies are established with public money and dominate the primary land market, acquiring and selling land of non-urban uses or existing urban land to developers. In addition, quasi-public development companies are often set up under names of independent business, as a safe way for government agencies to generate revenues without direct involvement in business activities (Zhang 2002). The common pro-development interest also has bonded local bureaucracy and real estate developers into an informal coalition. The combination of such thirst for land development, demand for buildings induced by economic growth, and limited supply of land has contributed to a drastic increase in land values in many cities (Zhu 2002). Rising land values would become a major driving force of the housing bubbles so pervasive in urban China in recent years, as discussed later in this chapter.

There remains inefficiency in land allocation, given the continuing lack of transparency and competition in the assignment of land use rights. In particular, the productivity level of industrial land (often measured by output value per unit of land) is low. As a result of fierce competition among municipalities to attract foreign investment, industrial land often is leased significantly underpriced. Provision of cheap or free land, for instance, is a common way for local agencies or state units to enter joint ventures with foreign investors. Another tactic is the creation of special development zones, which allow virtually autonomous rights of land subdivision and concession in land premium. This mentality of "if you build it, they (investors) will come" has led to idle industrial land and development zones (see Chapter 6). While engaged in industrial development, some firms also take advantage of the low prices to stockpile land. As a result, the average share of industrial land use in Chinese cities runs at about 21 percent and may be about 10 percentage points higher than the norm in Western cities. An official estimate in 2004 also put underutilized land at about 8 percent of total urban construction land in China (Association of Mayors of China 2009).

The footprint of urban land use also is expanding more than necessary as a result of fragmented spatial development in urban fringes. Encroachment into farmland is of increasing concern for the central government, given that arable land is a scarce, diminishing resource in China. Previously, urban fringes were often characterized by mixed land use patterns, where about half of the land would be for non-agricultural uses and about half of the agricultural land was for vegetable plantation. Alarmed by a steady loss of arable land, the central government issued a series of policies for land preservation and instituted a system of annual quotas and hierarchical review and approval (depending on the size of converted farmland) in the 1980s. But this system failed to force local governments to curtail farmland conversion. In the 1990s, the central government replaced the approval system with a more realistic rule of "no reduction." The results remain mixed, but the overall pace of farmland conversion has slowed in some areas (Lin 2009).

Such encroachment infringes upon farmers' interests and their collective stakes in farmland. It was reported that non-agricultural construction took away more than 1 million hectares of arable land in 2004 (Wu et al. 2007). While generally powerless against land conversion, farmers have nevertheless ratcheted up their resistance and sometimes expressed their anger through protests. Under pressure, the central government amended the Constitution in 2004 to create legitimacy for the state in acquisition of collective land for public interest. This implies that local governments retain the power to permanently change collectively owned land to state ownership. Such proceedings resemble the concept of eminent domain in the West – the power of government to take private property for a public use, without the owner's consent, on the payment of just compensation. But there are important differences. China's land, despite collective ownership in rural areas, is ultimately owned by the state. What is at stake is the user right and level of compensation. Also, how individual claims by farmers to collective land are decided vary from place to place. Even just compensation to a collective village may eventually filter down to affected farmers in an unjust fashion, because of corruption and conflicts of interests among village cadres.

The increasing awareness of property rights is echoed by urban residents facing redevelopment. If a site is planned for redevelopment, sitting tenants need to be compensated. There are three common methods of compensation. First, sitting tenants can choose to relocate to so-called "resettlement housing" built by local governments or with government subsidies. Second, they can choose to occupy a new unit built on site after paying the differentials between the new and old units. Third, they can opt for an all-cash package. The level of compensation is negotiated among existing residents, the developer, and the local government. It has to take into consideration not only the structure to be demolished, but also the market value of the site. As cities undergo physical transformation, local governments have initiated large-scale, often sweeping, demolition of inner-city neighborhoods to make way for new, commercial development (more in Chapter 8). Beginning in the early 1990s, affected residents began to protest unfair compensation and forced evictions. Such protests range from individual complaints to mobilization networks across neighborhoods (more on the latter in Chapter 14; Hsing 2010). The culmination of rights protest was the widely publicized "Wu Ping's nail house" in Chongqing (Box 10.1).

Box 10.1 Wu Ping's "nail house": one woman's quest for property rights in Chongqing

A "nail house" (*dingzihu*) is a term describing a home belonging to people who refuse to relocate and make room for redevelopment. In 2004, preparation for the construction of a six-story shopping mall began in the Jiulongpo district of Chongqing in southwest China. A total of 280 families had already been relocated from the site during the preceding 2 years, except for one two-story home owned by Wu Ping and her husband Yang Wu. Wu Ping had planned to open a restaurant in the home's ground floor. The developers cut their power and water, excavated a 10-meter-deep pit around their home, and requested administrative adjudication from the district Housing Administration Bureau in 2005 (Figure 10.3). The Bureau subsequently filed and was granted a request for forceful eviction with the district's general court in March 2007. Wu Ping, at the same time, issued interviews and frequent press releases and garnered unprecedented public support. She and her husband eventually settled with the developers in April 2007 (Lu 2007).

Figure 10.3 Wu Ping's "nail house."

Source: Wikimedia Commons, retrieved 3 January 2012 from http://commons.wikimedia.org/wiki/File:Chongqing_yangjiaping_2007.jpg (donated by Zola).

The timing of the resolution coincided with the passing of China's first modern private property law in March 2007. The law prohibits governments from land acquisition, except when it is in the public interest, but does not stipulate whether making room for private commercial development constitutes public interest. Wu Ping's "nail house" symbolizes the heightened public awareness of property rights protection.

Housing reforms and implications

Recapping housing reforms

The start of market reform in 1979 led to a drastic increase in government-allocated funds for housing construction. Throughout the 1980s, housing construction funds made up around a quarter of total investment funds and reached 7 percent of GNP (compared to about 1.5 percent prior to 1978: Wang 2003). In 1980, housing reforms began to take place. The first change was to share funding for housing construction between municipal authorities and enterprises. Some cities also began an experiment with commercialization by selling municipal public housing units to enterprises and individuals. By 1986–1988, more experiments went on nationwide to raise rents in the public housing sector and to introduce housing subsidies for state employees. However, these early efforts failed to solve housing shortages and maintenance problems caused by the highly subsidized housing provision, which subsequently limited the production and distribution of new units. In addition, the heavy burden of housing provision and social welfare contributed to substantial losses by many state enterprises.

Then, a turning point. In the early 1990s the central government required all cities to undergo housing reforms. The first step was to sell the existing public housing stock at very low prices and on a large scale, mostly to sitting tenants. Second, a gradual shift began, from the system of public housing provision to a paid, market-based distribution system. Largely based on the acclaimed experience of Singapore in increasing home ownership, a new housing savings system was established and continued to the present. This involves a housing provident fund through which every employer and employee makes a contribution to the employee's housing savings. The savings can only be used to purchase housing or for housing repairs, but also will be available as a supplement to the employee's pension upon retirement. The rates of saving vary from place to place, between 5 and 10 percent of an employee's monthly salary in the public sector. Some private firms pay a higher rate, while some small businesses and loss-making state enterprises do not participate in the system or pay a very low (2 percent) contribution (Wang and Murie 2000).

A parallel reform effort called for the establishment of housing insurance, finance and loan systems that would enable market-based housing development and distribution. At the beginning, only select banks designated by municipal authorities were authorized to supply mortgages to qualified home-buyers, drawing from the housing provident funds. By the late 1990s, all banks were cooperating with the funds to offer different types of housing finance, including regular mortgages, one-time withdrawal for home purchases, individual housing construction (by any resident with a rural spouse), and rent payments for low-income residents (Wang 2001). Some municipalities also injected public money into such funds to boost local property markets. In addition to helping with individual home finance, housing provident funds have provided finance for short-term loans to work units, housing cooperatives, and developers for building affordable housing. To facilitate housing finance, China also is developing a credit-rating system. In January 2006, the People's Bank of China launched a central credit history database to track payments of over 500 million people on their bills, taxes, credit cards, taxes, and loans.

The public sector, particularly work units, continued to be involved in housing provision at this point. The construction of the new, so-called commercial housing since the early 1990s involved a variety of public-owned development companies, quasi-public development companies, and, only recently, independent private companies. Once constructed, commercial housing could be sold to work units or directly to urban residents. Work units then could sell it to their employees at discounted prices. Employees buying at discounted prices often did not receive full right to their housing since the work units retained an interest. Employees in non-state sectors, such as foreign firms and private businesses, were not eligible for discounted housing and housing subsidies. Some state work units, using a different approach, provided cash payments to help their employees purchase fully-priced commercial housing. The amount of payments was determined by criteria such as rank, seniority, and merit. As a result, a significant amount of the so-called commercial housing remained a partially redistributive good rather than a true commodity. For instance, in Beijing, more than 90 percent of commercial housing was bought by work units in 1992 (Zhang 2000), and about 46 percent of housing investment was made by work units between 1992 and 1997 (Wang 2001).

Another turning point occurred around 1999. The involvement of work units in commercial housing ceased. This marked the end of the provision of all welfare housing (through both municipal and work unit distribution). Sitting tenants could choose to buy out the property rights of their public housing. Individuals buying at market prices can enjoy a full ownership right, whereas those paying cost prices obtain a limited ownership share or only a use right. As a result of these reform measures, there is a near-complete withdrawal of public investment in housing (with the exception of a limited number of affordable housing programs). Housing is no longer a free public good for urban residents (Box 10.2).

Along with commercial housing also comes another new element of housing reforms – property management. Most of the commercial housing has been constructed in the form of comprehensive development or large residential

---

***Box 10.2* Equitable housing reform scheme in Guiyang**

By 1999, housing no longer was a form of social welfare. The role of work units diminished, limited to cash contribution to housing provident funds. This withdrawal of public provision was difficult for many urban residents: they could not afford to buy or rent on the commercial market. To make matters worse, some state enterprises were unable to fulfill the cash obligation initially because of funding shortfalls stemming from declining profits and budgetary allocation by their supervisory government agencies. Guiyang, the capital city of Guizhou province, adopted a unique approach to implementing the cash contribution scheme that since has been used as a demonstration model for other localities (Lee and Zhu 2006).

Guiyang first revaluated the housing stock occupied by state employees, taking into account the location and physical conditions of each unit. This was followed by an assessment of housing entitlements of individual employees according to their salary levels and seniority. They were then assigned a housing allowance account. Those with housing areas beyond their entitlements would reimburse the municipal housing bureau, while others with under-entitlements would receive compensation grants. Once the differences were settled, sitting tenants were given full property rights. This model considered differential land rent as an essential component of housing benefits and costs. It also stipulated that once the price-to-income ratio (or affordability ratio) reached beyond 4, work units would need to provide cash allowance to employees (Guo 2000).

development projects undertaken and sold by real estate companies. This has allowed municipal housing bureaus and work units to retreat from housing provision, as well as subsequent maintenance and repairs. As such, housing management also becomes market-based. Residents in an estate pay an annual fee to support the operation of newly created property management companies that includes general maintenance, security and parking enforcement, landscaping, and waste collection (see Chapter 13). The fee varies by price levels of each estate and can be quite substantial in high-end housing developments.

Overall, housing reforms have produced some astounding results. There is a significant improvement in housing conditions. According to the 2000 census, housing space for urban households reached 25 square meters per person. This suggests a close to threefold increase in two decades. By 2005, the level rose to nearly 28 square meters, as indicated by that year's national One Percent Population Survey. Urban residents now have a broader range of housing choices. They can purchase new, commercial housing units, paying market prices. Many people have bought property rights to the homes they acquired under the old welfare housing system. If desired, they can put their homes on a secondary housing market and trade for better housing. They also can rent directly from homeowners with rental permits, as well as rent private housing inherited by urban families (a legacy of pre-1949 housing stock) or constructed by suburban farmers. However, reforms in urban housing provision have largely overlooked the needs of the migrant population in cities (see Chapter 5).

Home ownership and affordability

After nearly two decades of housing reforms, market mechanisms have largely replaced the old welfare-based system. Housing tenure choices, as a result, have changed. According to the most recent national survey (2005 One Percent Population Survey), home ownership now averages about 75.7 percent among urban households, up from 71.9 percent in 2000 (the first four columns together in Table 10.1). Ownership, however, is different from the practice in most other market economies. Since urban land is still publicly owned in China (by the state), homeowners can claim property rights only to the structure but not to the land underneath. Moreover, most urban housing is in the form of apartments, instead of single- or multi-family houses (Figure 10.4). Only in rural pockets of cities and in very high-end housing estates are there free-standing houses.

A common route to home ownership is through market mechanisms: urban residents purchase commercial housing, subsidized housing, or units formerly known as public housing. Commercial housing generally refers to new units, whereas used units are traded separately on a secondary housing market. A relatively new form, commercial housing only made up 27 percent of total housing investment in 1991 but doubled its share in just 3 years (Wang and Murie 1999). By 2005, about 16 percent of urban households owned such units (Table 10.1). For those households with higher income (above 10,000 yuan monthly), commercial housing was the overwhelming choice, indicating higher quality and better conditions. A substantial

Table 10.1 Types of urban housing tenure by income level, 2005 (percent of households)

Monthly income level (yuan)	Self-built housing	Commercial housing	Economic housing	Purchased public unit	Renting public unit	Renting private unit	Other housing	% of total households
Under 1,000	34.0	11.6	5.9	25.5	9.5	8.9	4.6	45.4
1,001–2,000	26.8	16.1	7.0	21.9	7.7	16.7	3.9	34.0
2,001–3,000	22.1	22.4	7.9	25.6	6.2	13.0	2.8	11.3
3,001–4,000	17.7	27.6	7.2	28.3	6.0	10.7	2.4	4.5
4,001–6,000	14.2	33.9	5.7	26.6	5.8	11.7	2.2	3.0
6,001–8,000	10.1	41.4	4.9	25.2	5.1	11.3	2.0	0.9
8,001–10,000	8.1	47.0	4.7	20.4	4.6	13.1	2.2	0.4
10,001–15,000	9.1	51.8	4.9	18.1	3.6	11.0	1.5	0.3
15,001–20,000	9.5	57.7	3.5	11.8	2.9	13.3	1.3	0.1
Over 20,000	14.5	53.9	2.7	9.5	3.0	14.9	1.5	0.0
Total	28.5	16.3	6.5	24.4	8.1	12.2	3.9	100.0
2000 Census	26.8	9.2	6.5	29.4	16.3	6.9	4.8	100.0

Source: Results from 2005 One Percent Population Survey (retrieved 21 June 2010 from http://www.stats.gov.cn/tjsj/ndsj/renkou/2005/renkou.htm); National Bureau of Statistics of China 2004.

Note: The total number of urban households surveyed was 1,528,461 in 2005. The housing information for 2000 was drawn from 8,154,917 urban households who filled out the long census form. Other housing may include staying with other urban residents, dormitories in factories or on construction sites, and informal housing built without official approval.

Figure 10.4 A typical commercial housing estate.

Source: Weiping Wu.

number of families (about 24.4 percent) remained in public housing units that went on sale in the 1990s, even though they have the option to trade older housing units through the secondary housing market. Some types of public housing, such as university faculty housing, are not allowed to be traded on the secondary market.

Some urban households (about 6.5 percent in 2005) benefited from the national Comfortable Housing Project, launched in 1995 to create private-sector housing (with government support) for low-income urban families. Units were sold at cost to such families, especially those with inadequate (less than 4 square meters) or no accommodation (Hui and Seabrooke 2000; Li 2000). In 1998, this project was revised and given a new name – Economic and Comfortable Housing (*jingji shi-yong fang*). The new emphasis is on developing housing for low- and middle-income groups. The key factors that reduce the price of such housing include free land allocation, a regulated profit level for developers, smaller housing size, and reduced government charges during the development and sale process.

There remains a large share of private housing (28.5 percent in 2005). This reflects both a historical legacy and fluidity in the urban–rural continuum. Private housing in cities generally consists of two kinds: older housing units passed on from one generation to the next within the family, and self-constructed housing in suburban areas by farmers or former farmers on land allotted by their villages. The first kind was constructed before 1949 and subsequently confiscated by municipal authorities and turned into public housing. Returned to original owners in the 1980s, today such housing is not common in most cities, confined mainly to downtown areas. The majority of private housing likely is the second type. As urbanization proceeds, land use in suburban areas or urban–rural transitional areas has turned from agriculture to manufacturing or services. But a substantial amount of the old village housing remains, accommodating both the original owners and renters. This is another example of the complexity in studying the Chinese city: urban and rural intermingle in places officially designated as urban.

Clearly, housing markets in urban China are segmented. There are different factors underscoring different types of owner-occupied housing. In addition, housing stocks vary across different cities. In places where government agencies and work units were involved heavily, there would be more formerly public housing units (e.g., Beijing and Chongqing). Elsewhere, particularly in smaller cities, private and commercial housing may be more prevalent (Huang 2004). Within cities, housing tenure now is correlated with household income. Higher-income households are more likely to reside in commercial housing, whereas owners of private housing and formerly public units tend to have much lower income levels (Table 10.1). Across different spatial locations, housing quality also varies. Patterns begin to resemble how urban space was differentiated into upper and lower ends before 1949 (see Chapter 8).

Housing prices in urban China are high, compared to worldwide norms where a reasonable housing price-to-income ratio is between 6 and 7. According to real estate estimates, it costs an average couple 20 years' salary to buy an apartment around the Fifth Ring Road in Beijing, 18 years of a Shanghai couple's salary to buy a flat outside the Outer Ring, and 16 years' income of a couple buying in Shenzhen's outskirts (Gerson Lehrman Group 2010). Also, housing mortgages require at least a 30 percent down payment, assuming a buyer is qualified. There is evidence that, in Guangzhou and Shanghai, slightly less than one-third of commercial housing buyers have resorted to mortgages (Li 2010). Personal savings and parental contributions are major paths to home ownership (Box 10.3).

Rental housing is a more affordable option. A typical rental housing unit is the second type of private housing, located in urban–rural transitional areas (Figure 10.5). With ample land to build private living quarters, an average rural household of four people tends to have at least eight to ten rooms. Hence, there is much space for rental. To a limited number of poor urban families, subsidized municipal housing is provided at below-market rent – a practice that began in mid-2001. There is evidence that rents correspond more closely to the size and quality of housing (Zhou and Logan 1996). However, the rental market is still immature, with thousands of intermediate rental agencies operating under limited regulatory oversight. A large number of deleterious building and rental activity continues, largely in the form of unauthorized construction and leasing of unsafe dwellings. This problem is particularly serious in urban–rural transitional areas. There, land is more readily available, the migrant population is more concentrated, and local residents have more incentive to rent out rooms (due to the loss of agricultural income). Even when regulations on rental housing take effect in some cities, concerns for adequate housing conditions and rental rights tend to be secondary.

Box 10.3 Bubble and burst: affording a home in urban China

Imagine someone making an equivalent of US$18,000 a year (like what a shift manager makes at McDonald's in the USA) in a position to buy a US$250,000 house on his own. This is just what many young urban Chinese are doing today. So how do they manage? They save like mad (to continue a tradition of frugality and self-reliance), borrow from parents and relatives, or prepay mortgage debts (for high-income borrowers). China has one of the highest personal saving rates in the world, at about 25 percent. But, income levels remain low. The country's per capita annual nominal income is less than US$2,000 (about US$7,500 after adjusting for purchasing power parity).

The story of Mark Li, a software engineer in Shanghai, is a case in point. His monthly salary is about US$1,500, an enviable income for any young Chinese. Yet the steady rise of property prices outpaced his personal savings. To settle down, he had to borrow from his frugal parents to pay the down payment of US$75,000 – effectively their life savings. That helped him close on a US$250,000 three-bedroom apartment in August 2011, still in the midst of the housing bubble. But weeks later, the market began to slow: by November 2011 home prices nationwide had declined for the third straight month. Responding to this sign of burst, Li's developer slashed prices by a quarter on identical units in an effort to attract more buyers. This outraged Li and many others who paid full price. They confronted the developer and demanded refunds. The alternative would not be an option: they would lose the US$75,000 down payment. Similar demonstrations occurred in Beijing as well (Pierson 2011).

For millions of urban Chinese, home ownership is part of the new, middle-class dream. The building and inhabitation of new, commercial housing have effectively created new forms of identity for residents, as well as new forms of governance for communities (Zhang 2010; see Chapter 13). The property sector is a huge employer, accounting for about one-fifth of the country's economic output. The quest for housing also has become part of the national psyche, played out in television soap operas, songs, and new slang. For instance, debt-strapped home buyers have been dubbed house slaves (*fangnu*) (Pierson 2011). Much is riding on China's ability to avoid the kind of housing bubble burst seen in the recent global economic downturn.

Property boom and housing bubbles

Today, housing has become a major form of expenditure for urban households. Over 80 percent of homes are now privately owned. China's housing investments exceeded 8 percent of GDP in 2005, a key driving force of urban economic expansion. Furthermore, home loans have grown rapidly since 2000 and mortgage

Figure 10.5 Private housing for rental in an urban–rural transitional area.

Source: Weiping Wu.

Note: The sign in the picture indicates availability of rental housing.

portfolios accounted for 12 percent of China's GDP in 2005 (Gibson 2009). Given such magnitudes, fluctuations in housing prices would affect the financial well-being of households. They also would trigger or reinforce fluctuations in the economy on the macro scale.

Housing prices have shot up. Some housing markets, such as in Shenzhen, have recently stagnated or even declined. But housing prices in other major cities are still growing by as much as 10 percent a year (Gibson 2009). The National Bureau of Statistics reported that housing prices on average increased by 15 percent nationwide in 2009. Since 2002, many government officials and economists have claimed that some cities should be wary of signs of housing bubbles – when real estate prices are not in line with variations of macroeconomic variables (e.g., growth of disposable income, consumption expenditure). There is evidence that Shanghai already showed signs of housing bubbles in 2003 (Hui and Yue 2006).

Such bubbles come alongside a highly distorted supply structure. In recent years, the proportion of ordinary dwellings has fallen, giving way to the development of villas and luxury apartments. In addition, speculative home purchases are increasing, after the sharp downturn in stock prices in late October 2007. For some urban households, the housing market could be taken as a substitute for the stock market. Foreign investors also are playing a key role in the housing boom.

Learning from the devastating impact of Japan's real estate market collapse in the 1990s on that nation's economy, the Chinese government has promulgated a series of administrative actions to cool the housing market. Since 2005, the People's Bank of China has repeatedly raised the benchmark loan rate in hopes of tempering the housing boom. Individuals applying for mortgages on apartments larger than 90 square meters must now supply at least a 30 percent down payment. The government has implemented various taxes on real estate transactions (Gibson 2009). Credits also were suspended to buyers of third and more homes. Another so-called 70/90 policy, unveiled in 2006, stipulated that 70 percent of commercial housing under construction in any given city should be earmarked for housing units with area of 90 square meters or less (the effectiveness of this policy, however, is in doubt).

But cooling the market has proved to be more difficult than anticipated. Even the recent global economic recession (2008–2009) failed to make a significant dent in China's property boom (Figure 10.6). Housing prices dropped sharply in mid-2008 but went back up within a year. They rose 10–15 percent in the first quarter of 2010 in cities such as Beijing, Hangzhou, Shanghai, and Shenzhen (Lan 2010). It was not until late 2011 that there were signs of the "housing bubble losing air" (Box 10.3). On the one hand, national-level regulating agencies lack effective clout – political as well as economic – over local interests. On the other

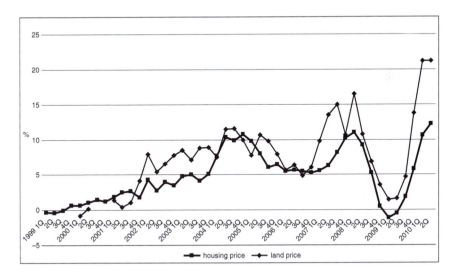

Figure 10.6 Growth rates of urban land and housing prices, 1999–2010.

Source: Su and Tao 2010.

Note: The data are drawn from the National Bureau of Statistics of China. Before the fourth quarter of 2006, they are based on 35 large and medium-sized cities. After that time, they are based on 70 cities.

hand, local governments have a powerful incentive to place their local banks' excess liquidity into real estate, which would expand the local housing market and their bottom lines. They have a direct interest in seeing land and construction prices rise, to increase government revenues. As much as 40 percent of local government revenues is tied to property development (Gibson 2009).

Conclusion

Marketization has been a prominent feature of urban land and housing development in China. Reforms in both areas have been gradual, in contrast to the rapid, wholesale privatization programs implemented in other former socialist countries. Nonetheless, within a matter of two decades, housing morphed from a form of social welfare to a private good largely based on market principles. Land reform, on the other hand, has been through a more treacherous path. Collusion between local governments and developers, price distortions, and social discontent remain large. Rising land prices likely have played a part in pushing up housing prices in large cities, making home ownership less affordable for ordinary residents. In the aftermath of the recent global economic downturn, concerns about housing bubbles and their subsequent burst remain high.

Bibliography

Association of Mayors of China. 2009. *China Urban Development Report 2008*. Beijing: China City Press.

Bertaud, Alain and Renaud, Bertrand. 1997. "Socialist Cities without Land Markets." *Journal of Urban Economics* 41(1): 137–151.

Chen, Cheryl. 2011. "Income for Local Officials in China Relies on Land, and More Land." Epoch Times (retrieved on 3 January 2012 from http://www.theepochtimes.com/n2/china-news/income-for-local-officials-in-china-relies-on-land-and-more-land-62535.html [based on data from http://photo20.hexun.com/p/2011/0917/454408/b_vip_38692D80562CE80A0A28D2D78355E8FC.jpg]).

Gerson Lehrman Group. 2010. "Housing Affordability in China." Published on 9 March at www.chinadaily.com.cn (retrieved on 10 July 2010).

Gibson, Neil. 2009. "The Privatization of Urban Housing in China and its Contribution to Financial System Development." *Journal of Contemporary China* 18(58): 175–184.

Guo, Shuqing. 2000. *Monetarization Reform of Housing Distribution in Guizhou [Guizhou zhufang fenpei huibi hua gaige]*. Beijing: China Finance and Economy Press [*zhongguo caizheng jinji chubanshe*].

Hsing, You-tien. 2010. *The Great Urban Transformation: Politics of Land and Property in China*. Oxford: Oxford University Press.
This book emphasizes the centrality of cities in China's ongoing transformation and is based on fieldwork in 24 Chinese cities between 1996 and 2007. The author advances two theoretical tenets. One, urbanization of the local state, is a process of state power building entailing an accumulation regime based on the commodification of state-owned land, the consolidation and legitimization of territorial authority through construction projects, and a policy discourse dominated by notions of urban modernity. The second,

civic territoriality, encompasses the politics of distribution engendered by urban expansionism, and social actors' territorial strategies toward self-protection.

Huang, Youqin. 2004. "Housing Markets, Government Behaviors, and House Choice: A Case Study of Three Cities in China." *Environment and Planning A* 36: 45–68.

Hui, Eddie Chi-Man and Seabrooke, Bill. 2000. "The Housing Allowance Scheme in Guangzhou." *Habitat International* 24(1): 19–29.

Hui, Eddie Chi-Man and Yue, Shen. 2006. "Housing Price Bubbles in Hong Kong, Beijing and Shanghai: A Comparative Study." *Journal of Real Estate Finance Economy* 33: 299–327.

Lai, Lawrence Wai Chung. 1995. "Land Use Rights Reform in China: Some Theoretical Issues." *Land Use Policy* 12(4): 281–289.

Lan, Xinzhen. 2010. "Toil and Real Estate Trouble." *Beijing Review* April 15: 28–29.

Lee, James and Zhu, Yapeng. 2006. "Urban Governance, Neoliberalism and Housing Reform in China." *The Pacific Review* 19(1): 39–61.

Li, Si-Ming. 2000. "Housing Consumption in Urban China: A Comparative Study of Beijing and Guangzhou." *Environment and Planning A* 32: 1115–1134.

Li, Si-Ming. 2010. "Mortgage Loan as a Means of Home Finance in Urban China: A Comparative Study of Guangzhou and Shanghai." *Housing Studies* 25(6): 857–876.

Lin, George C. S. 2009. *Developing China: Land, Politics, and Social Conditions*. London: Routledge.

This is a systematic documentation of the pattern and processes of land development taking place in post-reform China. The author goes beyond the privatization debate to probe directly into the social and political origins of land development. The rural–urban interface is shown to be the most significant and contentious locus of land development where competition for land has been intensified and social conflicts frequently erupted.

Lin, George C. S. and Ho, Samuel P. S. 2005. "The State, Land System, and Land Development Process in Contemporary China." *Annals of the Association of American Geographers* 95(2): 411–435.

Logan, John R. and Bian, Yanjie. 1993. "Inequalities in Access to Community Resources in a Chinese City." *Social Forces* 72(2): 555–577.

Lu, Shangqin. 2007. "Analysis of Chongqing 'Nail House' Event from the View of Property Law [*Chongqing 'dizihu' shijian de wuquanfa fengxi*]." *Journal of Anhui University of Technology (Social Sciences)* 24(4): 33–34 and 53.

National Bureau of Statistics of China. 2004–2008. *Statistical Yearbook of China's Land Resources*. Beijing: China Statistics Press.

National Bureau of Statistics of China. 2004. *China Population Statistical Yearbook 2004*. Beijing: China Statistics Press.

Pierson, David. 2011. "China's Housing Bubble is Losing Air." *Los Angeles Times*, December 13 (retrieved on 17 January 2012 from http://articles.latimes.com/2011/dec/13/business/la-fi-china-housing-bubble-20111213).

Su, Fubing and Tao, Ran. 2010. "Growth Coalition in China's Rush to Urbanization: Stimulation and Stabilization in China's Real Estate Markets, 2008–2010." Paper presented at the American Political Science Association annual meeting, Washington, DC, September 2–5.

Wang, Feng. 2003. "Housing Improvement and Distribution in Urban China: Initial Evidence from China's 2000 Census." *The China Review* 3(2): 121–143.

Wang, Yaping. 2001. "Urban Housing Reform and Finance in China: A Case Study of Beijing." *Urban Affairs Review* 36(5): 620–645.

Wang, Yaping and Murie, Alan. 1999. *Housing Policy and Practice in China*. London: MacMillan Press.

Wang, Yaping and Murie, Alan. 2000. "Social and Spatial Implications of Housing Reform in China." *International Journal of Urban and Regional Research* 24(2): 397–417.

Wong, Christine P.W. and Bird, Richard. 2004. "China's Fiscal System: A Work in Progress." Paper presented to the Conference on China's Economic Transition, Pittsburgh, November.

World Bank. 1993. *China: Urban Land Management in an Emerging Market Economy.* Washington, DC: The World Bank.

Wu, Fulong. 1996. "Changes in the Structure of Public Housing Provision in Urban China." *Urban Studies* 33(9): 1601–1627.

Wu, Fulong, Xu, Jiang, and Gar-On Yeh, Anthony. 2007. *Urban Development in Post-Reform China: State, Market, and Space.* London: Routledge.

This is an in-depth and thorough study of the urban development process in China's transition era, focusing primarily on land and housing development. It shows how the market has been "created" under post-reform urban conditions, highlighting how changing urban governance towards local entrepreneurial state facilitates market formation. It provides a wealth of conceptual ideas and empirical details to help specialists rethink the relationships between marketization, state action, and the transformation of urban spaces.

Zhang, Li. 2010. *In Search of Paradise: Middle-Class Living in a Chinese Metropolis.* Ithaca: Cornell University Press.

This book is an informed account, based on anthropological research in Kunming, of how the rise of private home ownership is reconfiguring urban space, class subjects, and ways of life in the reform era. A rising middle class hopes to find material comfort and social distinction in newly constructed and, often gated, communities, profoundly transforming the physical and social landscapes of urban China. The privatization of property and urban living also has engendered a movement of public engagement among homeowners as they confront the power of the developers.

Zhang, Tingwei. 2002. "Urban Development and a Socialist Pro-Growth Coalition in Shanghai." *Urban Affairs Review* 37(4): 475–499.

Zhang, Xingquan. 2000. "Privatization and the Chinese Housing Model." *International Planning Studies* 5(2): 191–204.

Zhou, Min and Logan, John R. 1996. "Market Transition and the Commodification of Housing in Urban China." *International Journal of Urban and Regional Research* 20(3): 400–421.

Zhu, Jieming. 1994. "Changing Land Policy and its Impact on Local Growth: The Experience of the Shenzhen Special Economic Zone, China, in the 1980s." *Urban Studies* 31(10): 1611–1623.

Zhu, Jieming. 2002. "Urban Development under Ambiguous Property Rights: A Case of China's Transitional Economy." *International Journal of Urban and Regional Research* 26(1): 41–57.

Part IV
Urban life

11 Environmental quality and sustainability

The environment is one of the key issues in China's contemporary urbanization, urban development, and planning. Pan Yue, vice-minister of then China's State Environmental Protection Administration (renamed as the Ministry of Environmental Protection in 2008), has warned: "[t]he [economic] miracle will end soon because the environment can no longer keep pace" (Economy 2007). Yet, paradoxically, China also is home to some of the most environmentally innovative urban planning in the world. Chinese cities have begun a remarkable range of efforts to mitigate the environmental impacts of rapid urbanization – from the introduction of hybrid buses to the imposition of stringent air quality standards for industries. Nonetheless, these efforts are offset by rapid increases in pollution-generating activities as the Chinese economy surges forward.

This chapter examines the challenges facing China's urban environment and the new approaches to planning and urban management designed to address them. China's primary urban environmental problems are in air quality, water quality, solid waste management, and sewage treatment. This chapter provides an overview of each of these issues today, followed by a discussion of efforts to improve the urban environment. The following questions should help guide the reading and discussion of the materials in this chapter:

- Why are China's air quality and water quality so problematic today?
- What are the linkages between China's rapid economic development and its growing solid waste problems?
- How is China trying to address its environmental problems in cities?
- How effective are the different strategies for mitigating urban environmental problems?
- How do different interests (urban residents, business managers, government officials) seem to perceive urban environmental problems?
- Which mitigation programs have the greatest possibility of success?

State of the urban environment

Acid rain is falling on one-third of the Chinese territory, half of the water in our seven largest rivers is completely useless, while one-fourth of our citizens does not have access to clean drinking water. One-third of the urban population is breathing polluted air, and less than 20 percent of the trash in cities is treated and processed in an environmentally sustainable manner.

(Pan Yue, vice-minister of the former State Environmental Protection Administration, 2005; cited in Lorenz 2005)

Challenges to improving China's urban environment are twofold: first, new environmental standards and procedures need to be established; second, those standards need to be enforced. As Elizabeth Economy (2007: 43) explains, there are "plenty of laws and regulations designed to ensure clean water [but] factory owners and local officials do not enforce them."

There is hope for rapidly developing cities, however. As Ajay Chhibber, UN Assistant Secretary General and the United Nations Development Program's Regional Director for Asia and the Pacific, has noted, much of China's urban area is in new development, and this will continue as cities grow over the next few decades. Rather than retrofitting old polluting factories and trying to convert outmoded infrastructure, China is increasingly building from scratch. This new construction can be guided by environmental priorities from the outset (Chhibber 2011). This potential for stricter guidelines on new development is reflected in China's most recent (2008) national planning law, which includes guidelines on environmental planning for new areas.

While high-profile cities such as Beijing and Shanghai, with the aid of high-profile events such as the Olympics and the World Expo, massive foreign invest-ment funds, and the most talented college graduates in the country, can make great and innovative strides in environmental planning, it is the mid- and small-sized cities whose environments suffer the most from China's rapid development.

As China has become an increasingly large participant in the global economy, its impacts on the world environment have grown. As a result, much world atten-tion has been focused on the state of the Chinese environment and environmental planning and mitigation in China. In 2007, the Worldwatch Institute devoted its annual *State of the World* report to cities. Both the problems and the possibilities for the Chinese urban environment played a central role in this publication.

Air quality

During the 1970s, black smoke from stacks became the characteristic of Chinese industrial cities; in the 1980s, many southern cities began to suffer serious acid rain pollution; and recently, the air quality in large cities has deteriorated due to nitrous oxides (NOx), carbon monoxide (CO), and photo-chemical smog, which are typical of vehicle pollution.

(He et al. 2002)

Air quality has been an issue in all Chinese cities. In 1997, a World Bank report estimated that 178,000 people per year suffered premature deaths due to pollution. In 2005, more than 3 million urban residents in China died of chronic or acute air pollution-related conditions (Matus et al. 2011). Nearly half of China's cities are plagued with dangerous particulate pollution (Lorenz 2005). The primary source of urban air pollution was the burning of coal in power plants, industries, domestic heating, and even in kitchens until the late 1990s. But recently, vehicle emissions have become a significant component of air pollution as well (Figure 11.1). Another significant source of urban pollution, particularly in northern China, is dust blown in from degraded rural areas.

China relies on coal for about 70 percent of its energy needs, and doubled its consumption of coal during the first decade of the twenty-first century. Coal contributes to particulate pollution in the atmosphere that leads to respiratory hazards, sulfur dioxide emissions, the main cause of acid rain, and carbon dioxide emissions – one of the main contributors to global warming (Economy 2007). China's carbon dioxide emissions rose 10.4 percent in 2010, nearly twice the world growth rate in carbon dioxide emissions (Chestney 2011). The more developed cities in eastern China have been working to diversify and clean their energy sources. In Shanghai, for example, carbon emissions attributed to coal now account for about 15 percent of the city's total carbon; automobile exhaust accounts for about 50 percent (Chan and Yao 2008).

More than half of China's cities experience acid rain; the problem is severe in about half of those cities – especially in the southeastern region south of the Yangtze River. China emits more sulfur dioxide than any other country in the world. At times these emissions are so extreme that they generate dangerous conditions not only for the health of plant life, but for people as well. In winter 2007, for

Figure 11.1 Air quality in downtown Shanghai, 2011.

Source: Piper Gaubatz.

example, the city of Guangzhou experienced 100 percent acid rain (pH level 3.8), an acidity level so high that it could contribute to the spread of infectious diseases (Ma and Schmitt 2008). There has been a slight improvement in acid rain conditions in Chinese cities during the past few years, but this problem will not be improved until the air pollution has been significantly reduced (MEP 2010a).

Automobile emissions have become one of the main sources of air pollution in large and medium-sized cities. In 1980, there were only about 680,000 cars in China. Most urban dwellers rode buses or bicycles; China was held up as a model of "bicycling culture" (more in Chapter 8). But transportation became increasingly diverse, and included both private cars and rising numbers of taxis (Box 11.1).

***Box 11.1* What happened to the "bread loaf" taxis?**

In 1990, Beijing had only one subway line, public buses were overcrowded, there were fewer than 14,000 taxis, and the city was preparing to apply to host the 2000 Olympics – an event that eventually included more than 10,000 athletes and 6 million spectators. Clearly the city's public transportation network was in need of an upgrade. Moreover, taxis were too expensive for most Beijing residents, and only served popular destinations. In order to apply a "quick fix," in 1991 the municipal government introduced *miandi* (little bread loaves) – tiny yellow five-passenger vans with motorcycle engines and very low fares. Taxi transport became affordable and convenient within a matter of months. A survey of 1,000 Beijing residents conducted by the China Social and Economic Study Center in December 1993 found that 60 percent had used a taxi at least once, and 10 percent used them regularly (He 1994). Legislative changes allowed any of the city's work units to operate taxis. The numbers of taxis in the city rose from 14,000 in 1991 to 28,962 in 1992, 56,684 in 1995, and 60,000 in 1997 (Gaubatz 1999).

Yet by 1998, the proportion of *miandi* in the city's traffic had dropped. Not only were there more vehicles of other types to dilute the sea of yellow that once characterized the *miandi*'s dominance of Beijing's traffic, but a large proportion of *miandi* had by this time been taken out of service due to mechanical problems or replacement with more reliable and comfortable sedans. Flooding the city with thousands of two-stroke engines had generated a significant increase in air pollution. Beijing officials estimated that the *miandi* were responsible for about 15 percent of Beijing's air pollution in the mid-1990s, though they were only 1.4 percent of the city's traffic. Meanwhile, Beijing was preparing to bid for the 2008 Olympics as "Green Beijing." The municipal government ordered the *miandi* destroyed. In 1999, all of the remaining *miandi* in the city were taken to the Shougang Iron and Steel complex in Beijing's western suburbs, torn apart, and crushed. Ironically, the Shougang Iron and Steel complex itself was later moved farther away from Beijing in order to mitigate its own pollution in preparation for the Beijing 2008 Olympics.

By 2009, there were 170 million automobiles in China. It has passed increasingly strict emission standards for cars in 2000, 2005, 2007, and 2010. However, both the rapid rise in numbers of automobiles and the continuing operation of high-emission cars and trucks continue to have a devastating effect. The Ministry of Environmental Protection estimates that although 28 percent of the vehicles on the road are rated as "high-emission," they account for about 75 percent of all vehicle-based pollution (MEP 2010b). During the Beijing Olympics, about half the private vehicles on the roads (taxis were exempt) were removed as part of a stringent effort to reduce the city's air pollution. The measure was considered successful. In April 2009, Beijing implemented a more limited license plate-based control on driving that keeps about 20 percent of the city's cars off the roads each day. A similar scheme was used in Shanghai, about a decade earlier, to ward off congestion in the downtown area. China's cities also have introduced hybrid and natural-gas buses to curb pollution generated by the public transport system.

Another source of air pollution in northern Chinese cities, in particular, is fine, yellow loess dust blown in from the long-deforested upper reaches of the loess plateau (see the discussion of Taiyuan, a city that experiences these dust storms on an annual basis, in Box 1.1). Loess is particularly fine-grained material which travels easily on the wind. In fact, the deep loess deposits of northern China are the result of loess being blown across much of Eurasia in the distant past. Due to degradation such as deforestation which occurred long ago, vast areas of the loess plateau in Shanxi province, Inner Mongolia Autonomous region, Shaanxi province, and Gansu province are covered in bare loess. This loess is picked up each spring by the prevailing winds and deposited in cities such as Taiyuan and Beijing. This regional transport of atmospheric particles generates visible clouds of yellow, swirling dust. Local residents are urged to stay indoors or use dust masks during these events. In recent years, there have been efforts to mitigate this problem through reforestation, but progress is slow.

Water quality

Water is a significant issue in Chinese cities, in terms of both supply and quality. About two-thirds of China's cities have less water than they need, and about 16 percent of cities have serious water shortages. The water supply problem stems from both growing demand and wasteful or inefficient use of existing water resources. Water consumption in cities has risen dramatically as the rapidly growing industrial sector consumes ever-increasing resources, especially water. Even domestic use, though much less significant than industrial use, is rising rapidly, as urban residents take showers and use washing machines, dishwashers, and other modern appliances. As for inefficiency, the Ministry of Housing and Urban–Rural Development estimates that as much as 20 percent of urban water may be lost underground due to leaky pipes (this is a problem in many cities around the world). The shortage of urban water is not only an inconvenience for urban residents, but it generates serious environmental issues as well. A number of eastern cities have overpumped their aquifers (in other words, they have used too much of their groundwater reserves), which has caused land subsidence (by more than 2 meters in Beijing and Tianjin) and threatens saltwater intrusion into the freshwater aquifer as subsurface coastal saltwater seeps in to replace the freshwater that has been pumped out (Economy 2007).

Water pollution is another serious issue. The Ministry of Environmental Protection estimates that about 74 percent of urban well water in China is unfit for drinking, and groundwater is of poor quality in many cities. Ninety percent of urban waterways are polluted (State Council 2005). For example, Lake Dian, which lies just south of the city of Kunming, remains heavily polluted with nitrogen and phosphate despite years of mitigation efforts, due to the fact that about 62 percent of the rivers flowing into the lake are severally polluted (MEP 2010a).

The problem of drinking water quality in Chinese cities also has generated interest in the privatization of water supply. Initially this could be seen in the rising consumption of bottled water in Chinese cities during the 1990s. This led, in a few large cities, to the development of small-scale piped high-quality water networks (see Chapter 9). Initial costs for such networks were high, but delivery of high-quality water with this method was ultimately less expensive than bottled water. Nonetheless, early efforts of providing Dual Water Systems, in which piped water was differentiated by quality, had a distinct socioeconomic component, by offering better-quality water at a premium to those who could afford it. Most of the companies operating these systems were joint ventures between private companies and public water authorities, further complicating the justification for the limited offering of potable water (Boland 2007).

Nonetheless, the trend in most Chinese cities today is to improve water quality within the existing systems. The Ministry of Environmental Protection and four other ministries released the Urban Drinking Water Sources Protection Plan 2008–2020 in 2010 to support the removal of pollution sources near public drinking-water supplies, such as drain outlets and factories.

Solid-waste management

One of the major byproducts of rapid urban and economic development is a sharp rise in garbage. Whereas urban residents in the lean years of the 1960s may have intensively recycled and reused in order to stretch meager resources, the far more affluent urbanites today are faced with increasing volumes of disposables. The total volume of solid waste in Chinese cities has increased more than six times since 1979 (Li 2010). The Association of Environmental Protection Industries predicts that China's municipal solid-waste output will continue to rise to 179 million tons in 2015, and 210 million tons in 2020 (MSW 2011).

Collecting, processing, and storing the growing piles of urban waste is a challenge for urban governments. In the classic "waste hierarchy," the most favored way to handle municipal waste is to reduce the amount of waste and to reuse as much as possible, followed by recycling, recovery (recovering energy), and finally, the least environmentally, socially, and economically sound method: relegating it to a landfill. In Chinese cities today, reduction and reuse are negligible, recycling accounts for about 3 percent of municipal waste, and recovery (incineration) accounts for about 15 percent of municipal waste. The remaining 82 percent of waste is packed into growing landfills (Li 2010). In March 2011, the Ministry of Environmental Protection announced plans to improve garbage separation and collection, recycling and reducing package waste in cities (Figure 11.2).

Solid-waste management is particularly a problem for mid- and small-sized cities, which lack the resources for effective processing. In the mid-sized city of Tieli, Heilongjiang (population 400,000), for example,

> there are only 15 sets of dump trucks, 2 back fighting machines and 1 bull-dozer in the sanitation department, MSW [municipal solid waste] collection is based on human tricycles, and there is almost no waste sorting, which would be piled up outside or landfilled in the only one simple equipped landfill site. Waste relevant equipment and instruments in the city now are inadequate and not well organized and transport efficiency is low. The facilities of waste collection, transportation are primitive. Smell, and drift material and exudates will cause secondary pollution.
>
> (Li 2010: 42)

In fact, there is great variation in waste generation among cities. All Chinese cities produce a large proportion of organic kitchen waste – about 60 percent

Figure 11.2 Solid-waste management containers fortified for high wind conditions, Shenzhen, 2009.

Source: Piper Gaubatz.

nationally – but the composition of the waste does vary. For example, Beijing produces about 64 percent organic garbage, 11 percent paper, and 13 percent plastic. In contrast, Shanghai produces 67 percent organic garbage, only 5 percent paper, and 20 percent plastic (Zhang et al. 2010a). Variations usually reflect economic differences (higher-income households tend to generate more inorganic waste). In the case of Beijing and Shanghai, the differences may also stem from variations in local recycling practices. This emphasizes the need for procedures to be developed to address different local needs. There also can be significant differences in the characteristics of solid waste collection in different neighborhoods within cities. In Beijing, for example, waste collection is adequate in the high-rise central districts, but many suburban areas are underserved (Zhang et al. 2010a).

Recycling in China is relatively haphazard. Formal systems for the collection of recycling are not well developed. In spring 2011 a system was implemented in several suburbs of Guangzhou in which residents who do not properly sort their trash into recycling containers will be fined, but this is only a pilot project at this time and has not been widely implemented (Law 2011). What recycling takes place is largely left to the informal sector. As one observer notes:

> In major cities like Beijing, Shanghai, and Guangzhou, recycling bins are a common sight, but most serve as spare trash cans. There's no consistent pickup service by municipal workers. Mostly, scavengers looking to make some pocket change do the job.
>
> (Law 2011)

As Chinese cities become more affluent, their trash has become more valuable. As is the case in other developing countries, scavengers have learned to make a living off picking through garbage dumps for items of value, including recyclable materials. In a number of cities, small migrant communities have arisen adjacent to dumps in order to engage in scavenging. For example, in Beijing, migrants from Henan established a settlement near one of the city's large garbage dumps and support themselves through trash picking. While scavenging plays a vital role in the environmental functions of cities in China, it is also an important "last-resort" occupation for retirees, the unemployed, and recently arrived migrants as they struggle to cope with the rising cost of living in urban China. Scrap sorting and selling are, in fact, highly developed sectors of the informal economy. There are specialists in a wide range of materials, from rare earths to string, and buyers are available to purchase and reuse the materials. Like many informal activities, the materials and labor value which scavenging, as an informal occupation, generate eventually end up in the formal economy. The buying and selling of recycled materials, especially metals, are big business in China. Formal companies purchase scavenged materials for sorting and resale in an industry which, while reliant on informal labor (a typical scrap sorter might bring in less than US$100/ month), generates millions in profit (Minter 2009).

While scavenging and trash sorting are, at one level, solutions to the growing environmental challenges of the increase in solid waste which comes with rapid development, they also represent the first step in a process of relocating, rather than solving, environmental issues. Trash generated in the cities eventually ends up somewhere else. One of the most notorious examples of this process in China is the electronics dump at Guiyu, a small city northeast of Hong Kong. The electronic waste collected by scavengers in cities throughout China is eventually sent to Guiyu, where it is dismantled, sorted, and resold. Guiyu has been so successful in this operation that they take in international e-waste as well. In fact, about 80 percent of the waste processed at Guiyu comes from overseas. The city's 5,500 waste-processing businesses dispose of more than 1.5 million computers each year (Johnson 2006; Walsh 2009). Unfortunately, the manual labor and low-tech approach to dismantling the computers expose residents of Guiyu to high levels of toxins. As an article in *Time Magazine* explains:

> A lot of exported e-waste ends up in Guiyu, China, a recycling hub where peasants heat circuit boards over coal fires to recover lead, while others use acid to burn off bits of gold. According to reports from nearby Shantou University, Guiyu has the highest level of cancer-causing dioxins in the world and elevated rates of miscarriages.
>
> (Walsh 2009)

Wastewater treatment

In 2005, only 23 percent of the factories in 509 cities properly treated sewage before disposing of it; one-third of all industrial wastewater in China is released untreated. The situation for domestic waster is even worse – two-thirds of household wastewater is released untreated (Economy 2007). There has been significant improvement in sewage treatment in Chinese cities over the past few years (more in Chapter 9). Nonetheless, the Ministry of Environmental Protection reports that 48 percent of urban sewage was improperly treated in 2007. Ironically, China was known historically for its careful processing and reuse of human waste. In the traditional system, human waste was collected and treated (usually by being buried and mixed with other ingredients), then used as fertilizer for peri-urban agriculture. When properly processed, this traditional fertilizer was relatively effective; improper processing could contribute to health problems (Xue 2005).

Many cities in China have used financial incentives to encourage industries, in particular, to manage their wastewater better. Between 2000 and 2008, charges for wastewater disposal in China's larger cities rose by about 300 percent (Mol 2009). Cities also have begun to impose stricter enforcement on sewage treatment. For example, in 2010, the city of Xining, Qinghai province, established a sewage remediation company which has worked to fine or shut down companies which fail to comply with treatment regulations.

Strategies for environmental sustainability

Cities, as places that concentrate both environmental pollution and the poor and disadvantaged populations particularly vulnerable to environmental pollution, are a primary focus of environmental planning in China. The central government has assessed urban environmental quality yearly since 1989. Several types of urban-focused programs have been initiated in order to promote and support improvements in urban environments, especially the National Model City program, the "key cities" program, and the eco-city movement. The Chinese Academy of Social Sciences prepares a "blue book on urban competitiveness" on a regular basis, which includes environmental rankings for cities.

Model City programs

In 1997, China launched a National Model City for Environmental Protection (NMC) program in order to identify cities that meet strict environmental standards in energy supply, waste management, and green space. Theoretically, cities named in the NMC program are supposed to be able to attract more foreign investment on the basis of their environmental quality. Between 1997 and 2007, 72 cities achieved this status. Some cities have carried out genuine efforts in order to achieve this status (Shenyang, for example, increased its good-air-quality days by 84 percent in 2003 and achieved NMC status in 2004; this may have had a significant impact on Shenyang's successful 2007 bid for hosting the International Horticultural Exposition). But there is concern that some cities have achieved this designation by relocating, rather than mitigating, environmental problems. For example, the city of Zhongshan is believed to have improved its environment for this program largely by relocating polluting industries to outside the boundaries of the assessment area (MEP 2010a; Bremer 2011). Moreover, these 72 cities still represent only about 11 percent of China's cities – the rest have yet to meet these environmental standards.

In all, 113 Chinese cities have been named as "key cities" for environmental monitoring and protection. This is different from the Model City program discussed above. In this case, these cities are closely monitored each year; their condition shapes the annual State of the Environment report published by the Ministry of Environmental Protection. There is no implication that the cities are particularly "good" or "bad" in environmental protection; rather, they are major cities, distributed throughout the country, which provide a benchmark for environmental monitoring.

Eco-cities and green urbanism

There has been a remarkable growth in the planning of eco-cities – new development centered on ecological principles and "green urbanism" retrofitting of existing cities in China. The goal for most such cities is to make as much use as possible of sustainable environmental practices ranging from wastewater recycling to use of solar energy. There are more than one hundred different eco-cities in various stages of planning, development, and construction in China today (Liu 2011). These projects vary in size from high-profile, large cities designed by prestigious international architectural firms to much smaller local efforts. China has attracted numerous innovative design firms in recent years, as its rapid development and large-scale vision provide many opportunities for experimentation and creativity. However, some of the most well-known experiments in building entirely new eco-cities have failed in recent years. These failures have been due primarily to failures in finance and administration. For example, Shanghai's Dongtan Eco-City Project, an ambitious effort to create a relatively large sustainable city on an island in the Yangzte River, was cancelled after its financing and political support collapsed (Box 11.2).

Box 11.2 Dongtan: rise and fall of the "world's first eco-city"

In 2005, the Chinese government and the British engineering design firm ARUP announced plans for the world's first true, purpose-built "eco-city." The prestigious consulting firm McKinsey had first conceived the project. The city was to be built on Chongming, a sparsely inhabited island at the mouth of the Yangtze River. The 1,041 km² island, which falls under the administration of the Shanghai City government, is the largest alluvial island in the world. It boasts large tidal mudflats that attract migratory birds, wetlands and a flat expanse of scattered villages, orchards, and rice paddies. The ARUP plan, to be completed in time for the 2010 Shanghai World Expo, included an 8,400 ha compact, low-rise, high-density city for 500,000 inhabitants (50,000 by 2010; the remaining 450,000 to settle in the city during subsequent development phases) on the eastern end of Chongming Island. The ARUP plan included a bridge to link the island to the mainland, greenbelt buffer zones to protect the wildlife and wetlands, the production of energy from renewable resources, zero-emission transport, water recycling, a "near-zero" landfill, and landscaping designed to preserve biodiversity (Wood 2007; Normile 2008; Brenhouse 2010). The Dongtan Eco-City Project attracted worldwide attention from admirers and skeptics alike. What set the Dongtan plan apart from other efforts to create green cities was not just the combination of green technologies, most of which can be found elsewhere, but also the total ban on older technologies, from petroleum-powered vehicles to its requirement that most construction be carried out by local labor (Cherry 2007).

> In theory, Dongtan will be self-sufficient in energy, food, and water, with close to zero carbon emissions from transportation. If this is accomplished, each person living in Dongtan will exert much less pressure on nature than a New Yorker does today . . . The Dongtan eco-city project is one of the latest attempts to design an urban form that brings the needs of people in line with the needs of the environment . . . this project comes at a time when humanity needs new models for urban development.
>
> (Lee 2007: 3–4)

But China was not ready for Dongtan. In 2009, the 25 km bridge and tunnel to the mainland were completed, and staggering numbers of tourists

(600,000 in the first few days) flocked to the island despite its lack of completed facilities. Although much of the skepticism expressed in the press had centered on whether or not the project could meet its environmental goals, in the end, the project was indefinitely put on hold in 2010, due to a lack of political support (its principal support had come from a Shanghai Communist Party leader who was jailed in 2008 on corruption charges) and disagreements over financing (Brenhouse 2010). The ecological showpiece was transformed from a model of forward-thinking development to a lesson in the pitfalls of poor management. Long before the project actually failed, one journalist wrote:

> In a country overloaded with environmental challenges, Dongtan is a symbol of political overreach that straddles nearby Shanghai and Britain, the home base of ARUP, the firm that dreamed up Dongtan. Its failings show the limits to getting bold ideas off the drawing board, even in China's top-down political culture, where outsized schemes get traction.
>
> (Montlake 2008)

Yet the fact that the Dongtan Eco-City Project was even attempted in China is a credit to the forward-thinking that continues in Chinese cities, despite the rampant, environmentally damaging development that overshadows the best intentions.

A few eco-cities seem on track for at least some success in terms of environmental sustainability. Most notable, perhaps, is the Sino-Singapore Tianjin Eco-City, a joint venture launched in 2007. The project, which is aimed at housing 350,000 residents by 2020, has been developing at a steady pace so far with a wide range of facilities, housing, and an industrial park constructed thus far (Liu 2011). The plan for this eco-city, located on the coast 40 km east of central Tianjin, includes nine goals: (1) energy efficiency and the use of renewable energy resources for at least 20 percent of all energy use in the city; (2) compliance with "green building" standards for all structures; (3) "green transportation," in which 90 percent of all trips would be on foot, by bicycle, or with public transportation; (4) an ecologically friendly design which preserves wetlands and other wildlife habitats and builds new green spaces; (5) water management and conservation, including the provision of potable faucet water and the use/recycling of gray water and desalinized sea water; (6) integrated waste management ("reduce, reuse, recycle"); (7) economic vibrancy, as an educational and research and development center for environmental technologies; (8) social harmony, especially provision of public facilities, care for the elderly, and accessibility for the handicapped; and (9) heritage conservation, including preservation of a 1,000-year-old canal which crosses the site (SST 2011).

Another kind of eco-city program links "green development" with economic development. An experiment is underway in Beijing entitled the Capital

Eco-Economic Zones. This is part of a development program based on the "economic zones" approach. The first "eco-economic zone" was formally established in Shandong in 2008 as part of the Yellow River Delta Efficient Eco-Economic Zone. Jiangxi province, the Three Gorges Reservoir area, and other regions have begun development of eco-economic zones. Beijing is distinctive in this context, however, as it ties the eco-economic planning approach more directly to a large city than the other programs. The program is designed to accelerate economic development while protecting ecological resources in the mountainous areas surrounding the city proper. Activities associated with the zone range from tourism development and the establishment of eco-towns to high-tech industrial and agricultural ventures (Zhang et al. 2010b). The development of such ecological zones may begin to resemble the "zone fever" of the 1990s, as cities throughout China attempt to use planning for "green urbanism" to attract investment.

Global events as impetus for change

Ever since the Chinese sent a delegation to the First United Nations Conference on Human Environment in 1972, care for the environment has played an increasingly important role in China's efforts to manage its global image. The impacts of these efforts, intensifying since the 1990s, should not be underestimated. In 1994, China was an early signatory of Agenda 21, the broad-reaching United Nations action plan developed for the Rio environment summit of 1992. Agenda 21-based environmental programs have been established throughout China. Twenty-seven urban districts and cities were identified as National Sustainable Experimental Cities in the mid-1990s in order to achieve Agenda 21 goals. For example, Huairou county, within Beijing's administrative area, was chosen under this program to experiment with accelerated urban development while protecting the environment. Huairou continues to participate in a variety of sustainability and green development programs, including Beijing's Capital Eco-Economic Zones.

But the most high-profile green planning has been that surrounding the staging of international events in China's cities. The staging of the Kunming World Horticultural Exposition (May–October 1999) marked a shift in the discourse of mega-event planning in China from development aimed at building society and achieving economic progress to building society and improving the environment. The theme for the Kunming Expo, Man and Nature: Marching into the 21st Century, boldly brought environmental themes forward. Kunming implemented new urban environmental measures as part of its bid to host the 1999 Expo. The city proposed a major cleanup of Lake Dian (a large nearby lake), substantial upgrading of urban infrastructure, and a major increase in the amount of "green area" throughout the city. US$2 billion was spent in air pollution abatement and clean-up of the lake, including the costs of relocating factories to the urban outskirts to curb particulate and other pollution from the use of coal. This was the first high-profile instance in China of using urban environmental quality as a component of urban (and national) "image marketing." It won Kunming recognition in 2000 as a "national pilot city" for environmental protection and management. Despite continuing difficulties with pollution in Lake Dian, this "green" event had contributed to improvement in Kunming's environment, particularly through relocating polluting factories and creating new green space within the city.

Beijing's bid for the 2008 Olympics, and Shanghai's bid for the 2010 World Expo, also were centered around environmental themes and led to massive new environmental initiatives. Beijing set a new standard for using the Olympics to achieve fundamental environmental change (Box 11.3) – a standard that will be difficult for subsequent Olympic cities to meet. Despite the much-publicized air quality issues during the 2008 Olympics (due as much to the weather as to human-generated pollutants), the Olympics was a true catalyst for change in Beijing. Similarly, the Shanghai World Expo (2010), with its slogan, Better City, Better Life, emphasized ecology, technology, and culture. Three of the five large themed exhibition venues at the Expo were devoted to exhibits on urban themes with explicitly environmental foci: one housed exhibits on cities that won an "urban best practices" competition held by the Expo organizers, another was devoted to urban innovation, and a third was devoted to urban origins, development, and futures. The Expo plan explicitly focused on the interconnections between urban life, the environment, and environmentalism. Shanghai was awarded the Expo in 2002, and, like Beijing's preparation for the Olympics, implemented a rapid US$45 billion facelift for the event. This included displacing an outmoded, heavily polluting shipyard, about 272 other companies and about 18,000 households (Huang 2010).

Box 11.3 Green Beijing, Green Olympics

Beijing was awarded the 2008 Olympics in 2001 with a plan centered on three strategic concepts: green Olympics, high-tech Olympics, and people's Olympics. The goals set forth in the initial plan included building "the basic framework of a modernized international metropolis." Beijing's Olympic bid argued that "hosting of the Games in Beijing will serve as a 'catalyst' for environmental improvement and help promote sustainable development in Beijing and China."

The city's bid, titled New Beijing, Great Olympics, outlined a complete restructuring of the city's basic infrastructure and land use with an emphasis on environmental planning and management. The action plan for "Green Beijing" included major environmental reforms such as dramatic reductions in automobile emissions; substitution of natural gas for coal use; use of solar and geothermal energy; increased public bus and subway transit capacity; extensive new subway and light rail lines; replacement of public buses and taxis with natural gas-powered, electric, and hybrid vehicles; improvements in the city's water supply; improved wastewater treatment;

solid waste reduction; large-scale afforestation efforts; planting of trees and new green areas throughout the city; and the relocation of 200 central city industries to the urban periphery. After winning the right to host the games in 2001, the Beijing city government, with the assistance of the national government, embarked upon a massive restructuring of the city's urban form and infrastructure. The US$50 billion preparation for the games thus provided unprecedented national government support for Beijing's urban development, a means to address the city's most pressing environmental issues and an opportunity to revise and implement the existing comprehensive plans meant to shape the city's development through 2010. As the opening paragraph of the Beijing 2004–2020 Comprehensive Plan explains, the combination of the recent fast pace of urban development and the specific requirements of the Olympics generated a need to create a new plan before the 1991–2010 plan had run its course.

The Olympic Plan accelerated the pace of progress toward environmental goals laid out in the early 1990s, and introduced substantial new targets in all areas (see, for example, Jiang 2003). The Olympic Plan also lent power to planning goals that challenged entrenched power structures. For example, the Shougang Iron and Steel complex, located just 11 km from Tian'anmen Square, was long known to be one of the worst polluters in Beijing. But it was also one of the most powerful state enterprises; previous attempts to relocate it had failed. With the power of the Olympics at hand, planners were able to insist that Shougang move out of Beijing. The company relocated more than 100 km away. However, because they could not keep to the fast-paced schedule required by the Olympic Plan, a portion of the factory had not yet moved when the Olympics started. Shougang had explained as early as 2005 that they would not be able to meet the goal until at least 2010. Nonetheless, only one blast furnace remained by the time of the Olympics, and that one ceased operation during the Olympics.

While many of the environmental efforts associated with the Beijing Olympic Plan were far-reaching – from improved water systems to natural gas buses – others were ephemeral and specific to the time of the Olympics, such as temporarily banning construction activities, closing factories, and limiting traffic. In the days leading up to the opening of the Olympics in August 2008, workers could be seen in many Beijing neighborhoods carefully arranging massive displays of potted flowers and shrubs. This only temporarily increased the "green" surface of the city.

There is no doubt that the Olympics contributed substantially to Beijing's character as a leader in a number of approaches to sustainable urbanism. In focusing on the "green Olympics," Beijing also set an agenda for urban environmental planning that will reach every province in China.

Financial incentives

A variety of financial strategies also have been planned or undertaken to encourage more sustainable development. Pilot programs are under way to offer tax incentives to enterprises engaging in five areas, including municipal wastewater treatment, municipal garbage treatment, comprehensive development and utilization of methane gas, energy conservation upgrading and emissions reduction technologies, and seawater desalination. A "green credit" policy was put in place by the Ministry of Environmental Protection, the People's Bank of China, and the China Banking Regulatory Commission in 2007 in order to encourage banks to deny credit to polluting enterprises, and to offer preferential lending to environmentally friendly ventures. Several of China's major banks now collect environmental data as part of the commercial loan application process, deny loans to polluting businesses, and promote more "green" development (Aizawa and Yang 2010).

Moreover, nine cities, such as Shenyang and Shanghai, have developed policies advocating or requiring the purchase of "green insurance," which covers the liability of firms involved in industrial pollution, to help cover the costs of industrial pollution events (MEP 2010a). The green insurance program originates in a set of guidelines produced jointly by the Ministry of Environmental Protection and the China Insurance Regulatory Commission in 2007. In 2008, the first compensation was paid out under this program when the China Ping An Insurance Company paid out on a claim by residents of Zhuzhou City (Hunan province) for damage caused by hydrogen chloride leaks. This program thus provides a mechanism for polluting industries to pay damages and mitigate severe environmental problems (Aizawa and Yang 2010).

Environmental management

Environmental governance

China's environmental protection is overseen by agencies at all levels of government. Although the first nature reserves were established in 1956, concerns over the environment were generally set aside in favor of economic growth throughout the 1950s and 1960s. But in 1972, as part of China's early push to re-establish its international relations, it sent a delegation to the First United Nations Conference on Human Environment in Stockholm. Subsequently, the country's first national environmental protection organization was established in 1973 (Edmonds 1999).

This organization evolved into a subministry called the National Environmental Protection Agency. This agency was upgraded to the status of a full ministry in 1998 and renamed the State Environmental Protection Administration (SEPA). In 2008, SEPA was reorganized and renamed the Ministry of Environmental Protection. Other agencies, such as the Ministry of Water Resources, also have authority over China's environment. The national government reports spending about 3 percent of its gross domestic product on mitigating environmental pollution (Orchison 2009).

Cities vary in their environmental oversight but most large cities have a similar structure for governing the environment. Beijing, for example, has several municipal bureaus dedicated to environmental and resource protection, including the Municipal Bureau of Environmental Protection, the Municipal Bureau of Forestry and Parks, and the Beijing Water Authority. Xining, capital of Qinghai province, has an Environmental Protection Bureau, a Water Conservation Bureau, and a Forest Bureau. Many critics have expressed concern, however, that China's local environmental governance still suffers from a lack of transparency and official accountability, and that environmental laws are weakly enforced.

Legal instruments

Municipal environmental management has been incorporated into the local urban planning framework. China's urban planning is governed by a national urban planning law. The 1989 national urban planning law, which was in effect from 1990 to 2008, encouraged cities to plan for the environment (see added italics below):

> **Article 6** The compilation of the plan for a city shall be based on the plan for national economic and social development *as well as the natural environment.*
>
> (City Planning Law of the People's Republic of China 1989: http://www.china.org.cn/english/environment/34354.htm)

> **Article 14** In the compilation of the plan for a city, attention shall be paid *to the protection and improvement of the city's ecological environment, the prevention of pollution and other public hazards, the development of greenery and afforestation, the improvement of the appearance and environmental sanitation of urban areas.*
>
> (City Planning Law of the People's Republic of China 1989: http://www.china.org.cn/english/environment/34354.htm)

The current urban and rural planning law (2008+), which supersedes the 1989 law, pays somewhat more attention to environmental and conservation planning, and has more forceful language, in the "mandatory" requirement for plans to address watersheds, afforestation areas, and environmental protection areas, and in the call for the planning of new areas to consider the ecological environment.

> **Article 4** In making and implementing urban and rural plans, attention shall be paid to following the principles of overall planning for urban and rural areas, rational layout, *conservation of land*, intensive development and planning before construction; *to improving the ecological environment*; *promoting conservation and comprehensive utilization of resources and energy*; to preserving cultivated land and other natural resources and historical and cultural heritage; to maintaining the local and ethnic features and traditional cityscape, *to preventing pollution and other public hazards*; and to meeting

the need of regional population development, national defense construction, disaster prevention and alleviation, and public health and safety.

> (Law of the People's Republic of China on Urban and Rural Planning
> 2007: http://www.china.org.cn/china/LegislationsForm2001-2010/
> 2011-02/11/content_21899292.htm)

Article 17: The following shall be made *mandatory* for the overall plan of a city or town to include: the area covered by the plan, the scale of the land used for construction in the area covered by the plan, the land used for infrastructure and public service facilities, *the watershed sites and water system*, capital farmland *and land used for afforestation, environmental protection*, preservation of natural and historical and cultural heritage, disaster prevention and alleviation, etc.

> (Law of the People's Republic of China on Urban and Rural Planning
> 2007: http://www.china.org.cn/china/LegislationsForm2001-2010/
> 2011-02/11/content_21899292.htm)

Article 30 In the development and construction of the new areas in a city, attention shall be paid to rational determination of the scale and schedule of construction, to the full use of the existing urban infrastructures and public service facilities, *careful preservation of the natural resources and ecological environment* and embodying of the local features.

> (Law of the People's Republic of China on Urban and Rural Planning
> 2007: http://www.china.org.cn/china/LegislationsForm2001-2010/
> 2011-02/11/content_21899292.htm)

China also has an overarching national environmental law as well as a number of specific national environmental laws that have been strengthened in recent years, both in their specificity and their enforcement, such as the *Cleaner Production Promotion Law* (2002) and the *Environmental Information Disclosure Decree* (2008). There are also new environmental policy instruments, such as emission trading. However, implementation of these strategies is sometimes weak, as fines are relatively low and "promotion" is emphasized over regulation (Mol 2009).

The increasing devolution of authority from the central state to local governments has meant more local control over environmental concerns in cities. At the same time, national legislation such as the 2003 *Environmental Impact Assessment Act* has generated opportunities for more public participation in the care of local environments (Mol 2009). Most cities now have a mechanism, such as an online complaint forum, for the public to communicate their environmental concerns directly with local agencies responsible for environmental conservation and management. Although this form of commentary may not have much influence on local officials, it is the start of a public dialogue. Recently, the city of Beijing came under considerable pressure from both the media and citizens when it was revealed that the city has not been reporting data on fine particulate air pollution (PM 2.5 pollution – that is, pollution made up of particles less than 2.5 microns in size). Beijing has had increasing difficulties over the past few years with particulate pollution, which has led to flight cancellations, dark skies, and health hazards. A Beijing citizen mounted an intense internet-based campaign in December 2011 for more accurate public reporting of pollution data. In January 2012, the municipal government agreed to change its reporting practices (Wong 2011; Barboza 2012).

Environmental justice

"Environmental justice" is the concept that there should be a fair spatial distribution of both the benefits and burdens of environmental conditions. In urban areas, this is usually understood to mean that urban residents should have equal access to beneficial aspects of the environment such as clean air and water, or green areas, and that no neighborhood or segment of the population should be disproportionately exposed to environmental hazards, such as toxic waste or untreated sewage. Chinese urban residents are increasingly aware of and concerned about environmental health hazards in their cities. An increasing proportion of the civil unrest in China seems to be related to environmental issues. Most commonly, these concerns center on the location of polluting factories or other institutions in close proximity to residential areas. In May 2008, for example, 400–500 protesters in Chengdu, capital of Sichuan province, staged a peaceful march to protest the construction of an ethylene (plastic) plant and oil refinery just outside the city, citing concerns about environmental health risks (Wong 2008). Another high-profile example comes from the town of Huashui, a village within the administrative area of Dongyang City in central Zhejiang province. A development zone for chemical factories was opened in 2001. Although the new zone was initially welcomed for its contributions to the local economy, 4 years later local residents identified a 10-km-wide "death zone" around the factories where trees and crops had withered and died, and there was an increasingly high rate of serious birth defects and still-births. In 2005, a 3-week-long mass protest and road blockade, which ended in a violent conflict with police, ultimately resulted in the closure of the area's 13 factories (Ma and Schmitt 2008).

There are tens of thousands of environmental protests each year in China. In 2005 alone, there were about 51,000 pollution-related protests in China; complaints about the environment to government officials via letters and hotlines recently numbered near half a million per year (Economy 2007). These protests both point to serious problems – the "tip of the iceberg" – as yet more pollution issues must exist for which there are no protests, and demonstrate that the Chinese population is becoming aware of environmental issues and has some confidence in the potential efficacy of lodging complaints about them. Although many have expressed skepticism over the impact of such protests, and it is likely most protests yield few, if any, results, nonetheless the protestors are not alone in their concern over the dangerous environmental side-effects of China's rapid development. In an interview with the German magazine *Der Spiegel*, Pan Yue, vice-minister of the former State Environmental Protection Administration, noted that "In Beijing alone, 70–80 percent of all deadly cancer cases are related to the environment. Lung cancer has emerged as the No. 1 cause of death" (Lorenz 2005).

The business of the environment

Environmental management in China, as in much of the world, has become increasingly privatized. In 1993, the China Association of Environmental Protection Industry (CAEPI) was established to coordinate environmental efforts across different municipalities, regions, government agencies and, eventually, environmental management businesses. As such it coordinates the efforts of state agencies with private firms. Initially a domestic organization, it has grown to engage in a wide range of international activities. Most recently, the CAEPI has hosted yearly business-oriented conventions, such as the China Municipal Solid Waste Summit and the China International Environmental Protection Exhibition and Conference, which provide a venue for information sharing, networking, inviting international investment in China's environmental management, and selling Chinese technologies abroad. The organization also sends delegations to international environmental industry conventions. At the 2011 Municipal Solid Waste Summit, for example, there was a strong focus on new technologies for efficient incineration of urban garbage.

During the reform era, China has seen the growth of environmental advocacy not only from coalitions of state agencies and private business, but also through the development of non-profit organizations. These include government-organized non-governmental organizations (GONGOs), such as the China Women's Development Foundation, international non-governmental organizations (INGOs), such as the Worldwide Fund for Nature (WWF), and domestic non-governmental organizations (NGOs), such as the China Youth Climate Action Network (CYCAN). In Beijing, the Urban China Initiative has been established as an NGO which facilitates communication between government, business, and academia. They produce the Urban Sustainability Index, which provides an annual measure of social, economic, environmental, and resource sustainability for 112 cities across China (UCI 2012).

Conclusion

The rapid pace of urbanization and development in China clearly is harmful to the urban environment. Chinese cities today are plagued by bad air, unsafe water, water shortages, and garbage and sewage problems. Yet the cities are undertaking a remarkable range of projects designed to mitigate both short-term and long-term environmental issues, and to generate increased sustainability for future generations.

Bibliography

Aizawa, Motoko and Yang, Chaofei. 2010. "Green Credit, Green Stimulus, Green Revolution? China's Mobilization of Banks for Environmental Cleanup." *The Journal of Environment and Development* 19(2): 119–144.

Barboza, David. 2012. "China to Release more Data on Air Pollution in Beijing." *The New York Times*, January 6, 2012. (Retrieved on 19 January 2012 from http://www.nytimes.com/2012/01/07/world/asia/china-to-release-more-data-on-air-pollution-in-beijing.html?scp=1&sq=Beijing%20pollution&st=cse).

Boland, Alana. 2007. "The Trickle-Down Effect: Ideology and the Development of Premium Water Networks in China's Cities." *International Journal of Urban and Regional Research* 31(1): 21–40.

Bremer, Mark. 2011. "China National Model Cities for Environmental Protection." *Green Explored* February 20, 2011.

Brenhouse, Hilary. 2010. "Eco-City Plan Shrivels as China Expo Unfurls: Poor Consultation, a Backer Sent to Jail, and Confusion Over Funds Sabotaged Sustainable Town." *International Herald-Tribune* June 25, 2010.

Chan, Chak C. and Yao, Xiaohong. 2008. "Air Pollution in Megacities in China." *Atmospheric Environment* 42: 1–42.

Cherry, Steven. 2007. "How to Build a Green City: Shanghai Hopes to Build the World's First Truly Sustainable City." *IEEE Spectrum*, June 2007 (online).

Chestney, Nina. 2011. "China's CO2 Emissions Rise Sharply." Reuters News Service, June 9, 2011.

Chhibber, Ajay. 2011. "Green Cities for Blue Skies in China." *China Daily*, January 12, 2011.

Economy, Elizabeth C. 2007. "The Great Leap Backward? The Costs of China's Environmental Crisis." *Foreign Affairs* 86(5): 38–59.

Edmonds, Richard Louis. 1999. "The Environment in the People's Republic of China 50 Years On." *The China Quarterly* 159: 640–649.

Gaubatz, Piper. 1999. "China's Urban Transformation: Patterns and Processes of Morphological Change in Beijing, Shanghai, and Guangzhou." *Urban Studies* 36(9): 1495–1521.

He, Kebin, Huo, Hong, and Zhang, Qian. 2002. "Urban Air Pollution in China: Current Status, Statistics, and Prospects." *Annual Review of Energy and Environment* 27: 397–431.

He, Zong. 1994. "Life Going Well for Beijingers." *China Daily* 2–14–94: 6.

Huang, Carol. 2010. "China's Shanghai Expo 2010 – By the Numbers." *Christian Science Monitor*. April 29, 2010. (Retrieved on 19 January 2012 from http://www.csmonitor.com/World/Global-News/2010/0429/China-s-Shanghai-Expo-2010-by-the-numbers.)

Jiang, Yi (ed.) 2003. *Implementation Manual for Green Buildings for the Beijing Olympics* [*luoshi aoyun jianzhu shifan jienan*]. Beijing: China Architecture & Building Press [*zhongguo jianzhu gongye chubanshe*].

Johnson, Tim. 2006. "E-waste dump of the world." *The Seattle Times* April 9, 2006. (Retrieved on 19 January 2012 from http://seattletimes.nwsource.com/html/nationworld/2002920133_ewaste09.html.)

Law, Violet. 2011. "As China's Prosperity Grows, So Do Its Trash Piles." *Christian Science Monitor* July 28, 2011. http://www.csmonitor.com/World/Asia-Pacific/2011/0728/As-China-s-prosperity-grows-so-do-its-trash-piles

Lee, Kai. 2007. "An Urbanizing World." *State of the World 2007: Our Urban Future*. New York: WW Norton Company and the Worldwatch Institute.
The *State of the World* is an annual publication by the Worldwatch Institute. Each year features a different theme. In 2007, the volume focused on urbanization, and particularly upon China and its rapid urban transformation.

Li, R. 2010. "Municipal Solid Waste Management in China." Thesis, Roskilde University, Denmark.

Liu, Coco. 2011. "China's City of the Future Rises on a Wasteland." *Scientific American*, September 28, 2011. http://www.scientificamerican.com/article.cfm?id=chinas-city-of-the-future-tianjin-rises-on-wasteland

Lorenz, Andreas. 2005. "The Chinese Miracle Will End Soon. SPIEGEL Interview with China's Deputy Minister of the Environment." *Der Spiegel* March 7, 2005. http://www.spiegel.de/international/spiegel/0,1518,345694,00.html.

Ma, Li and Schmitt, Francois O. 2008. "Development and Environmental Conflicts in China." *China Perspectives* 2: 94–101.

Matus, Kira, Nam, Kyung-Min, Selin, Noelle E., Lamsal, Lok N., Reilly, John M., and

Paltsev, Sergei. 2011. "Health Damages from Air Pollution in China." Report #196. Cambridge, MA: The MIT Joint Program on the Science and Policy of Global Change.

Ministry of Environmental Protection (MEP). 2010a. *2009 Report of the State of the Environment in China.* (Retrieved on 1 May 2011 from http://www.mep.gov.cn/)

Ministry of Environmental Protection (MEP). 2010b. *China Annual Report on the Prevention and Control of Vehicle Pollution.* (Retrieved on 1 May 2011 from http://www.mep.gov.cn/)

Minter, Adam. 2009. "Reassessing and Regrouping." *Scrap Magazine* January/February.

Mol, Arthur P. J. 2009. "Urban Environmental Governance Innovations in China." *Current Opinion in Environmental Sustainability* 1: 96–100.

Montlake, Simon. 2008. "In China, Overambition Reins in Eco-city Plans." *Christian Science Monitor* December 24.

MSW. 2011. China Municipal Solid Waste Summit. (Retrieved on 3 March 2011 from http://www.mswforum.com.cn/index.asp.)

Normile, Dennis. 2008. "China's Living Laboratory in Urbanization." *Science* 319(5864): 740–743.

Orchison, Keith. 2009. "Green Economy Awaits Push – Special Report: Waste Management." *The Australian* January 23.

Roseland, Mark and Soots, Lena. 2007. "Strengthening Local Economies." *State of the World 2007: Our Urban Future.* New York, NY: WW Norton and the Worldwatch Institute.

Sino-Singapore Tianjin Eco City (SST). 2011. (Retrieved on 1 January 2011 from http://www.tianjinecocity.gov.sg/.)

State Council. 2005. 11th Five Year Plan – China National Environmental Protection Plan (2006–2010). (Retrieved on 10 February 2011 from: http://www.caep.org.cn/english/paper/China-National-Environmental-Protection-Plan-in-11th-Five-Years.pdf.)

Urban China Initiative (UCI). 2012. *2011 Urban Sustainability Index.* Beijing: McKinsey & Company.

United Nations Environment Program (UNEP). 2003. *Urban Environmental Management in China.* Osaka, Japan: The UNEP-International Environment Technology Centre.

Walsh, Bryan. 2009. "E-waste Not." *Time Magazine* January 9. (Retrieved on 18 January 2012 from http://www.time.com/time/magazine/article/0,9171,1870485,00.html.)

Wong, Edward. 2008. "In China March, Hints of a Movement: Protest of Pollution is Latest in a Series." *International Herald-Tribune* May 6.

Wong, Edward. 2011. "Outrage Grows Over Air Pollution and China's Response." *New York Times* December 6. (Retrieved on 19 January 2012 from: http://www.nytimes.com/2011/12/07/world/asia/beijing-journal-anger-grows-over-air-pollution-in-china.html.)

Wood, Roger. 2007. "Dongtan Eco-City, Shanghai." Powerpoint presentation. Online at: http://www.arup.com/_assets/_download/8CFDEE1A-CC3E-EA1A-25FD80B2315B50FD.pdf.

World Bank. 1997. *Clear Water, Blue Skies: China's Environment in the New Century.* Washington, DC: the World Bank.

Worldwatch Institute. 2007. *State of the World's Environment 2007: Our Urban Future.* Washington, DC: Worldwatch Institute.

Xue, Yong. 2005. "Treasure Nightsoil As If It Were Gold: Economic and Ecological Links between Urban and Rural Areas in Late Imperial Jiangnan." *Late Imperial China* 26(1): 41–71.

Zhang, Dong Qing, Tan, Soon Keat, and Gersberg, Richard. 2010a. "Municipal Solid Waste Management in China: Status, Problems, and Challenges." *Journal of Environmental Management* 19: 1623–1633.

Zhang, Yifeng, Lan, Tingting, Liu, Chunla, and Zhang, Hongye. 2010b. "Construction and Strategic Vision of the Capital Eco-economic Zones." *Journal of Resources and Ecology* 1(3): 274–283.

12 Lifestyle and social change

There have been fundamental changes in the lives of urban dwellers throughout China since the onset of economic reform in 1979, and nowhere are these changes more profound than in the large cities of the east coast. As the often gray and monotonous streets of the Maoist era have given way to the visual cacophony of contemporary city landscapes, so too, the day-to-day rhythms of life have changed. Home life, work, transportation, and leisure have all undergone remarkable transformations. Old neighborhood communities have been broken apart, while new communities are forming in new housing areas. Lunch at the work unit cafeteria is being replaced by grabbing a quick bite at a fast-food restaurant, new hobbies are emerging, such as home decoration or pet care, and, for many, shopping has become a recreational activity. As the physical patterns of the city change (more in Chapter 8), so, too, do these activities.

As China's cities are growing with influxes of capital, development projects, and migrants, urban life and culture are changing as well. What is it like to be an urban resident in China today? From sports events to art shows, urbanites are creating space and time for new leisure activities. Yet contemporary urban life also brings new challenges. This chapter surveys five aspects of life in China's changing cities: shopping, use of public spaces, growing old, crime, and communities of difference (gay and lesbian life, ethnic enclaves, and art communities). The following questions guide this chapter:

- How has daily life changed for urban Chinese in recent years?
- What types of spaces are generated by these changes in daily life?
- What challenges are generated by social change in Chinese cities?

Shopping and consumerism: changing landscapes and lifestyles

It is difficult for those who did not experience it to imagine the fundamental changes in urban life that have occurred over the lifespan of Chinese people born around the time of the 1949 revolution. As China's economy, society, and political system were repeatedly readjusted, urban life took on new everyday rhythms with surprising speed. A simple task, such as shopping, is a good example of an activity in which every urban family must engage and which is integral to the function of the city as a whole. During the war-torn years leading up to the Chinese revolution, shopping for food, clothing, and other daily necessities in urban China, for most residents, involved trips on foot or by trolley to many different markets – fresh food in neighborhood vegetable markets, clothing in the street of the tailors, books in narrow lanes of an urban district devoted to bookshops. Most purchases involved lively bargaining with the shopkeepers. There were a few upscale department stores and specialty boutiques downtown in the European-influenced, Treaty Port cities, but these were expensive and did not cater to the bulk of the population.

Shopping in Mao's China

After the establishment of the People's Republic of China, however, the task of obtaining basic food and clothing changed dramatically in Chinese cities. During the early 1950s, increased jobs and wages in the growing urban industrial sector widened the gap between urban and rural Chinese. While urban residents, for example, were consuming about 7 kg of meat per year, rural Chinese were consuming only half that amount (Huenemann 1966). At the same time, shortages developed in key subsistence items, such as grain and cloth. In the cities, urban residents had to form long lines to shop for limited quantities of basic items. In 1954, for example, lines began to form for sugar. Other commodities soon followed. This meant that, while some people were able to obtain these basic goods, others went without them. In response, the central government imposed rationing systems, starting with grain and cooking oil in 1953, cotton cloth in 1954, and tofu, sugar, meat, fish, and eggs in 1959. Ration coupons replaced currency for these purchases (Huenemann 1966; Rada 1983). Meanwhile, the state had taken over the retail outlets for these goods. Commercial shopping districts were significantly reduced in their extent and function. With the government controlling the supply and sale of basic commodities, "shopping" became more a matter of picking up the goods as needed rather than the more free-wheeling bargaining and comparison which had characterized earlier markets. Even restaurant meals required coupons, so that restaurant managers could use the grain coupons to pay their suppliers.

During the 1960s, rationing was extended for a time to light industrial goods such as pots and pans and bicycles. Thus, in general, subsistence and consumer goods were in relatively limited supply for much of the pre-reform era. Some goods and services were distributed directly by employers. By the 1970s, state-run stores carried a limited supply of goods. Even in the mid-1980s, many of these stores were relatively empty and offered few choices to customers. Given both the limited supply of goods and the uniformity of the offerings in the shops, there was little need for most Chinese to travel beyond their local neighborhood for shopping. Many urban residents lived in close proximity to their place of work in housing provided by their work unit (see Chapter 8), shopped in local markets and state-run stores, and lived out their daily lives largely within their own neighborhoods. Even entertainment and recreation, such as movies or recreation centers, were provided by employers at the workplace or in nearby housing complexes (Davis 2005).

Shopping in the reform era

During the reform era, however, this way of life has undergone a remarkable transformation. The introduction of a responsibility system in the rural sector during
the early 1980s drastically increased agricultural production and output. The grain
markets and other basic commodities were deregulated by the early 1990s to end
the rationing system. Rising industrial production has generated an increasingly
wide range of products. During the 1980s, the new development zones largely produced goods for the international market, and relatively few of these goods were
available to Chinese citizens. A transitional form called "free markets" developed,
where overruns of manufactured goods, fruits and vegetables, and other consumer
products were sold by peddlers from stalls and carts along back alleys or in empty
lots. Although some of these markets were eventually licensed and regulated, many
operated outside the state-controlled system (the so-called "dual-track" approach
in which both the plan and market coexisted). By the 1990s, however, manufacturers increasingly produced for both international and domestic markets. The nascent market-based economy became increasingly consumer-oriented. During the
1990s, Chinese cities experienced a "consumer revolution" (Davis 2005).

Specialization and commercial centers are being restored in Chinese cities,
but not to their pre-1949 patterns. District specialization is less by trades (though
trades are sometimes clustered) than by function – districts devoted to education,
or residential, or commercial functions are emerging as distinct areas of the cities. This generates a land use pattern which resembles that produced by zoning in
North America. Commercial activity at different scales seems to have permeated
every cell of urban tissue in China's large eastern cities (more in Chapter 8).

The most recent trend in retailing in Chinese cities is the construction of
enclosed shopping malls (containing multiple retailers) and hyper-markets (operated by single retailers). This newly emerging retail landscape also contributes
to the increased specialization of neighborhoods in Chinese cities into distinct
functional niches. Both new shopping districts and large downtown and suburban
shopping malls are proliferating. Six of the 25 largest shopping centers in the
world are located in Chinese cities. Most of these commercial spaces are conceived as foci for entertainment and other public activities, as well as the buying
and selling of goods (Gaubatz 2008a).

The changing retail landscape in Chinese cities has altered (and been changed
by) the way Chinese urban residents spend their time. Those with money to spend
have become "avid – and knowledgeable – consumers of transnational-branded
foodstuffs, pop-music videos and fashion" (Davis 2005: 692). Along with this new
consumerism comes changing shopping patterns. "Shopping" in Chinese cities, as
in many cities around the globe, has become a pastime in and of itself – "shoppertainment." Upscale shopping districts in many large cities now boast Starbucks,
McDonald's, and dozens of other international and domestic restaurants to attract
young workers to spend a Saturday afternoon or a night out. Like malls in the
rest of the world, these new "quasi-public" spaces are highly regulated. Although
people may perceive them as "public," they represent a highly controlled and sur-
veilled environment (Gaubatz 2008b). While enclosed malls are occupying an

increasingly important niche, many Chinese also are drawn toward modern spaces that mimic older forms, such as open-air markets. In Shanghai, Beijing, Kunming, and other large cities, the main shopping streets have been pedestrianized and decorated with benches, statuary, and summer-time outdoor cafés to enhance their function as public and entertainment space. In other cities, streets are temporarily blocked off at night to offer similar fair-weather entertainment space.

But these new globalized spaces cater to a growing class of upscale consumers while leaving others behind. Between 1998 and 2003, while "the official consumer price index for urban China barely changed, the income disparity between the richest 10 percent and the poorest 10 percent of urban residents effectively doubled" (Davis 2005: 694). There is a large segment of the urban population who cannot possibly afford to shop in the neon and glass downtown shopping districts. For recent migrants to the cities, who might spend most of their time working long hours for meager wages, the new shopping districts offer an unattainable glimpse of a different world (Pun 2003). More traditional markets (particularly grocery markets) do exist in Chinese cities, although they, too, are undergoing change. While the global firms compete for China's new elite, there are hundreds of domestic enterprises catering to more average Chinese citizens, and local markets continue to thrive as well. Business analysts refer to this as a dual system, in the sense that there is simultaneously convergence with global trends and lifestyles, and divergence, which in this case is the persistence of culturally specific approaches to urban life. Many urban neighborhoods support small businesses, from hairdressers to corner sundries stores (Uncles 2010).

Public spaces and entertainment

Chinese cities are densely built, crowded, and noisy. Although per capita housing area is increasing (see Chapter 10), apartments are often cramped. Urban residents spend significant portions of their leisure time in public spaces. These spaces range from informal and transient to elaborately landscaped parks and gardens. Similarly, the activities carried out in public places range from visiting with friends to participating in carefully planned events.

Informal public space

Any open space in urban China is likely to become a public space. For example, as cities have undergone massive redevelopment in recent years, the process of clearing older housing in favor of new development often means that bare sites appear for short periods of time – from weeks to months – during the relocation and construction process. Usually these spaces are only walled off from the public when the construction starts. Until then, these open spaces quickly become neighborhood space, as local residents carry chairs out to sit and visit in the evenings, children fly kites or dig in the dirt, or entrepreneurs erect temporary markets, cafés, or entertainments (Figure 12.1). Sidewalks sometimes perform a similar function, with neighborhood residents setting up chairs and tables in the evenings

Figure 12.1 An informal market in a temporarily vacant lot, Kunming, Yunnan province.
Source: Piper Gaubatz.

to engage in "people watching," and sometimes for street vendors selling snacks and nightcaps. Unused spaces under overpasses, at the center of traffic circles, or adjacent to structures also are used as public space, and may become the temporary venue for activities from dancing lessons to chess clubs.

New activities in urban parks

More formal public space includes parks and public squares. In traditional China, public, landscaped, park-like areas could be found in the grounds of temples, which were often open to those wanting to stroll or contemplate fragments of nature within the crowded cities. "Public" parks were introduced during the Treaty Port era but often were closed to most of the Chinese public or access-limited through entrance fees (when these parks were located in foreign-controlled areas). Since 1949, many public parks have been developed either through the opening of historic sites (such as temples or former palaces) to the public or directly through the design and construction of new public landscape spaces. Beijing's 12 major parks, for example, are all on historic sites of temples and palaces. Since the 1990s, the dual drives to improve China's urban environment and to produce international-standard cities has led to a marked increase in park area, as smaller parks have been developed along urban canals and in the construction of new and redeveloped urban districts. For example, in Xining, the cleanup of an urban river whose banks had been lined by dilapidated shacks has resulted in a chain of small parks that draw regular multi-generational crowds to sing, play games, drink tea, or jog along the long paved paths which line the river (Gaubatz 2008a). Many cities, such as Beijing, Xi'an, Xining, and Höhhot, have restored sections of old city walls or canals and created linear parks along them.

Public parks have become the venues for a wide range of recreational activities. Although some parks provide formal spaces for recreation, such as basketball courts or exercise courses, most simply provide the space for residents to use as they like. Public parks are used for band rehearsals and performances, singing clubs, dance activities, and story telling (Box 12.1).

Box 12.1 Dancing in the streets: informal dance craze in Beijing

Underneath a freeway overpass, on a vacant concrete traffic island in the middle of bustling Beijing, 40 Chinese women in their sixties and seventies, dressed in silk brocaded jackets and padded silk pants, slowly waved lime-green handkerchiefs and fluttered hot pink, white, and green striped fans above their heads . . . It was 7 a.m. in deep December in northern China. Mingguang Qiao, a concrete pad underneath a freeway overpass, belonged to *yangge* dancers for the moment, despite the passing rumbling buses, honking taxis, and sweeping street cleaners.

(Chen 2010: 21)

. . . a new fad has arisen among the early risers [in Beijing] – ballroom dancing. Yes, it's Fred Astaire and Ginger Rogers, except there's no black tie or sequins. And here the ballroom is a park, and the dancers are positively proletarian. Bundle into heavy leather coats, don wool mittens, turn on the cassette player: Shall we dance?

(Marquand 2000: 1)

Dancing, whether the Chinese *yangge* folk dance, the Argentine tango, or ballroom dancing, has become a popular form of recreation in Chinese cities. In the dense quarters of urban residential districts, dancers often congregate in parks, on sidewalks, underneath overpasses, or in parking lots. They might be young or old, talented or just starting out. In the early reform years, visitors to China were often fascinated by glimpses of groups of silent people moving through tai chi (*taiji*) exercises in the early-morning mist. But as the reform era has progressed, the practice of group exercise in public spaces has widened to include a range of dance and martial arts activities. Some of these gatherings are informal clubs, while others might be lessons that come at a price. Other than tai chi, the most popular of these dances include *yangge*, a folk dance performed with fans or colorful handkerchiefs, Western-style ballroom dancing, especially waltzes and tangos, and ribbon and fan dances, which are hybrids of traditional Chinese dance forms and modern dance (Figure 12.2).

Figure 12.2 Ribbon dancers at the Temple of Heaven, Beijing.

Source: Piper Gaubatz.

Ballroom dancing was popular in China during the 1950s and early 1960s, a legacy of the Treaty Port era, but was banned during the Cultural Revolution. Ballroom dancing experienced a resurgence during the 1980s. Today it attracts young and old alike. In 2007, the *Beijing Daily* estimated that there were about 500 non-commercial ballroom dancing sites in Beijing (Lu and Lu 2008).

> Morning in Beijing arrives with the soft surrealism of sibilant danc-
> ing feet. First one couple, then another and another arrive at Beijing's
> numerous parks and courtyards with the dawn's light, paring off and
> floating in waltzing embraces. These public spaces are their momen-
> tary haven from the pressures and frustrations of Beijing life.
>
> (Evans 1993)

Many of these activities have in common aspects of performance, skills practice, and community building. On a fair-weather Saturday afternoon in Beijing's Tiantan (Temple of Heaven) Park, for example, it is sometimes difficult to walk through the crowds of onlookers that surround people engaged in activities as varied as "water calligraphy," in which people write poetry in water on the pavement with massive brushes, then watch the poetry fade as the water dries, or band practice for small bands devoted to performing marches by John Philip Sousa. The music groups, from Sousa to karaoke, are just barely spread far enough apart that the sound of one does not bleed to the next.

Urban recreation and public squares

A new kind of public open space was introduced after 1949 in the form of large public squares originally modeled after Moscow's Red Square. The original purpose of these squares was to stage mass patriotic demonstrations. Many of the squares, most notably Tian'anmen Square in Beijing, also had symbolic importance, as they replaced the closed spaces of imperial China with "people's" space at the heart of the city. Tian'anmen Square itself replaced the government offices that had flanked the road leading to the gates of the imperial palace ("the Forbidden City") with a public gathering place. Chairman Mao declared that Tian'anmen Square should be built "big enough to hold an assembly of one billion" (Wu 2005: 23). In fact, the Square began as a hastily widened space to serve as the venue for a gathering of about 70,000 people for the declaration of the establishment of the People's Republic of China. By 1959 it had been expanded to hold about 400,000 people; in 1977 it was expanded again, with the addition of Chairman Mao's mausoleum, to hold about 600,000 people. Most cities throughout China built similar, if somewhat smaller squares, during the early years of the People's Republic of China. Before the 1980s, these squares were used primarily for occasional public gatherings, and as shortcuts for bicycles crossing the city. Some became gathering places on hot summer nights, but their bare nature (no landscaping, benches, or other amenities) limited their utility. Beginning in the 1980s, however, and especially during the 1990s, many squares were redesigned with landscaping, benches, and art to create more useful public space (Box 12.2). Today they function as parks, concert venues, and gathering places (Gaubatz 2008b).

***Box 12.2* Riding tricycles in Höhhot: the evolution of public squares and recreation**

New China Square, the large public square in Höhhot, capital of Inner Mongolia, was built in 1953 as a location for mass patriotic demonstrations. Originally, this square was simply a large open area of bare ground occupying a significant portion of the space between Höhhot's twin traditional cities (one built by the Manchu, the other by Chinese, Mongol, and Hui). It was created by bulldozing existing housing and shops. The square was paved in the early 1960s. Some might find this stark expanse of concrete a void within the fabric of the city, yet the square has always been a popular recreational space in Höhhot. On hot summer nights, in particular, despite dim lighting and a lack of landscaping or benches, Höhhot's citizenry flocked to this massive open space to escape the heat; during the day, it became a popular local venue for kite flying. The New China Square was expanded and renovated in the mid-1980s, when some flowerbeds were added around the periphery, cement tiles were replaced with textured non-slip tiles, and a podium was erected on the south side. The recreational use of the square became more diversified, with entrepreneurs offering children rides on tricycles and toy cars, and concessionaires offering snacks from mobile carts at night. The square continued to function both as an "official" space for ceremonies and mass events, especially those commemorating the People's Republic of China, and as an informal recreational space where local citizens could gather. Still primarily a vast expanse of pavement, the square remained largely empty during the days, but came to life at night.

In 2002, the city completed a third expansion and renovation of the square. In its third life, New China Square was transformed from its bleak origins into a dynamic and multifaceted public space. Not only was the square enlarged, but it was completely redesigned. The dais that once served as a podium for political rallies became a contemporary, state-of-the-art stage augmented with lighting, sound, and other equipment for staging concerts and plays. The flat expanse of the pavement was broken up by a new design that includes sunken and raised areas, trees, benches, and flowerbeds. Most spectacular, perhaps, is an elaborate computer-controlled, lighted fountain system. Much of the open expanse at the center of the square is covered in hundreds of recessed fountains programmed to turn off and on in different patterns and timings, lit by thousands of lights.

During the summer months, Höhhot's residents flock by the thousands to the square. Young people dare each other to run through the pulsating fountains; others stroll paths through trees and flowers, or relax on benches. Most Friday and Saturday nights during the summer, the square becomes quite crowded as bands play on the stage, lights and fountains put on a show, and fireworks are set off overhead. Like public squares in most cities, Höhhot's New China Square has evolved from a venue for political gatherings to the city's most popular public park. In 2006, Höhhot's New China Square won an award in a competition entered by 137 cities for the best public square in China (Figure 12.3).

Figure 12.3 New China Square, Höhhot, with fountains.

Source: Piper Gaubatz.

Growing old in urban China

China's population is aging. As the standard of living increases, a larger proportion of the population is surviving into old age. Life expectancy more than
doubled since 1949, from 35 to just over 70 years of age. In 2005, more than 10
percent of the population – about 143 million people – were over the age of 60.
Some estimate that, by 2050, as much as 25 percent of the population – about
400 million people – might be over 60 years old – a number of people larger than
the combined current populations of France, Germany, Italy, Japan, and the UK
(Zhang and Goza 2007). At the same time, the social and cultural mechanisms for
supporting the elderly have been seriously eroded. In traditional China, elderly
people were cared for by large extended families. Today, especially with smaller
urban family size, less flexible housing, and working women, the family can no
longer be relied upon as the primary safety net for the elderly in urban China.
International studies show that modernization tends to erode family support for
elders; a recent survey suggests that this is true in China as well. Higher levels of
urban development are correlated with lessening family support (both financial
and in kind) for the elderly (Cheung and Kwan 2009). Increasing numbers of
senior citizens are living alone. In Tianjin, for example, 54 percent of the elderly
lived alone in 1997; by 2002, this figure had risen to 62.5 percent; by 2012, it is
estimated to reach 90 percent (Zhang and Goza 2007). This generates significant

challenges and hardships for older adults as they age and become less capable of caring for themselves. Elderly people today, who were born in a society where large extended families were the norm, but are now living in increasingly isolated situations in cities, are increasingly experiencing loneliness and other discontents (Yang and Victor 2008).

Chinese cities are experimenting with new ways of accommodating and caring for the elderly in residential settings. Three of these new institutions – "elder-care institutions," which provide residential care outside the family, "institutions for paying respect to older adults," and "institutions for providing care for older adults" – are designed for low-income elderly residents as a form of social welfare. Two other institutional types – "senior apartments" and "nursing homes" – target wealthier elderly. But the "modernity" of these new institutions is often at odds with the expectations of the elderly, who often do not wish to abandon their homes in favor of an institutional setting. In a recent nationwide survey, only 20 percent of the urban elderly were interested in living in an institutional setting. Yet there is a substantial disjunction between this clear disinterest and state and local government policy, which prioritizes institutionalization as its main plan for accommodating the elderly (Chou 2010).

Crime and urban neighborhoods

One of the costs of China's opening to the world and liberalization of its economy has been a marked rise in crime. Between 1978 and 2006, "the homicide rate more than doubled; assault increased 7.9 times; robbery grew by 4.7 times; and larceny rose 8.7 times" (Zhang et al. 2009). In the cities, this has a number of manifestations, including the emergence of areas local residents avoid because they are reputed to be dangerous, and an increasing concern about pickpockets, break-ins, and more serious crimes. Fortress-like spaces have developed, from routinely affixing bars on the windows of lower floors to reduce break-ins, to exclusive "gated communities" (see Chapter 8).

The common public perception is that the rising crime rate is directly correlated with the growing presence of migrants in the cities (Nielsen and Smyth 2008; Zhang et al. 2009). Statistics are not readily available to confirm this perception, but one study in the 1990s suggested that as many as 70–80 percent of urban theft might be carried out by migrants (Yu 1993); in Beijing in the mid-1990s, there were clear spatial correlations between the number of crimes committed and migrant neighborhoods (Ma 2001). Certainly a key component of this correlation is that migrants make up a disproportionate share of the urban poor; the poor are more likely to turn toward petty crime (Cao and Dai 2001). Migrants also are disproportionately male and young, in comparison with urban residents; young males are more likely than other cohorts to be engaged in crime.

The public perceptions of spatial patterns of danger and public safety are changing the experience of living in urban China. Not only are urban neighborhoods becoming increasingly differentiated on the basis of wealth (see Chapter 8), but these differentiations extend to safety as well. Whereas cities under state socialism had relatively undifferentiated public safety landscapes, contemporary Chinese cities are developing "good neighborhoods" and "bad neighborhoods." A recent survey in the city of Tianjin found that Chinese urban residents, not unlike those in other countries, feel safer if they are living in neighborhoods where they have strong ties with their neighbors (Zhang et al. 2009). This is a particular social challenge for China. The reform-era restructuring of Chinese cities has moved large portions of the population of every city. It is possible that most urban Chinese live in situations where they have known their neighbors for less than 3–5 years.

Authorities are struggling with the spatial dimensions of urban crime. The vice-party secretary of Guangzhou, Guifang Zhang, has remarked that there has been a major crackdown on crime in "important" areas of the city. He explained to the *South China Morning Post* that "in the city center, after we stepped up action, criminals moved to the fringes. These two areas account for 50 per cent of crime, so if we can control it in both places, we are hopeful the situation will improve" (SCMP 2007). This type of spatial strategy has two potentially unwanted consequences: it may well move crime to different areas rather than actually reducing it, and it may contribute to increasing differentiation of property values and safety in the city.

Communities of difference: sexual preference, ethnicity, and art

For all that there are substantial differences in urban lifestyles between migrants living quasi-urban/quasi-rural lives in informal settlements and young, upwardly mobile Chinese living in high-rise luxury apartments complete with servant's quarters and 24-hour doormen, another kind of difference can be found in growing communities of urbanites whose difference is not economic, but rather is grounded in identity, such as sexual preference, ethnicity, religion, or even art. The social norms of the pre-reform era were highly conformist; not only was difference not tolerated, but often it was criminalized or actively purged. Contemporary Chinese society is gradually acknowledging the diversity in urban society.

Gay and lesbian communities

Homosexuality was usually considered illegal (under a law banning "hooliganism") in the People's Republic of China until 1997, and remained on an official list of mental illnesses until 2001 (Balzano 2007; Bartram 2010). The reform era and its liberalization have opened possibilities for the development of gay and lesbian spaces in cities. Larger cities in particular have seen the growth of small clusters of gay- and lesbian-friendly bars, tea houses, and cafés. These clusters, though sometimes ephemeral, have generated spaces for emerging gay and lesbian communities. In fact, since the late 1990s, increasingly visible, though still small, gay and lesbian spaces have developed in nearly every city (Ho 2008, 2009; Wu 2003). In Chengdu, for example, the gay community is centered around Hua Xin, a small neighborhood on the northeast edge of the downtown area. This area has emerged since the 1990s as a cluster of bars, clothing shops, beauty parlors, and video stores catering to a gay clientele (Wei 2007). It seems, however, that the small gay neighborhoods that are emerging in Chinese cities are still primarily commercial, rather than residential, in character.

At the same time, urban gay and lesbian communities are becoming increasingly complex and socially stratified. Differences and hierarchies have developed between long-term urban residents and recent migrants, between those with and without access to the international internet-connected gay community, and between people in different economic classes (Ho 2008, 2009). This developing stratification suggests a maturing of the gay and lesbian community as it becomes more established in the cities. Nonetheless, homosexuals continue to experience ostracization by their families and society, and overt homosexuality is far from commonplace. In early 2010, what was to be China's first gay pageant, to be held in Beijing, was cancelled by officials hours before it was scheduled to begin (Bartram 2010).

Ethnic-origin communities

Ethnicity in the contemporary Chinese city, as in the past, may be constituted not only through the overt cultural distinctions between different peoples, such as Tibetans, Uygur, or Koreans, but within the Han Chinese community through place of origin. Recent migrants in particular tend to congregate with others from similar places of origin (see Chapter 5 on migrants and migrant villages). On the outskirts of Beijing, for example, migrants have formed "villages" dominated by single places of origin, such as Zhejiang Village, Henan Village, Anhui Village, and Xinjiang Village. These areas form cultural clusters where distinct dialect, occupations, and social networks are maintained through daily practice and performance. These neighborhoods tend to be underserved by basic infrastructure, such as water, power, or schools, and often come under attack by local officials seeking to move illegal migrants or illegitimate businesses. Zhejiang Village, for example, formed early in the 1990s as a mecca for migrants from the Wenzhou area of Zhejiang (Box 5.1). Both Zhejiang Village and Xinjiang Village were largely bulldozed to the ground by local governments in the late 1990s, but the former has been rebuilt.

Most large cities had small Muslim enclaves centered around mosques in Imperial and Republican China. Hui, ethnically Chinese Muslims, congregated in these neighborhoods within the traditional cities. After the revolution, however, many of China's mosques were closed and these neighborhoods became more heterogeneous. Some contemporary cities retain their Muslim neighborhoods, which in many cases have been expanded or revitalized in recent years as Chinese Muslims gain both the economic and political ability to restore and expand community institutions such as mosques and markets (Gaubatz 2002). Beijing, for example, has two traditional Muslim districts that are still somewhat intact: Niujie, in the southwestern quadrant of the former Qing city of Beijing, and Madian, which lies just northwest of the path of the Qing city walls, due north of Niujie. Niujie began its life as a Muslim neighborhood in the twelfth century; Madian was formed during the seventeenth century. Madian, however, lost much of its Muslim character when it became a center for new housing development during the early years of the People's Republic of China. In the 1950s, its population was 90 percent Muslim; it is 25 percent Muslim today. Niujie, by contrast, has maintained its integrity with the support of the district and municipal governments. The neighborhood was renovated in the late 1990s – most of the 3,000 or so families residing there moved out temporarily. When given the choice to return to the newly renovated neighborhood, nearly all of the Muslim families returned, while the non-Muslims tended to go elsewhere. Niujie today serves as a center for Beijing's Muslim community, with Muslim-oriented hospital, schools, social services, shops, cafés, and restaurants (Wang et al. 2002).

While Niujie and Madian are traditional Muslim enclaves, there also has been a notable new Muslim enclave in Beijing – Xinjiang Village, a migrant settlement that began its development early in the reform era in northwestern Beijing. Here Uygur migrants who have come to do business in Beijing have congregated in increasing numbers, and have established new mosques and other Muslim social institutions. Although this neighborhood started as a ramshackle collection of shacks, it gradually developed a character of its own, with restaurants decorated with onion domes and minaret-like towers, grape arbors shading cafés, and numerous stalls and shops selling kebabs, flat breads, and Central Asian goods. The neighborhood was bulldozed in 1999 and the appellation "Xinjiang Village" shifted to a nearby neighborhood. That neighborhood, in turn, was bulldozed in 2003 and again in 2005. Today there are only a few Muslim restaurants left, and the Muslim presence has been much reduced (Rayila 2011).

Although in most Chinese cities minority populations typically are so small that they are virtually invisible, this is less true in the frontier regions which are the homelands of distinct non-Han Chinese peoples such as the Uygur or the Tibetans, or in cities where there is a long tradition of cultural diversity. Nonetheless, the status and visibility of different peoples in contemporary Chinese frontier cities vary widely. For example, in Urumqi, capital city of the Xinjiang Uygur Autonomous region, despite the fact that the majority of the city's population is Han, the traditional Muslim neighborhood, which formed just south of the city's southern gate when the city was developed in the Qing dynasty, remains today. This neighborhood is marked by the spires of mosques and large outdoor markets. In Xining, capital of Qinghai province, the traditional Muslim neighborhood centered on the city's great mosque has flourished in recent years. Its streets are lined with shops selling clothing for several different local Muslim peoples, books in Arabic, and halal foods. But other peoples have had very different interactions with the city. Tibetans founded this outpost high on the Tibetan plateau, and once controlled a vast area from its walls. Kumbum (Chinese: Ta'ersi) monastery, one of the great monasteries of Tibetan Buddhism, lies within the city's administrative boundaries. Yet Tibetans and their distinctive architecture are little evident in central Xining today. Most of the local Tibetan community lives in the surrounding countryside or in villages close to the Kumbum monastery.

Artists' colonies

Artists have created new urban spaces in recent years, which nurture urban China's burgeoning interest in modern art. There is a continuing tension between the avant garde and mainstream aspects of the art community. Artists in China provide a lively and critical commentary on contemporary life; thus they live on a sometimes dangerous edge between legitimacy and incarceration. Art communities in Chinese cities have developed in the spaces abandoned by the rapidly growing economy, such as outmoded factories, warehouses, and shops. They are mostly found along the loose boundary between the urbanized and rural spaces of the city, where land use controls are relatively fluid (Ren and Sun 2012).

In Beijing, for example, there are about 20 artists' colonies located on the edges of the traditional urbanized area. Although most of these are on the northeastern side of the city, the first was located in the northwest near the ruins of the ancient Yuanmingyuan Palace. This village of about 300 artists began in 1984 and lasted until 1995, when the Haidian district government closed it down (Ren and Sun 2012). Dashanzi, a new artist's colony, emerged soon after on the northeastern edge of the city. This colony formed where there had been an industrial zone during the 1950s–1970s. By the 1990s, however, the factories had been abandoned. Artists began to make use of the vast, empty halls of the former 798 munitions factory for studio and exhibition space. In the early years, the artists of 798 were often harassed by city authorities for essentially squatting in these abandoned buildings. But the district continued to grow, and by 2006 was legitimated when the local Chaoyang district government officially recognized its management company as an authorized "creative industry" venture (Zhuang 2009). Today Dashanzi maintains a professional-looking website, invites government officials to participate in art workshops, and serves as a massive cluster of art galleries, studios, and exhibitions catering to domestic and international art aficionados. As Dashanzi has become more "legitimate," however, some artists have chosen to take their art, with its social and political commentary, to new locations that are less mainstream. Some art exhibitions have been mounted, for example, in the basements of new high-rise buildings – a new and growing urban space that is largely underutilized (Visser 2010).

Similarly, Shanghai's Moganshan Road art district (often referred to as M50) began in 2000 when a few artists began to make use of some old warehouses dating from the 1930s for studio space. The district started out in relatively derelict structures along Suzhou Creek, as well-known artists, such as Shanghai's Xue Song, moved their studios there. Local authorities attempted to close down this art colony and evict the artists in 2004, but the artists were able to fend this off with the aid of local publicity (Kong 2009; Visser 2010). Since that time, domestic and foreign investment has upgraded the spaces, and the area quickly became well-known, trendy, and high-priced (Kong 2009). Although it currently serves as a center for Shanghai's avant garde art community, other art salons and studios have also developed. As M50 becomes increasingly expensive and well known, it is likely that some artists will seek more economical and less well-known spaces.

Conclusion

Life in China's cities is becoming increasingly diverse and exciting. Opportunities abound for new activities and new spaces are emerging which reflect the exuberance of these new pursuits. There are new ways to spend money, new forms of entertainment, and many opportunities for social activities to bring people together in a rich variety of new communities of affinity. While many of these new aspects of Chinese urban life are positive, some have generated or exacerbated social problems, such as care for the elderly or street crime. Discrimination against those who are different persists, though there are signs of increasing tolerance.

Bibliography

Balzano, John. 2007. "Toward a Gay-Friendly China? Legal Implications of Transition for Gays and Lesbians." *Law and Sexuality* 16: 1–43.

Bartram, David. 2010. "China's Gay Rights Revolution." *The Guardian*. Online edition. January 19, 2010.

Cao, Liqun and Dai, Yisheng. 2001. "Inequality and Crime in China," in Messner, Steven F. *Crime and Social Control in a Changing China*. Westport, CT: Greenwood Press.

Chen, Caroline. 2010. "Dancing in the Streets of Beijing: Improvised Uses Within the Urban System," in Hou, Jeffrey (ed.) *Insurgent Public Space: Guerilla Urbanism and the Remaking of Contemporary Cities*. Oxford and New York: Routledge.

Cheung, Chau-Kiu and Kwan, Alex Yui-Huen. 2009. "The Erosion of Filial Piety by Modernisation in Chinese Cities." *Ageing and Society* 29: 179–198.

Chou, Rita Jing-Ann. 2010. "Willingness to Live in Eldercare Institutions Among Older Adults in Urban and Rural China: A Nationwide Study." *Ageing and Society* 30: 583–608.

Davis, Deborah. 2005. "Urban Consumer Culture." *The China Quarterly* 183: 692–709.

Evans, Martin C. 1993. "China Exercising By Dawn's Early Light." *Seattle Times*, September 5, 1993 http://community.seattletimes.nwsource.com/archive/?date=19930905&slug=1719461.

Gaubatz, Piper. 2002. "Looking West Toward Mecca: Islamic Enclaves in Chinese Frontier Cities." *Built Environment* 28(3): 231–248.

Gaubatz, Piper. 2008a. "Commercial Redevelopment and Regional Inequality in Urban China: Xining's Wangfujing?" *Eurasian Geography and Economics* 49(2): 180–199.

Gaubatz, Piper. 2008b. "New Public Space in Urban China: Fewer Walls, More Malls in Beijing, Shanghai and Xining." *China Perspectives* 76(4): 72–83.

Gu, Chaolin, Chan, Roger C. K., Liu, Jinyuan, and Kesteloot, Christian. 2006. "Beijing's Socio-spatial Restructuring: Immigration and Social Transformation in the Epoch of National Economic Reformation." *Progress in Planning* 66: 249–310.

Ho, Loretta Wing Wah. 2008. "Speaking of Same-Sex Subjects in China." *Asian Studies Review* 32: 491–509.

Ho, Loretta Wing Wah. 2009. *Gay and Lesbian Subculture in Urban China*. London: Routledge.

Huenemann, Ralph. 1966. "Urban Rationing in Communist China." *The China Quarterly* 26: 44–57.

Kong, Lily. 2009. "Making Sustainable Creative/Cultural Space in Shanghai and Singapore." *Geographical Review* 99: 1–22.

Lu, Hongmei and Lu, Dongmei. 2008. *Beijing at Play: Two Sisters' Journey into the Park Culture of Beijing*. Beijing: Intercontinental Press.

Ma, Guoan. 2001. "Population Migration and Crime in Beijing, China," in Messner, Steven F. (ed.) *Crime and Social Control in a Changing China*. Westport, CT: Greenwood Press.

Marquand, Robert. 2000. "It's 6 a.m. Shall we dance?" *Christian Science Monitor* 93(7): 1.

Nielsen, Ingrid and Smyth, Russell. 2008. "Who Wants Safer Cities? Perceptions of Public Safety and Attitudes to Migrants Among China's Urban Population." *International Review of Law and Economics* 28(1): 46–55.

Pun, Ngai. 2003. "Subsumption or Consumption? The Phantom of Consumer Revolution in Globalizing China." *Cultural Anthropology* 18(4): 469–492.

Rada, Edward L. 1983. "Food Policy in China: Recent Efforts to Balance Supplies and Consumption Requirements." *Asian Survey* 23(4): 518–535.

Rayila, M. 2011. "The Pain of a Nation: The Invisibility of Uyghurs in China Proper." *The Equal Rights Review* 6: 44–67.

Ren, Xuefei and Sun, Meng. 2012. "Artistic Urbanization: Creative Industries and Creative Control in Beijing." *International Journal of Urban and Regional Research* 36(3): 504–521.

South China Morning Post (SCMP) Staff Reporter. 2007. "Crime Still Serious But We Will Reduce It, Guangzhou Deputy Pledges." *South China Morning Post* May 23.

Uncles, Mark D. 2010. "Retail Change in China: Retrospect and Prospects." *The International Review of Retail, Distribution, and Consumer Research* 20(1): 69–84.

Visser, Robin. 2010. *Cities Surround the Countryside: Urban Aesthetics in Postsocialist China*. Durham, NC: Duke University Press.
This book offers a fresh lens on the rapid changes taking place in Chinese cities and societies. Building on a base of scholarship grounded in more conventional social science, urban planning and historical perspectives, Visser presents the Chinese city as it is portrayed in contemporary Chinese art, cinema, and fiction. She uses interviews with artists to weave their life experiences and perceptions together with the art they produce to paint a compelling alternative vision of the rapid changes which are taking place in urban China.

Wang, Wenfei, Zhou, Shangyi, and Fan, C. Cindy. 2002. "Growth and Decline of Muslim Hui Enclaves in Beijing." *Eurasian Geography and Economics* 43(2): 104–122.

Wei, Wei. 2007. "'Wandering Men' No Longer Wander Around: The Production and Transformation of Local Homosexual Identities in Contemporary Chengdu, China." *Inter-Asia Cultural Studies* 8(4): 572–588.

Wu, Hung. 2005. *Remaking Beijing: Tiananmen Square and the Creation of a Political Space*. Chicago, IL: University of Chicago Press.

Wu, Jin. 2003. "From 'Long Yang' and 'Dui Shi' to Tongzhi: Homosexuality in China." *Journal of Gay and Lesbian Psychotherapy* 7(1/2): 117–143.

Xiang, Biao. 2005. *Transcending Boundaries: Zhejiangcun: The Story of a Migrant Village in Beijing*. Leiden: Brill.

Yang, Keming and Victor, Christina R. 2008. "The Prevalence of and Risk Factors for Loneliness among Older People in China." *Ageing and Society* 28: 305–327.

Yu, Lei. 1993. *Crime Research in Today's China*. Beijing: China People's Public Security University Press.

Zhang, Yuanting and Goza, Franklin. 2007. "Who Will Care for the Elderly in China?" Working Paper, Center for Family and Demographic Research. Bowling Green, OH: Bowling Green State University.

Zhang, Lening, Messner, Steven F., Liu, Jianhong, and Zhuo, Yue Angela. 2009. "Guanxi and Fear of Crime in Contemporary Urban China." *British Journal of Criminology* 49: 472–490.

Zhuang, Jiayuun. 2009. "Factory 798: The Site of Nostalgia and its Incontinent Dweller. " *Extensions* 5. Retrieved from http://www.extensionsjournal.org.

13 Urban governance and civil society

The state, both central and local, remains strong, even under increasing globalization and marketization. But its function and scope have shifted. Because China's political system remains centralized, there continue to be direct political, administrative, and fiscal relationships between the central government and provincial and local governments. Notwithstanding, there has been a steady decentralization of administrative and fiscal powers from central to local levels. Today, municipal governments increasingly resemble local developmental states: more interested in economic growth than ideological pursuits.

Situated in the large context of changing central–local relations, this chapter outlines the transformation of governance institutions, the urban administrative hierarchy, and the state–society relationship. Cities now are facing challenges of developing effective institutions of local governance and giving non-state actors a voice in decision making. The following questions should help guide the reading and discussion of the materials in this chapter:

- What hierarchy does China's urban administrative system follow?
- The increasing number of cities in China partially stems from administrative changes. How so?
- China's largest city, Shanghai, has experimented with a new model of municipal administration: two-level governments and three-level governance. How does this model work?
- In what ways have municipal governments become local developmental states?
- Under changing state–society relations, how is urban society organized? What is the primary function of neighborhood committees today?
- An increasing range of private interests and civil society institutions is filling the vacuum of political representation. What are they and what are their roles?

Reorganization of the state

Urban administrative hierarchy

Under state socialism, the central government was the only key actor and major beneficiary of state development. Local administrative units, such as provinces, prefectures, cities, counties, and communes, served as links in the administrative hierarchy to implement centrally designed national plans of social and economic development (the so-called "Five-Year Plans"). Today, the chain of command is different: local governments have more autonomy in designing place-based policies, more power in seeking new sources of revenue, and more responsibilities in providing public services. This indicates a different form of state intervention at lower spatial scales. The administrative hierarchy, on the other hand, remains more or less intact in terms of official appointments and accountability. That is, decentralization has not happened on the political front. Local politicians and cadres remain accountable to their superiors.

Broadly, China's administrative hierarchy has five main levels (Figure 13.1). At the highest level is the central government in the capital city, Beijing. The second level includes 23 provinces, four centrally controlled municipalities (Beijing, Shanghai, Tianjin, and Chongqing) with the same status as provinces (also known as provincial-level cities), and five autonomous regions (populated by concentration of ethnic minority groups), as well as two special administrative regions (Hong Kong and Macau). The next three levels are local and begin with prefectures and prefecture-level cities, followed by counties and county-level cities, and then by townships and towns. Each local level is represented by its own government. Counties and townships belong to rural administration, separate from urban administration. Regular staff in urban administration belong to the state nomenklatura system (now known as the civil service). This entitles them to welfare benefits such as pension, health care, and housing subsidies. But most rural administrative staff are not part of the state nomenklatura system and have limited

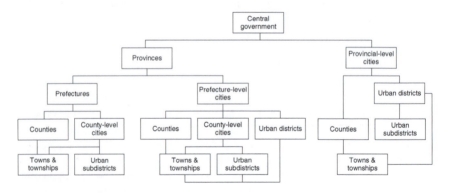

Figure 13.1 China's administrative hierarchy.

Source: Based on Ma 2005.

Notes:

a There are autonomous regions with concentration of minor ethnic groups and special administrative regions (Hong Kong and Macao) at the provincial level, autonomous areas at the prefecture level, and autonomous counties at the county level.

b A small number of prefecture-level cities (about 3 percent) do not have subordinate counties, because of either limited rural hinterland or conversion of counties into county-level cities or urban districts (Ma 2005).

career mobility in the political system. The hierarchy is complex, particularly at local levels. Counties and townships may be enveloped within cities (i.e., rural administration within urban administration).

The urban administrative system follows a similar hierarchy: provincial-level cities, subprovincial-level cities (often the capitals of their sitting provinces), other prefecture-level cities, county-level cities, and officially designated towns. Cities of the first three types together form a category called prefecture-level cities. As higher-level urban units, they oversee the governance of lower-order ones. In the strictest sense of the term, a prefecture-level city is not an equivalent of a city or municipality. Instead, it is an administrative unit comprising both an urban core (a city as in the North American context) and surrounding rural or less-urban-ized areas. A prefecture-level city may contain multiple urban districts, counties, county-level cities, or other similar units (Figure 13.1). As a result, it can be con-sidered as the Chinese version of the term "metropolitan area," as used in many other countries. In general, cities higher in the hierarchy are favored in important ways: greater autonomy in decision making, more public finance resources, and greater access to transport infrastructure.

The central government has made adjustments to the urban administrative sys-tem for the benefit of overall economic growth and urban expansion. A policy of "cities leading counties" in the early 1990s linked cities to their hinterlands to spur rural development. As such, rural administration, in hundreds of counties annually in the immediate years of the new policy, came under the economic directives of cities. Today more than 1,200 counties are under the jurisdiction of about 280 prefecture-level cities (Cartier 2005; Chung 2007). Under this arrange-ment, a prefecture-level city becomes an administrative body managing a number of counties. County-level cities, previously governed by provincial governments, also are placed under the jurisdiction of prefecture-level cities. This change has created a spatial arrangement that is closer to city-centered regions than a true city. From a positive viewpoint, this bridges the production systems and resource allocations between rural and urban areas and extends the horizontal reach of cit-ies (Figure 13.2). Before market reform, administrative barriers prohibited direct interactions between urban and rural areas. On the other hand, the change has inevitably led to physical expansion of cities by encroaching upon farmland, with serious implications for food production and environmental quality. There are also cases of illegal land grabs and subsequent rural unrest (see Chapter 10). Under the new relationship, subordinate counties may lose their autonomy over certain administrative and budgetary issues, often against their will and to their detriment. As such, it is not uncommon to see cities and counties still existing more or less as independent units with different interests, even after the intended integration.

Another series of policy changes reclassified a larger number of counties as cit-ies (in 1983, 1986, and 1993), giving formerly rural areas urban power that entails more administrative and fiscal autonomy. The latest criteria indicate that a rural township can be designated as a county-level city with a minimum nonagricul-tural population of 60,000 and GDP of 200 million yuan. County-level cities with

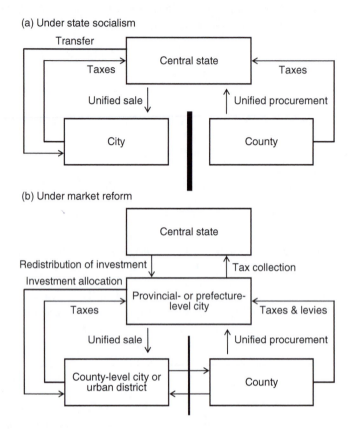

Figure 13.2 Resource allocations between urban and rural areas.

Source: Based on Chung 2007.

non-agricultural populations of 200,000 and GDP of 2.5 billion yuan (in which 35 percent is in the tertiary sector) could be promoted to prefecture-level cities (Chung 2007). All of this underscores the steady rise in the number of cities, particularly prefecture-level cities that enjoy greater administrative power (Table 13.1). Suburban counties outside large cities, in other cases, have been converted into urban districts. Overall, the reorganization of urban administrative hierarchy goes beyond simply changing the ranks of different territorial units: it involves a restructuring of power from rural to urban areas and a shift of economic decision making from vertical authorities based on central ministries to horizontal authorities based on local units. To be sure, a place's administrative rank continues to affect its political and economic relations with other places. There is evidence of a correlation between a city's rank and its size – larger cities tend to be at higher administrative levels than smaller cities (Ma 2005).

Table 13.1 Cities in the administrative hierarchy, 1949–2009

Year	Provincial-level cities	Prefecture-level cities	County-level cities	Total
1949	2	56	61	119
1950	2	64	66	132
1955	3	83	78	164
1960	3	88	109	200
1965	3	76	90	169
1970	3	79	95	177
1975	3	96	86	185
1979	3	104	109	216
1980	3	107	113	223
1985	3	162	159	324
1990	3	185	279	467
1995	3	210	427	640
2000	4	259	400	663
2005	4	283	374	661
2010	4	283	370	657

Source: National Bureau of Statistics of China 2009, 2011.

Municipal administration

Within a city above the prefecture level, there are two levels of administration – municipal government and district government. For provincial-level cities, districts are equivalent in rank to prefecture-level cities. Districts in prefecture-level cities are equivalent to county-level cities (Figure 13.1). County-level cities, on the other hand, do not have urban districts. For civil service staff working in urban administration, such classification is important because it determines grade and pay levels. District governments also establish Subdistrict Offices (also known as Street Offices) to act in their interests. Between 50,000 and 100,000 residents live in each subdistrict, whose size in population census resembles that of census tracts in the USA. There is no People's Congress at the subdistrict level and thus, according to China's Constitution, the subdistrict office does not constitute a level of government. Rather, its divisions are responsible for tasks delegated by the respective departments in the municipal or district government. Mostly as a branch office of the district government, its resources and functions are limited.

There have been recent experiments with new ways of organizing local administration. The best known is the Shanghai model: two-level governments and three-level governance. Launched in the late 1990s, this model allows more powers for district governments and expands the functions of subdistrict offices. First of all, there are additional lines of professional staff in these offices, as well as an increased share of district government budget. Then, a local police station and a commerce department are put in place in each subdistrict. As such, subdistrict offices have become the grassroots level of administration in Shanghai. Clearly, governmental responsibilities and services are decentralized. This is partially in response to the declining role of work units and their traditional role in providing social services. Given the enormous size of the Shanghai municipality (about 18 million in population) and its provincial-level status, this decentralization process seems a sensible move to make government units more responsive to public needs.

This model of changing governance structure is primarily driven by the public sector, with substantial resource demand on municipal finance. Most other Chinese cities have not followed the same path. Some have experimented in different ways, expanding the role of resident engagement (see the Shenyang model later in this chapter). The Shanghai model also has had unintended consequences. District governments' over-zeal to pump up land transfer revenues contributed to the over-building of commercial real estate and luxury residences. The accumulation of excess capacity resulted in a sharp drop in rents during 1996–1999. Consequently, in mid-2000 the city enacted a new regulation depriving all district governments of the approval right of land leasing. Instead, the Shanghai Municipal Housing and Land Administration would have the sole authority, and could stop the leasing approval of land for projects such as shopping malls, entertainment centers, golf courses, and grade A villas and office buildings (Yusuf and Wu 2002).

Developmental state in the making

While the central government dictates critical policies determining economic openness, municipal authorities have a large role to play in devising and implementing strategic local objectives. They have a firm hand in determining the timing, pace, and configuration of local development. Current municipal leaders are generally better educated, technically trained managers and professionals. More of them are native to the cities instead of outsiders directly appointed by the central or provincial government. With increasing local autonomy and a rising cadre of younger technocrats, local leadership showcases the rise of a new developmental state intent on being fully engaged in the global market and associated economic growth. This is different from how local governments, say in the USA, behave. Although business-oriented, American local governments can best be regarded as an enabling state and rarely give the strong guidance that is key to state developmentalism. Some even argue that local governments in China do not simply try to reproduce or follow development efforts initiated by the central government. Instead, they are often the actual architects of growth and development policies that are conducive to local institutional frameworks and specific development needs (Segal and Thun 2001). Nowhere are the imprints of this local developmental state more clearly reflected than in the process of economic planning and urban development.

Local economic planning

Under reform, China has undergone a gradual process of decentralization in both decision making and fiscal responsibility (more in Chapter 9 for the latter). Cities have gained more power to approve foreign investment projects, control land lease and transfer, and annex or acquire agricultural land for urban expansion.

Working out a development strategy that can stimulate growth is an essential goal for local governments. At least two considerations motivate them to do so: exhibiting achievements to the central government and promoting economic development to serve local interests (Zhang 2002; Zhu 2000). Some call this "local state corporatism": a system in which local governments "treat enterprises within their administrative purview as one component of a larger corporate whole" (Oi 1995: 1132).

At the local level, political leadership has played an important role in attracting global capital. The central state grants different degrees of power to different levels of local government in the approval process. With increasing competition among cities, localities have adopted markedly different methodologies for attracting FDI and cultivating technological capabilities. They offer a broad array of incentives in financing, taxation, intellectual property protection, commercialization, and support services. As a result, patterns of growth and development diverge markedly across the country, in both urban and rural sectors (see Chapter 7). Such local or regional variation, however, is not unique to China; it also is seen in many developing countries undergoing rapid economic and social transformation.

A key distinction lies in the extent to which municipal authorities exercise control over the corporate sector. Some cities retain a tighter rein over firms, while others are more hands-off with a less hierarchical system. The best case in point is the contrast between Shanghai and Beijing, China's two largest cities. In Shanghai, power originated from the top of the municipal bureaucracy and filtered down to firms through a tightly organized hierarchical structure. Top municipal officials controlled the head offices of corporate groups, which in turn controlled the individual firms within the group. This bureaucratic structure allowed the municipal government to provide investment capital to firms directly, compensating for underdeveloped financial networks. In addition, large-scale corporate groups were able to coordinate internal activities, structure internal incentives properly, and monitor the performance of subordinate firms (Segal and Thun 2001). Early on, municipal officials funneled the majority of local investment and FDI to large state-owned corporate groups or firms. Such large groups tended to dominate the local economy, a legacy of Shanghai's historical role as the "industrial workhorse" of the planned economy. The emphasis on large enterprises, in particularly, helped launch the city's successful automobile sector, with several champion firms such as Shanghai Volkswagen and General Motors' joint venture.

Such a local economic hierarchy, however, marginalized smaller, more flexible non-state firms. For instance, it was not until the early 1990s that the development of private technology enterprises came into focus in Shanghai, after such enterprises in other cities (particularly Beijing) had achieved critical fame. The focus on government guidance in Shanghai also is reflected in the balance of power among municipal institutions. Agencies traditionally active in the state plan, such as the planning and economic commissions and industrial bureaus, dominate local politics. Other bureaus, mostly non-industrial, occupied lower positions on the ladder.

Beijing, on the other hand, took a more hands-off approach to economic planning, particularly in corporate control. This was in part a result of the city's role as the national capital: municipal officials were reluctant to promote local over national interests. Firms tended to be smaller and were not as likely to be wholly owned branches of a larger corporate structure. Local officials promoted a wide range of ownership structures in the information technology sector, directed capital to smaller non-state firms, and helped scientists build a network to facilitate information and personnel flows. Branches of the municipal government also stepped in to arrange funding for small technology firms from third sources (Segal and Thun 2001). This proved to be quite conducive to high-tech development, as seen in the rise of China's foremost science park – Zhongguancun (more in Chapter 6).

Beijing also used infrastructure projects and the local science budget to support small, non-state firms. These investments were not firm-specific and were not based on political or administrative connections: benefits were either public goods or distributed by competitive means. Unlike in Shanghai, the majority of research funds in Beijing went to independent scientific organizations, not to labs within state-owned enterprises. In the high-tech sector, most of the earliest technology enterprises drew their personnel, technology, equipment, and investment capital from the Chinese Academy of Sciences, a branch of the central government over which municipal officials had little control (Segal and Thun 2001). Overall, Beijing sought to implement development policies that did not require extensive government supervision and that would turn the strength of non-state firms into an advantage.

This contrast in local economic planning also mirrors how artistic and cultural activities have developed in the two cities. Unlike Beijing, where ministries overseeing cultural activities are perhaps more preoccupied with running the country, Shanghai cultural authorities keep a tight rein on the arts. They play a more active guiding role than those in Beijing, and local officials often use cultural institutions as vehicles for personal ego trips. The Bureau of Broadcasting, Film, and Television approves all documentaries and other film-related activities. The Cultural Bureau reigns over the performing arts, making sure to attend all rehearsals. As a result, there are fewer loopholes for artists to exploit, and the city lacks the creative passions that fuel an underground culture in Beijing. In particular, painters have taken advantage of the looser political environment of the 1980s and formed artistic colonies in Beijing. Artists, ironically, enjoy more freedom to create in the nation's political center than in its cosmopolitan commercial hub. The increasingly lively creative atmosphere, coupled with more abundant foreign buyers and consumers, attracts even more artists to Beijing, generating a virtuous cycle. The congregation of talent ultimately makes it the country's artistic capital (Esaki-Smith 2001; Wu 2004).

Changing processes of urban development

Under decentralization, the central government has significantly reduced its direct involvement in local development projects. Urban development is now shaped by local governments, contrasting sharply with the pre-reform era. But the central government still holds some critical power, such as the appointment of key local officials. It also requires review and revision of comprehensive plans of all municipalities. In addition, it periodically issues directives in the form of national urban policy, indirectly affecting local development issues. For example, the cultivated land preservation policy of 1998 has forced local governments to curtail converting farmland to urban uses and to emphasize redeveloping existing built areas (see Chapter 10).

Expanding the local revenue base has been at the top of municipal agendas. One important mechanism is through engagement in land development. Public-owned development companies are established by either municipal or district governments with public money. They dominate the primary land market, acquiring and selling land to developers. State work units also have joined the real estate business, partially to retain the development rights of their existing land. Some have set up or funded development companies to redevelop previous factory sites. On the other hand, local governments would induce state enterprises to give up land use rights of premium locations by providing assistance for factory relocation (Wu 1999; Zhu 2002).

The involvement of state work units and public-funded companies complicates the development process. Even real estate development companies are connected with various government branches through formal institutional linkages and/or informal personal contacts. The close relationship between municipal authorities and the development industry has prevented the industry from becoming truly market-based. Developers also receive benefits in the form of tax reductions and land subsidies. Imposing ad hoc local fees and extending government responsibilities to developers, such as providing public goods, also are common practices (Zhu 2000). As such, local bureaucracy and developers often form an informal coalition, to promote a common pro-development agenda.

The constantly shifting balance of power between the government and economic interests also has complicated the implementation of urban plans. The City Planning Bureau traditionally examines and approves the location and boundary, provides conditions for designing, and issues licenses for construction. But efforts of planners are often blocked by ill-defined enforcement procedures and numerous concessions made to high-profile projects. Planning tends to assume a passive role, following rather than leading the pattern of land development. Often, planners are under great pressure from local politicians to play an active role in competition with other jurisdictions for capital and industries.

The lack of effective planning and regulations has contributed to the "property bubble" (see Chapter 10) and the "development zone boom" (see Chapter 6) in many cities. In addition, development projects in strategic locations boosting local images get more attention from municipal leaders, as do those with potential economic payoffs such as luxury hotels and shopping centers, grade A offices, and high-end housing (more in Chapter 8). Other more urgent and tougher issues, such as affordable housing, remain low on the local development agenda. The land leasehold system implemented in China offers a unique opportunity for municipal governments to set aside land for social housing. But this would mean reduced revenues from land leasing and offering other types of public subsidies (as in the case of social housing in Singapore). As such, it was not until 2010 that affordable housing gradually received more official attention, with pressure from the central government.

Evolving state–society relations

Organizing the urban society

Under state socialism, urban society was organized through both the workplace and residence – the former by state work units, the latter by a network of neighborhood committees (*juweihui*). Given that the majority of urban workers were employed by state work units, these units were primary agents of administrative and social control. As such, the society was tightly managed by the state system through workplace affiliation. But with the drastic downsizing of state-owned enterprises during market transition (see Chapter 7), the role of work units has declined. Neighborhood committees, on the other hand, remain pervasive. By some estimate, there are about 90,000 of them nationwide (Read and Chen 2008). This network covers the bulk of the urban society, except in some new housing estates and peri-urban areas. Equally prevalent in the rural society is a network of village committees. Such state-sponsored institutions of neighborhood administration also can be found in other Asian societies (such as Korea, Japan, and Taiwan) and have historical roots. At least in the Qing dynasty (1644–1911), the state designated headmen for groups of urban households with collective responsibilities (Bray 2005; see Chapter 3).

Today, the primary role of neighborhood committees is to facilitate governance. Their key staff are appointed and paid by the state (as part of the civil service), while others work as contract workers or volunteers. They are the eyes and ears for local governments and the police force. Directly reporting to subdistrict offices, they monitor family-planning compliance and maintain household registration records. With the rising influx of migrants, they also take on the responsibility of initial intake of requests for temporary residency filed by migrants. In some neighborhoods, they even manage the permit process of rental housing (see Chapter 5). In addition, they provide some services to residents, including collecting complaints, facilitating welfare programs (for elderly and disabled residents), organizing recreational activities, mediating disputes among residents, and sometimes carrying out neighborhood watch programs.

Experimenting with an expanded role for neighborhood committees, the city of Shenyang has moved governance reform beyond the public sector. Hard hit by economic restructuring and state-owned enterprise downsizing since the mid-1990s (see Chapter 7), the city was in no position to follow the so-called Shanghai model of decentralized administration. Public resources were tight, and state work units shrank. In response, Shenyang municipal government relied more heavily on neighborhood committees to deliver community services. The first step was to combine two to three of them to cover a larger population, reducing their total number from 2,700 to 1,277 citywide (Lu 2003). Many were renamed as community (*shequ*) committees, reflecting the change in geographic coverage. In addition, a community council, made up of resident representatives, becomes an important player in decision making, while a community negotiation committee plays the role of advising and monitoring. The goal of such reorganization was to encourage more resident engagement in self-governance, with less government resources and involvement.

Increasing residential mobility and rapid urban development have affected the functioning of neighborhood committees. It can take several years for municipal and district governments to set up a committee in a new neighborhood. When peri-urban areas transition from rural to urban administration, village committees give way to neighborhood committees. Such administrative reorganization often leads to neglect or incapacity. Even after staff are hired, it takes time to build the necessary ties with residents to lubricate their operation. In older neighborhoods, some committee staff are long-time residents themselves, and enjoy closer connections with their constituents. Overall, these grassroots organizations have slowly redirected more of their function as providers of services and organizers of social activities. Their monitoring role, a legacy of the Communist state's grip on urban society, has become secondary.

Private interests in community governance

With housing commodification (see Chapter 10), a range of private institutions and resident groups have become important players, especially in new neighborhoods. Such non-state interests, while in no way replacing the state network of social control, open up more space for civic engagement. Through them, some public responsibilities are privatized and transferred to the private sector and community. The first such institution is the private property management companies, and the second is a type of resident group commonly known as homeowners' associations (HOA). In many new private housing developments, the two together take over some of the functions traditionally served by neighborhood committees.

The first property management company was formed in Shenzhen in 1981, and the commercial industry known as property management began developing in the early 1990s. A central regulation, issued by the then Ministry of Construction in 1994, called for the professionalization of residential management. By the turn of the twenty-first century, there were tens of thousands of property management companies nationwide (Zhang 2010). Their primary function is to provide basic services to both the community (demarcated by the development) and to residents, including facility maintenance and repair, security surveillance, ground maintenance and sanitation, and, increasingly commonly, parking management. Residents pay annual fees to them for such services, and have access to an on-site office every day of the week. Some companies even get involved in neighborhood organizing activities, such as helping with census data collection and mediating disputes among residents.

As private entities, property management companies lack formal regulatory power, but have wide latitude to set their own rules and fees. Conflicts with residents regarding fee scales and quality of services are inevitable. They also are a form of local authority in maintaining social order and controlling access to the community (particularly in keeping out the unwanted). They hire security guards to take charge of safety issues. The guards can question people with suspicious behavior, detain intruders, and then hand them over to local police. Such privatization of maintaining public order has helped create an increasingly differentiated or even segregated urban space.

HOAs are a very recent phenomenon in urban China. In the 1990s, when massive construction of commercial housing went on, there was no clear state policy officially sanctioning them. *The Property Management Regulation*, issued by the State Council in 2003, was the first to grant legal status to HOAs (Zhang 2010). But the conditions for such legal status are very strict and border on unrealistic – requiring the election of a steering committee with the approval of at least half of homeowners in a given community. As such, the presence of HOAs is uneven across different types of residential development and across localities.

The nature of HOAs varies, from those dominated by developers to others with leaders democratically elected by members. Some large developers with more resources actively initiate the establishment of HOAs, aiming to create friendly relations with residents and ultimately safeguarding the company's own reputation. In neighborhoods where homeowners are dissatisfied with property management, HOAs become a vehicle for residents to protect their interests against corporate power. Such contrasting relations also characterize the interactions between HOAs and neighborhood committees. While national policy guidelines stipulate that neighborhood committees overrule HOAs, the relations between the two occupy a wide spectrum – cooperative in some places and antagonistic in others.

Because of the investment and prestige associated with purchasing a home in private housing compounds, residents have become more vocal in defending their perceived rights. In fact, rights protection is becoming a new trend of activism and a powerful social force today, similar to those involved in protesting against forceful eviction in redevelopment areas (see Chapter 10). The most drastic opposition has come about when residents feel that the quality of their living environment is infringed upon. Often, it is against developers using (or misusing) common land to build more housing units at a higher density than promised, at the expense of green space, or at too close a distance to allow for adequate sun exposure. The latter is a unique feature in Chinese architectural practice and consumer preference. With a stronger reliance on natural lighting and ventilation, construction convention requires that each unit gets at least an hour of sunlight on the shortest day of the year (December 21), which then determines how far apart buildings need to be. As disputes over the use of communal space for new construction escalate to public protests, lawsuits, or even violence, involved residents become highly organized. With the state's retreat from direct provision of community services, residents feel the need to defend their self-interest against corporate greed. This power relation adds a new layer of complexity to urban governance.

Space for public participation

While the central state has diffused its power to local levels, it continues to hold a tight rein over civic life. In addition to its reluctance to engage in political reform, there is limited space for public participation. On the surface, new laws and policies, especially at the national level, have been developed since the 1990s to promote, confer, and formalize the rights of public involvement in planning and managing environmental issues related to urban development. In particular, the *Environmental Impact Assessment Act* of 2003 specifically encourages the involvement of work units, experts, and the general public in environmental assessment (Tang et al. 2008). The development of the legal framework for environmental public participation also reflects an evolving orientation from non-binding advisory mechanisms to binding obligations grounded in citizen rights.

In reality, however, barriers are multiple. Public engagement goes against the long-entrenched bureaucratic culture of China's authoritarian state. Little information on procedures and timing for public participation is available. Nationwide, there is no systematic information disclosure system, although several large cities have introduced "open government information" initiatives (most notably in Guangzhou and Shanghai). As such, people often do not fully understand policies. Where public participation does exist, it is often on inequitable terms or does not provide adequate opportunity for public input. At the local level, the benefit of public participation in development decisions often is not fully recognized by officials. With no election power to leverage, community groups are the weakest among different players in urban development. They are never members of the local growth coalition. Last, in terms of institutionalized participation, China does not have a long tradition of civil society organizations. Most such organizations are engaged in delivering services, from legal aid to environmental protection. Environmental non-governmental organizations were the first civil society groups to emerge in the mid-1990s (including such successful ones as Friends of Nature, Global Village of Beijing, and Green Earth Volunteers). The political situation has shifted to allow the growth of similar groups in the health sector, particularly groups focusing on HIV/AIDS (CEF 2006). Nonetheless, the relationship of these groups with the bureaucracy remains uneasy.

There are signs of progress, however slow. In Beijing, for instance, now there is an annual poll initiated by the municipal government. Inviting all residents to comment, the internet poll provides information for assessing and ranking the performance of municipal agencies. Such public polling, either electronically or otherwise, has been used in Shanghai for government-initiated redevelopment projects to gauge residents' willingness to relocate. Some cities, including Beijing, Shenzhen, and Shanghai, have begun putting some regulations on the internet and soliciting comments electronically. Public hearings also become more common, especially for pricing issues, local environmental licenses, and environmental plans.

A more institutionalized form of participation, participatory budgeting, was first introduced to China in the late 1990s. Implemented fully as in the successful case of Porto Alegre (a Brazilian city), participatory budgeting enables citizens to play an important role in budgetary decision-making processes. It is now more commonly practiced in China's rural areas, in villages and townships. A small number of urban subdistrict offices (about 20) and a handful of municipal governments have experimented. One of the earliest experiences was in Jiaozhuo city, Hunan province, in 1999. Under the supervision of the Ministry of Finance, the city introduced a series of public budgeting reforms as part of a World Bank project. It has established a number of procedures in achieving balanced budgets, monitoring budgeting implementation, and opening up budgets to citizens and members of the local People's Congress for discussion (He 2011).

In recent years, a growing number of pollution accidents and environmental protests have helped fuel a growing wave of mass mobilization. Through both official channels and unauthorized protests, citizens voice a sense of injustice over the sacrifice of environmental and human health in the name of economic development (see Chapter 11). Some organized opposition also rallies against unwanted uses either on or off site – the so-called "not in my backyard" phenomenon (NIMBYism). In June 2007, thousands of protesters massed on the streets of Xiamen (Fujian province), forcing the government to delay construction of a US$1.4 billion chemical plant. The demonstrators organized themselves by mobile phone text messages and put photos and video of the marches on the internet. Several months later, residents in Shanghai took to the streets to protest against a proposed expansion of the maglev train line and succeeded in derailing the project altogether (Box 13.1). In both cases, cautious responses from the municipal governments underscore how delicately authorities must tread in the face of residents demanding a say in protecting the homes and other amenities afforded by their rising living standards.

Box 13.1 Street democracy: protesting against proposed maglev line in Shanghai

One of the largest, and perhaps most effective, urban protests happened in Shanghai. Angry over the proposed extension of a high-tech train line and potential health risks, hundreds of people defied bans on public demonstrations and staged mass protests over a period of 2 years. The proposed project was high-profile, using costly German-made technology of magnetic levitation. Opened in 2004 as the only commercial operation of maglev technology in the world, the current line runs from Shanghai's new Pudong International Airport to a suburban subway stop, a distance of 32 km (20 miles). Plans to extend the line, drawn up before the protests, were shelved after residents mounted a letter-writing campaign and hung banners in opposition. Revised plans to take the train line through different neighborhoods 29 km (18 miles) to the old Hongqiao airport in western suburbs prompted the prolonged protests.

While the perceived health risk of radiation may not be based on scientific evidence, it touched the nerves of middle-class urban Chinese. An environmental assessment report released by the Shanghai Academy of Environmental Sciences, compiled for the municipal government after resident complaints, declared the extension plan safe. The project also would have low impact on air and water quality, according to the report. But a greenbelt buffer zone around the track would only be 22.5 meters wide. An original blueprint showed a buffer zone of 150 meters (German specifications required 300 meters) on each side. Spokespeople for the Shanghai Environmental Bureau stated that the radiation effects would extend only 3–5 meters from the train and that noise could be lessened by slowing the train in residential areas. But residents remained unconvinced. Such public protests helped bring residents' concerns to light. Under pressure, the municipal government began rethinking the route. It also set up an email account to collect public comments. To date, the proposed extension project has yet to break ground.

Conclusion

While the state has retained its central role in governing the market and society (down to neighborhood level), the relationship is increasingly complex between the state and society, and between the state and market. The bureaucracy has decentralized, to allow for more local autonomy and initiatives. The urban scale also has become the focal point in the administrative hierarchy, as cities take on the role of growth engines nationwide and across different regions. Hence, municipal governments increasingly resemble local developmental states, nested in a web of connections with business and private interests. Such a growth coalition, however, often overlooks the environmental and social costs associated with economic development. It is in these areas that we begin to see the emergence of civil society organizations and meaningful forms of public participation. A more educated and well-to-do middle-class citizenry now demands a say in the decision-making process. This is, and will prove to be, a new challenge for the evolving state–society relationship.

Bibliography

Bray, David. 2005. *Social Space and Governance in Urban China: The Danwei System from Origins to Urban Reform*. Palo Alto, CA: Stanford University Press.

Cartier, Carolyn. 2005. "City-Space: Scale Relations and China's Spatial Administrative Hierarchy," in Ma, Laurence J. C. and Wu, Fulong (eds) *Restructuring the Chinese City: Changing Society, Economy and Space*. London: Routledge, pp. 21–38.

China Environment Forum (CEF). 2006. *China Environment Series*. Issue 6. Washington, DC: Woodrow Wilson International Center for Scholars.

Chung, Him. 2007. "The Change in China's State Governance and its Effects upon Urban Scale." *Environment and Planning A* 39: 789–809.

Esaki-Smith, A. 2001. "Still China's Second City." *Newsweek International* July 16: 28.

He, Baogang. 2011. "Civic Engagement Through Participatory Budgeting in China: Three Different Logics at Work." *Public Administration and Development* 31: 122–133.

Lu, Hanlong. 2003. *From Party and Government Administration to Community Governance.* Shanghai: Institute of Sociology, Shanghai Academy of Social Sciences.

Ma, Lawrence J. C. 2005. "Urban administrative restructuring, changing scale relations and local economic development in China." *Political Geography* 24: 477–497.

National Bureau of Statistics of China. 2009. *Statistical Summary of Sixty Years of New China 1949–2008.* Beijing: China Statistics Press.

National Bureau of Statistics of China. 2011. *China Statistical Abstract 2011.* Beijing: China Statistics Press.

Oi, Jean. 1995. "The Role of the Local State in China's Transitional Economy." *China Quarterly* 144: 1132–1149.

Read, Benjamin L. and Chun-Ming, Chen. 2008. "The State's Evolving Relationship with Urban Society: China's Neighborhood Organizations in Comparative Perspective," in Logan, John R. (ed.) *Urban China in Transition.* Malden, MA: Blackwell, pp. 315–335.

Segal, Adam and Thun, Eric. 2001. "Thinking Globally, Acting Locally: Local Governments, Industrial Sectors, and Development in China." *Politics and Society* 29(4): 557–588.

Tang, Bo-sin, Wong, Siu-wai, and Lau, Milton Chi-hong. 2008. "Social Impact Assessment and Public Participation in China: A Case Study of Land Requisition in Guangzhou." *Environmental Impact Assessment Review* 28: 57–72.

Wu, Fulong. 1999. "The 'Game' of Landed-Property Production and Capital Circulation in China's Transitional Economy, with Reference to Shanghai." *Environment and Planning A* 31: 1757–1771.

Wu, Fulong. 2002. "China's Changing Urban Governance in the Transition Towards a More Market-oriented Economy." *Urban Studies* 39(7): 1071–1093.

Wu, Weiping. 2004. "Cultural Strategies and Place Making in Shanghai: Regenerating a Cosmopolitan Culture in an Era of Globalization." *Progress in Planning* 61(3): 159–180.

Yusuf, Shahid and Wu, Weiping. 2002. "Pathways to a World City: Shanghai Rising in an Era of Globalization." *Urban Studies* 39(7): 1213–1240.

Zhang, Li. 2010. *In Search of Paradise: Middle-Class Living in a Chinese Metropolis.* Ithaca: Cornell University Press.

Zhang, Tingwei. 2002. "Urban Development and a Socialist Pro-Growth Coalition in Shanghai." *Urban Affairs Review* 37(4): 475–499.

Zhu, Jieming. 1999. "Local Growth Coalition: The Context and Implications of China's Gradualist Urban Land Reforms." *International Journal of Urban and Regional Research* 23: 534–548.

Zhu, Jieming. 2000. "Urban Physical Development in Transition to Market: The Case of China as a Transitional Economy." *Urban Affairs Review* 36(2): 178–196.

Zhu, Jieming. 2002. "Urban Development under Ambiguous Property Rights: A Case of China's Transitional Economy." *International Journal of Urban and Regional Research* 26(1): 41–57.

Conclusion
Looking toward the future

Today's urban China is a fascinating topic of study. The nature and scope of its transformation have already broadened our understanding of cities in general. Although China is undergoing an urban transition that is in some ways comparable to what Western industrialized countries have gone through in the past, the outcomes – particular patterns of development, the nature of urbanism, interactions between urban and rural, and so forth – necessarily are quite different. This has to do with history and context, as we have shown throughout the book. Some of the major domestic forces shaping China's urban transition are distinctive, particularly as related to marketization, decentralization, and migration. Into the future, cities will continue to experience unprecedented changes that characterize contemporary China.

Future urbanization

China is now predominantly urban. By 2030, the urbanization level will rise to 65 percent. This means, between now and then, another 300 million people will become urbanites (UN-HABITAT 2010). The dominant driver of future urbanization will be migration, as is the case in most developing countries. Preliminary results of the 2010 Census, for instance, indicate that 8 million of Shanghai's 22 million population are migrants. This share will continue to rise, potentially reaching a point at which migrants constitute half of the urban population in many large cities by 2030. *In situ* urbanization will be another major driver of growth, and will continue to eat away arable land surrounding existing urban areas. How China accommodates its increasingly urban population is critical not only directly to the well-being of these people but also more indirectly to its sustained economic development.

Chinese cities will grow in size. As a function of the country's enormous population, the urban system will have an unprecedented number of large cities. There will be hundreds of cities with more than 1 million inhabitants. Shanghai, Beijing, and Tianjin are already mega-cities, each with more than 10 million in population. Several others are likely to join rank. The number of mid-sized and small cities will expand too, as a result of *in situ* urbanization. They will likely be in close proximity to larger cities, reflecting the tendency for agglomeration. With increasing attention and investment in the interior, from the central government, the central and western regions also may be home to more large cities. In fact, cities such as Chengdu, Chongqing, and Wuhan may become mega-cities. Nonetheless, the coastal region will continue to dominate the urban system.

The urban economy will continue to generate the bulk of the country's GDP. About half of the annual GDP in 2005 was concentrated in the top 40 cities. The four key drivers of economic growth – investment, trade, industrial restructuring, and consumption – remain robust. In particular, investment from both domestic and foreign sources continues to pour into production and infrastructure. Moreover, leading cities retain a relatively healthy mix of economic activities in both manufacturing and service sectors. Even in the commercial hub of Shanghai, for instance, there is robust development in several manufacturing industries, such as automobiles, electronics and telecommunications equipment, petrochemicals, and home appliances. Such economic diversity is likely a viable insurance against the kind of deindustrialization and subsequent decline seen in many Western cities.

China's urban system may evolve into several likely scenarios. On the one extreme is a highly concentrated pattern, with two or three leading cities growing in both size and dominance. On the other extreme is a dispersed system of cities and towns (MGI 2009). Perhaps the most likely (and currently most popular among academic and policy circles) is a pattern with strategically located networks of cities or urban regions. Urban regions align two or more metropolitan areas together (this means two or more prefecture-level cities, or city regions, in the Chinese context) and spell overall benefits for the national economy – the advantage of scale. History, location, economies of scale, and policy preferences have contributed to the relative success of large cities. They attract more talent and investment. A network of cities surrounding such large cities reinforces agglomeration economies, as in the case of Lower Yangtze River Delta and Pearl River Delta regions. Other emerging urban regions, although far less integrated at present, could include the Bohai region, Chengdu-Chongqing region, Qingdao-Jinan region, and Shengyang-Dalian region, to name a few. Together, they will likely command a lion's share of China's urban economy.

Continuing urbanization will reinforce social changes and spatial transformation. Over time, the growth of business and service professions requires well-educated and white-collar workers. They will make up a growing middle class in urban areas, demanding a set of amenities similar to what is available in the West. We are already seeing the beginning of suburbanization and motorization in many large cities. With an already intense population-to-land ratio, following the path of suburbanization and urban sprawl, as seen in some industrialized countries, really is not an option for China's cities. Some scholars believe that the divergent development trajectory in China has laid the foundation for institutional and policy responses that may offer innovative solutions (McGee et al. 2007). This remains to be seen.

Challenges for China's cities

As urbanization accelerates, China's cities face a number of pressing issues. First and foremost is to achieve full benefits of agglomeration. Functioning as city regions, cities at and above the prefecture level are now the principal engine of the national economy, counting for about two-thirds of the country's GDP. But most of them remain underurbanized (under 70 percent on average) and hence have unrealized agglomeration economies. There is evidence that the larger and more urbanized the city region, the higher productivity is (Kamal-Chaoui et al. 2009). Many municipal governments, however, have promoted the so-called polycentric strategy of urban development, to reduce high population density in the central city and direct new development towards suburbs. Continuing in this fashion, China's city regions will become sprawling urban areas high in congestion and pollution costs but less in benefits from agglomeration economies.

Measures to ensure an adequate supply of skills are another challenge in sustaining urban economic dynamism. Today, rapidly growing economies depend more on the creation, acquisition, distribution, and use of knowledge. While China's migrant workers continue to fuel the growth of urban construction and labor-intense manufacturing, the demand for college graduates will increase. Experience in the West indicates that, increasingly, non-routine analytic and interactive skills such as problem solving, creativity, entrepreneurship, and leadership are in demand. While ahead of many developing countries in specialized training in science and engineering, China's higher-education sector lags behind its East Asian neighbors (Japan, South Korea, and Singapore) in competency and quality measures. Catching up will entail a shift within universities to emphasize generic skills, experiential learning, and outcome-based teaching methods. Sustaining the supply of knowledge workers also calls for more attention to quality of life and services, as well as to investment and entrepreneur environment.

As China further undergoes the plan-to-market transition, urban social–spatial differentiation will intensify. Increasing migrant concentration, given that most migrant housing is in much worse condition than local housing, may aggravate the situation. Settlement patterns will influence the future socioeconomic standing of migrants. The experience of other developing countries shows that severe shortages of affordable housing tend to force migrants to live in squatter settlements and slums; many of these migrants eventually become the urban underclass. With more tolerant migration policies, over time urban ties will surpass rural ties and many migrants will settle permanently in cities. Although some migrants may adjust to urban life well, others will have no choice but to become the first of an emerging group of poor in cities with increasing spatial segregation.

There will be pressure on cities to find sustained sources to finance urban growth. The present fiscal system is far from rational, and contributes to the lack of transparency in public spending and an increasing disparity among cities of different geographic location and size. With *in situ* urbanization and expansion, tension will intensify between the loss of arable land and cities' dependence on land sales for revenues. Also challenging will be to extend the provision of services to the expanding migrant population in cities. Failing to do so will further fuel the growth of an urban underclass of migrants as well as the jobless. To avoid the traps other countries have fallen into, urban land needs to be planned to promote more efficient use and to tie to the financing of urban infrastructure through the development of a functional urban land market and residential property tax system.

China's fragile human–environment relationship will be severely challenged by rising urbanization. Urban expansion, often in fragmented and sprawling fashion, will intensify the depletion of land, water, and other resources. Containing urban sprawl will be a major imperative. At the root of fragmented expansion is the increasingly market-driven development process that collides with government interests. The local state's weak planning capacity and hunger for revenues undermine its ability to exercise control over land use. In its current practice, land leasing and transfer also are driven by short-term interests of local governments, instead of long-term land management strategies. Furthermore, the ecological footprint of the urban population will enlarge with modernization, motorization, and heightened consumption. Congestion, already choking some cities, will become extreme. So will pollution, of air and water in particular, and waste materials. Perhaps the most serious environmental challenge urban China confronts is access to water. This stems from both growing demand and rapidly increasing levels of pollution. Water conservation and reuse will be a rational choice.

Given these challenges, the path to sustainable urbanization in China no doubt will require policy interventions. The strong state may be to its advantage, if past experience is any guide. For example, public investment has enabled cities to make great strides in providing infrastructure services and perhaps to perform better than most of China's counterparts with similar income levels. Municipal authorities have proven capable in managing the rapid pace of modernization. Moving forward, however, will entail new governance practice in at least two directions. The first is at the grassroots level, to allow for more meaningful stakeholder engagement and public participation in urban governance. Second will be at the level of urban region, to formulate and implement region-wide efforts in economic cooperation, transportation and infrastructure planning, and environmental management. The rapid emergence of urban regions is pushing existing forms of subprovincial governance beyond capacity.

Implications for the world

China's urban transformation does and will have global implications. The Chinese city will continue its critical role in global production and supply networks. The country's stunning turn as the world's factory stems from its true factor endowment (labor) and a new developmental state firmly in control. This combination has made many Chinese cities very competitive for footloose global capital in labor-intensive industries. But China's manufacturing is moving up the learning curve. At the same time, the central government has begun to encourage strategic investment in research and development (R&D), modifying the science and technology development strategy from a "master imitator" to indigenous innovation. With a competitive reservoir of educated and skilled personnel, Chinese cities are becoming attractive to multinational firms in knowledge-intensive industries, particularly R&D firms. This will begin to challenge the dominance of Western cities in the global economic system.

The Chinese city will be a gigantic consumer of a variety of goods on the global market. To power its growth, urban China will account for about 20 percent of global energy consumption. Given the low per capita resource endowment domestically, most of this will be imported. For hundreds of cities, the building of mass-transit systems and new roads will continue, combined with a steady rise in office and residential construction. All of these will create immense demand for oil, steel, and other commodities which will impact world supply and prices. On the other hand, the aggregate consumption power of urban Chinese will be unparalleled. As standards of living increase, they will become a new driving force for sustaining the global consumer markets, particularly for such specialized markets in tourism, higher education, and luxury items (Wu 2007).

Leading Chinese cities will become important nodes in the global marketplace, joining the rank of world cities. Several types of function are commonly associated with this status: finance, corporate headquarters functions, global services, transport, information, and cultural activities. In the East Asia region, Tokyo, Hong Kong, and Singapore lead the way with respect to these functions. Both Beijing and Shanghai have achieved respectable levels in a short space of time and are registering increasing global linkages. Shanghai is likely the one best positioned to emerge over the next decades as a major regional hub or even a world city comparable to Hong Kong and Singapore. According to the Global Financial Centers Index, Shanghai was ranked 35th in 2009. This is based on criteria factoring in the quality of a city's workforce, business environment, market access, infrastructure, and an index of the general economic competitiveness (cited in Yusuf and Nabeshima 2010). Using a composite index of R&D capabilities (including indicators such as international exposure, human capital resources, R&D network, and policy setting), researchers rank Shanghai above four other major Chinese cities (Beijing, Guangzhou, Chengdu, and Tianjin). Shanghai's R&D capabilities are in fact close to those in Seoul, South Korea (Jefferson and Zhong 2004).

The environmental impact of urbanization has a global dimension. The most severe will be pollution of coastal water by industries and untreated wastewater, cross-border and intercontinental air pollution, and emissions of greenhouse gas (Kamal-Chaoui et al. 2009). China is the second largest contributor to global warming in the world. The fundamental problem is the enormous dependence on coal for energy. Industries such as steel, cement, and chemicals are by far the largest users. But much of the country's environmental protection effort relies on initiatives by local officials, leading to a patchwork of environmental protection. Wealthier cities with highly proactive mayors and strong ties to the international community tend to invest more in absolute terms, as well as a greater percentage of their local revenues into environmental protection. Shanghai, Dalian, and Zhongshan exemplify such regional environmental activism (Naughton 2007).

Accelerated urbanization contributes to the steady reduction in arable land. This has direct implications for food security, a subject of growing concern. According to official estimates, newly built urban areas have grown by 50 percent since 2000, which partially accounts for the loss of 8.3 million hectares of arable land during the same period (*China Daily* 28 March 2011). While it may be subject to debate whether China can achieve food security given the large population, the central government has pursued this goal as a matter of national security. The challenge is to seek a balance between food security and urbanization of the country's vast rural population. If China were to face food shortages, the impact on world food supply and prices would be immense. At present, official policies overly emphasize grain production. As such, there is a large demand for imported soybeans, sugar, and other agricultural products. The case of soybeans is particularly contentious, as efforts to convert to biofuels have driven up demand and prices. China's growing interests in farmland elsewhere to grow soybeans have already caused unease for other developing countries (e.g., Brazil).

Last, but not least, China and its cities pose a unique challenge for scholars and researchers around the world. Theorizing the country's economic transition and urban changes has proven difficult because of the rapid speed and explosive scale at which events and policies take place. Interpreting the changes we see today in urban China also makes us wonder: To what do we compare China? Is China like any other country? While the Chinese city is undergoing economic, social, and spatial transformation resembling what we have seen elsewhere, parts of its trajectory clearly push the limits of contemporary urban theories and experience. As John Logan and Susan Fainstein appropriately point out, Chinese cities are *sui generis* in combining aspects of different urban types along with certain uniquely Chinese qualities (Logan and Fainstein 2007). Many large cities have transformed from relatively compact, low-rise centers to sprawling metropolises surrounded by suburban developments and mega-malls along the American model. Yet continued elite preferences for the urban core leaves a social–spatial pattern similar to that seen in European cities. All of these, however, cannot disguise the persistence of dire poverty and substandard living conditions in pockets of cities and among segments of urban population. This hybrid trajectory and landscape are what makes studying urban China such an intriguing and challenging undertaking.

Bibliography

Jefferson, Gary H. and Kaifeng, Zhong. 2004. "An Investigation of Firm-Level R&D Capabilities in Asia," in Yusuf, Shahid, Altaf, M. Anjum and Nabeshima, Kaoru (eds) *Global Production Networking and Technological Change in East Asia*. New York: Oxford University Press, for the World Bank, pp. 435–475.

Kamal-Chaoui, Lamia, Leman, Edward, and Zhang, Rufei. 2009. "Urban Trends and Policy in China." OECD Regional Development Working Papers, 2009/1. Paris: OECD Publishing.

Logan, John and Fainstein, Susan. 2008. "Introduction: Urban China in Comparative Perspective," in Logan, John (ed.) *Urban China in Transition*. Oxford, UK: Blackwell Publishing.

McGee, Terry G., Lin, George C. S., Marton, Andrew M., Wang, Mark Y. L., and Wu, Jiaping. 2007. *China's Urban Space: Development under Market Socialism*. New York, NY: Routledge.

A comprehensive study of China's urbanization by a team of well-established researchers, this book explores urban changes in the coastal region in the transition era. It uses a holistic approach, positioning Chinese urbanization in the context of global urbanization, to capture the social and political dimensions. A useful focus of the book is its attention to the overall urbanization process and the recent emergence of urban regions.

McKinsey Global Institute (MGI). 2009. *Preparing for China's Urban Billion*. Retrieved on 20 October 2010 from www.mckinsey.com/mgi.

Naughton, Barry. 2007. *The Chinese Economy: Transitions and Growth*. Cambridge, MA: MIT Press.

United Nations Center for Human Settlements (UN-HABITAT), with China Science Center for International Eurasian Academy of Sciences and China Association of Mayors. 2010. *State of China's Cities 2010/2011: Better City, Better Life*. Beijing: Foreign Languages Press.

Wu, Fulong. 2007. "Beyond Gradualism: China's Urban Revolution and Emerging Cities," in Wu, Fulong (ed.) *China's Emerging Cities: The Making of New Urbanism*. London: Routledge, pp. 3–25.

Yusuf, Shahid and Nabeshima, Kaoru. 2010. *Two Dragon Heads: Contrasting Development Paths for Beijing and Shanghai*. Washington, DC: The World Bank.

Index

acid rain 219, 220, 221
activism 203, 236, 237, 270, 272, 279
administration: municipal 263–4;
    rural 261; urban 66, 259–62;
    see also government; municipal
    government
administrative areas: of cities 73;
    conversion of rural to urban 76–7
administrative compounds 57–60
administrative staff, urban/rural 260–1
adult education institutions 143
Agenda 21 231
agglomeration of cities 275, 276–7
aging population 142, 250
agriculture 21–2; 6th/7th centuries 38;
    collective farming 94, 98; decline
    in 98; employment in 4, 21, 135;
    productivity of 98; in Shanxi 16; youth
    views of 101; *see also* arable land;
    farmers; farming
air cargo hubs 88
airports, overbuilding of 87–8
air quality, Beijing 232; *see also* pollution
Altan Khan 26, 43
American fast-food chains 122–3
Amoy *see* Xiamen
ancient period, cities 31
arable land 202, 277, 279; *see also*
    farmland
artistic/cultural activities,
    Beijing/Shanghai 266
artists' colonies 255, 266
Asian Development Bank (ADB) 191
Asian financial crisis 85
Association of Environmental Protection
    Industries 224
atlas, urban 62
automobiles: arrival of 65;
    emissions 222–3; ownership 161, 163,
    164

automobile sector 164, 265
autonomous regions 22, 26, 254, 260

Bai people 26
ballroom dancing 247
bank credit, infrastructure
    development 138, 186
bank lending 139
banks: Asian Development Bank
    (ADB) 191; and environmental
    data 234; foreign 124, 191; state-
    owned 134, 138; *see also* People's
    Bank of China
Baoji 79
Baotou 15
Beihai 113
Beijing: air pollution 236; annual
    poll 271; artistic/cultural activities 255,
    266; automobiles 163; cancer
    cases 237; as capital city 42, 58–60;
    Capital Eco-Economic Zones 230–1;
    central business district 158; and
    city regions 86; coal hill 51;
    corporate headquarters 119, 147;
    crime in 251; dance craze 246–7;
    economic planning 266; financial
    sector 125; Forbidden City 57–8;
    foreign banks 124; house prices 213;
    hutong tours 165; infrastructure
    development 188; as million city 35;
    as modern international city 135;
    municipal bureaus/environmental
    protection 235; Muslim districts 253–
    4; Olympics 84, 223, 232–3; R&D 89,
    126; redevelopment of 165; science
    parks 116; size of 72–3; solid
    waste 226; tertiary sector 135; Tiantan
    Park 247; Zhejiang Village 104–5
Beijing Water Authority 235
bell and drum towers 61

benchmark loan rate 213
Bengbu 79
Better City, Better Life 232
bicycling 161, 222
black market, land 200
blue-stamp *hukou* 97
Bohai Bay Region region: as city region 84; manufacturing 147
bond markets, and infrastructure financing 184
borrowing, infrastructure development 184, 186
BOT/BOO schemes 191–2
brain drain, rural China 101
bread loaf taxis 222
British trade with China 45, 64
Buddhism 12, 25, 26, 43
budgeting, participatory 271–2
business services 133, 276
bus systems, urban 178

canals, Suzhou 17
Canton *see* Guangzhou
Canton Tramways Syndicate 64
capital: attracting global 265; ethnic 118; small scale FDI 118
capital cities 36; Beijing 42, 58–60; *see also* Beijing
Capital Eco-Economic Zones, Beijing 230–1
capital markets, industrial finance 139
carbon dioxide emissions 221
car manufacturing *see* automobile sector
car ownership *see* automobiles; ownership
cellular structure, housing 153
census data 6, 73, 75
central business districts (CBD) 156, 158
central government, and infrastructure 177, 178, 183
central place theory 29
Chang'an 33–4, 36, 38, 56, 62–3
Changchun 46
Changzhou 42
Chengdu 36, 86, 191–2, 237, 252–3, 276
children, migrant 106–7
China Association of Environmental Protection Industry (CAEPI) 237–8
China Banking Regulatory Commission 234
China Insurance Regulatory Commission 234
China International Environmental Protection Exhibition and Conference 238

China Municipal Solid Waste Summit 238
China Ping An Insurance Company 234
China-Singapore Suzhou Industrial Park 120
"China's sorrow" 16
China Statistical Yearbook for Cities 7
China Urban Construction Statistical Report 8
China Women's Development Foundation 238
China Youth Climate Action Network (CYCAN) 238
Chinese Academy of Sciences 127, 266
Chinese Communist revolution 80
Chinese Far East Railway 46
Chinese national census 73
Chinese Revolution 21, 29
Chinese technology associations, in Silicon Valley 127
Chinese urban ideal 50–1
Chongqing 17, 77–8, 86, 117, 188, 276
Christian churches 66
cities: ancient period 31; county-level cities 261, 263; defining 5, 73–4; distribution by size 75; distribution of 74; double-walled 55, 57; early imperial period 31; early traditional period 30, 35–8; expansion of 261; fortified 42; as growth engines 111, 273; historical expansion of 36; historical significance of 28–9; large 85, 275; late imperial period 30, 32; and market reform 154–6; medieval period 32; mega-cities 275; middle period 30; mid-sized/small 85, 275; million cities/million+ 35, 85, 275; numbers of 72, 262; port cities 41; prefecture-level cities 261, 262, 263; provincial-level 260, 261, 263; in Republican era 30; restructuring/ specialization of 134; Silk Route 40; size of 33, 72–8; socialist 152–4; subprovincial-level 261; traditional Chinese 67; after World War II 29
cities leading counties policy 261
citizenship, rural/urban 94
city gates 55–6, 65
City God, temples to 61
City Government Civic Associations 66
city of Cathay, A 62
City Planning Bureau 196, 267
City Planning Law 235
city regions, eastern region 86
city walls 53–6, 64

civic engagement 269
civic life, state control of 271
Civil Aviation Administration 177
civil service 260, 263
civil society organizations 271, 273
civil unrest, environmental issues *see*
    protests
clan connections, trade 63
Cleaner Production Promotion Law
    2002 236
climate 20–1
coal: pollution caused by 220, 221, 279;
    Shanxi 16; and urban expansion 80
coastal development strategy 114
coastal open cities 84, 112, 113, 178
coastal provinces: domination of urban
    system 276; economic priority of 134;
    prosperity of 121; redevelopment
    of 147; revenue share 187
coastal waters, pollution 279
collective farming 94, 98
Comfortable Housing Project 209
commercial development 159–60
commercial housing 102, 205, 206, 207,
    209
commercialization of cities 39
commercial spaces, and entertainment/
    public activities 243
communications, and foreign
    investment 116
Communist Party: antiurban bias 2–3; and
    SOE managers 137; *see also* Chairman
    Mao; People's Republic of China
community groups, and urban
    development 271
community service delivery, and
    neighborhood committees 268–9
community (*shequ*) committees 269
commuting 163
Company Law 1994 136
compensation, for displacement 166, 203
competition: for foreign
    investment 202; inter-urban 88;
    regional/inter-regional 86–9
concession areas 64
Confucianism 25
congestion, traffic 181, 278
construction bonds 186
Construction Commission 196
consumer economies, moving
    toward 122–8
consumer goods 4, 111, 120, 122, 133,
    134, 156, 159, 242
consumerism 243

consumer market, Chinese 126
consumer revolution 160, 243
consumer services 133
consumption 71, 80, 134, 146, 160, 171
contract, employment 141
control: adminstrative/social 268; state/of
    civic life 271
conveyance, land 199–200, 201
co-operative health services 95
corporate headquarters, location of 119,
    147
corporate technology centers 126
cosmic cities 2, 48
cosmology 12, 51
cotton 42
counties: in administrative hierarchy 260;
    reclassification to cities 261
county-level cities 261, 263
county-seat cities 31–2
courtyards 57, 58
credit, and polluting industries 234
credit-rating system 205
crime: and migrants 106, 251;
    spatial dimensions of 252; urban
    neighborhoods 251–2
cultivated land preservation policy 267
cultural affinity, and globalization 118
Cultural Revolution: education in 142;
    forced migration 97; and Third
    Front 81–2
cycling 161, 222

Dachang plant 192
Dadu 36, 41, 59
Dalian 113, 279
dance craze, Beijing 246–7
Daoism 12, 25
Daqing oil fields 80
Dashanzi 255
data, urban population 6; *see also* census
    data; environmental data
Datong 80
Daxing 34
Dazhai 15–16
decentralization 3–4; administrative/
    fiscal powers 4, 259; political 260; of
    population 171
demand, for commodities 279
Deng Xiaoping 84, 113, 114, 124
department stores, foreign 160
deregulation, grain
    markets/commodities 243
desakota model 76
desert, northwestern 14

designated towns 5, 261
developers, and social–spatial
  reconfiguration 167
development: environmental
  guidelines 220; green 230–1;
  incentives for sustainable 234; and
  infrastructure 178, 180; local 4; pro-
  growth agenda 170; redistribution
  of 89; rural 261
development companies 201, 267
development funds, competition for 86
development policies 71, 82
development process, urban 267
development projects, state 86
developmental state, new 264–8
development strategies: geographically
  based 83; and growth 265; outward-
  oriented 114; urban bias in 94
development zones 83, 84, 88, 267
devolution 4; and environmental
  issues 236
differential land rent 206
dioxins 227
disability benefits, non-local workers 141
discrimination, against migrants 94
disposable incomes, and automobile
  ownership 164
district governments: role of 171;
  subdistrict offices 263
district specialization, by function 243
diversity, urban society 252–5
domestic wastewater 227
Dongguan 121–2
Dong river 19
Dongtan Eco-City Project 229–30
double-walled cities 55, 57
drinking water 219, 224; *see also* faucet
  water; water
dual track approach, shopping 243, 244
dual water systems 224

early imperial period, cities 31, 37–8
early traditional period, cities 35–8
East Asian economic crisis 99, 128
East Asian investment 117
East China Normal University 116
east coast cities, motorization rates 163
east coast, urbanization 128
East (Dong) River 19
eastern region 13; cost of investment
  in 89; domination of 89; economic/
  urban development 89; infrastructure
  development 188
eco-cities 228–31

economc growth, redistributing 145–9
Economic and Comfortable Housing 209
Economic and Technology Development
  Zones 115
economic base 21–3
economic development: environmental/
  social costs 273; foreign firms 22; and
  green development 230–1; Treaty Port
  cities 45
economic disparities 89, 147; *see also*
  inequality
economic restructuring 133–5;
  financing 138–9; and northeastern
  region 147
economy: diversification of 42, 276;
  economic performance, regional 146–9;
  economic planning, local 264–6;
  economic reform 111–14, 117;
  economic system, structural changes
  to 3; growth 98, 111, 276; moving to a
  market system 84; urban industrial 98
education 142–5; and employment 140;
  higher-education sector 142–3, 150,
  277; migrant children 106–7, 107–8;
  reforms 150; secondary 143, 144;
  technical/vocational 145; *see also*
  universities
elderly care 250–1
electricity 65–6
electronics industry 120
Electronics Street 127
Emeishan 12
eminent domain 202–3
employment: in agriculture 4, 21,
  135; contract 141; expansion into
  suburbs 168; informal work 137–8,
  226; in property sector 211; service/
  manufacturing 135; urban 136; urban/
  type of ownership 137; women 142;
  see also unemployment; work
enceintes 56
energy consumption, and growth 279
engineering graduates 145
engineers, Western-educated 126–7
enterprise sector 125, 138, 145, 187
entertainment, and public spaces 244–50
entrepreneurial approach, of cities 83–4
entrepreneurs 105, 126–7
environment: and global image 231;
  physical 11–21; state of urban 219–27
environmental activism *see* activism
environmental assessment, public
  involvement in 271
environmental data, and banks 234

environmental governance 234–5
environmental guidelines 220
Environmental Impact Assessment
    Law 236, 271
environmental impacts: traffic
    congestion 181, 278; transport
    projects 164
Environmental Information Disclosure
    Decree 2008 236
environmental justice 237
environmental management 234–8
environmental planning 227–8
Environmental Protection Bureau,
    Xining 235
environmental sustainability, strategies
    for 227–34
ethnic capital 118, 120
ethnic-origin communities 253–4
European districts, of cities 64–5
European Investment Bank 191
Experimental Zone for New Technology
    and Industrial Development 127
export processing 120–2
exports: global export markets 113;
    manufacturing 4
extra-budgetary funds 186–7

factories, relocation of 156, 158, 165, 232,
    267; *see also* industry; relocation of
factory of the world, China as 116–22
family migration 101
fan dances 247
farmers: and housing 197; loss of
    land 202
farming, collective 94, 98
farmland, loss of 202, 261, 277, 279
fast-food chains, American 122–3
faucet water 180–1; *see also* drinking
    water; water
fengshui, and urban design 51–2
Fengyi 36
Fen River 15
foreign-invested enterprises 140
finance: housing 205; and
    infrastructure 182–7; *see also* foreign
    direct investment; foreign investment
financial incentives, sustainable
    development 234
financial regulatory system 125
financial sector 124–5
financial services: centers for 123–5;
    state-owned banks 134
fine particulate air pollution 236
fiscal contracts 183

fiscal decentralization 4; and
    infrastructure 182–4, 187
Five-Year Plans 260; Sixth 83;
    Seventh 83; Eighth 85, 114; Ninth 85,
    89; Tenth 89, 90
floating population (*liudong renkou*) 98;
    see also migrants
flooding, Yellow River 16
food production, and land loss 261
food security 279–80
Forbidden City 57–8, 60
foreign capital: industrial
    restructuring 139; and infrastructure
    development 186
foreign direct investment (FDI) 5;
    1990s 84; coastal provinces 147;
    competition to attract 265; and east
    coast urbanization 128; main sources
    of 117; and real estate/property
    development 118; sources of 120; and
    technology transfer 112, 118; urban
    destinations of 119; *see also* foreign
    investment
foreign empires, Treaty Port cities 45–6
foreign investment 116–20;
    attracting 112; and cheap land 158;
    competition for 116, 202; early
    reform years 83; East Asian 117; and
    infrastructure 178; and National Model
    City status 228; uneven 89; and urban
    system 128; *see also* foreign direct
    investment
foreign residents 64
foreign trade: 1980s 83; Song dynasty 39
Forest Bureau, Xining 235
Founder 127
Four Modernizations program 82, 83
French–Chinese treaty 45
Fujian Delta, as Open Economic Zone 113
Fujian province 41–2, 124, 188
funds, for infrastructure development 184;
    see also finance
Fuzhou 45, 113

Gansu 80
gardens, Suzhou 18
garrison cities, Qing dynasty 44
gated communities 156, 168, 170
gay/lesbian communities 252–3
gender wage gap 142
geographically based development
    strategy 83
geography: cultural 25–6; physical 2, 12
geomancy, and urban design 51

girls, education of migrant 107
global consumer markets 279
global economic downturn 128, 146, 213
global export markets 113
Global Financial Centers Index 279
global image, and the environment 231
global intelligence corps 158
globalization 5, 117, 118, 126
globalized spaces 244
global mass retailers 160
global pharmaceutical industry 125
global warming 279
governance: changing 264, 270; local 4;
  and neighborhood committees 268;
  new 278; private interests in
  community 269–70; self-governance by
  residents 269
government: and infrastructure 177,
  178, 183; and land ownership/
  acquisition 196; role of district 171; *see
  also* administration; local government;
  municipal government
government-organized non-governmental
  organizations (GONGOs) 238
Go West policy 86, 89, 90, 147
graduates: demand for 277;
  unemployed 146
grain production 280
Grand Canal 17, 19, 21, 39
Grand Canal Chain 44
grants, for infrastructure 183
Great Leap Forward 81
Great Wall 42, 44
Green Beijing, Green Olympics 232–3
green credit policy 234
green development, and economic
  development 230–1
green insurance 234
green planning 231
green urbanism 228–31
grid pattern, of streets 56
gross domestic product (GDP) 13, 86, 87,
  117, 148, 149, 180, 234, 261, 276
gross national product (GDP) 73, 98, 198
growth: economic 98, 111, 276; and
  energy consumption 279; financing
  of 138–9, 277; and inequality 145;
  Pearl River Delta 121–2; redistributing
  economic 145–9
Guangdong 88, 113, 119, 124
Guangxi 188
Guangzhou: acid rain 221–2; as city
  region 86; coastal open city 113; crime
  in 252; as dual city 64; Hong Kong

investment in 113; labor reform 141;
  premium water networks 193;
  R&D 126; recycling 226; regional
  position of 87–8; trade share 124;
  transport network 179; as Treaty
  port 45; unduly large role of 81; urban
  expansion in 169
guidelines, new development 220
Guihua 43
guild halls 63
guilds, trade/native place-associated 63–4
Guiyang 206
Guiyu, electronics dump 227

Hainan Island 83, 113, 124, 188
Hakka people 94
Han dynasty: Chinese empire in 25;
  cities 31, 36, 37, 38, 39; population
  in 24; Silk Routes 40; tea trade 40;
  Western 33–4
Hangzhou: and city regions 86; corporate
  headquarters 147; and Grand Canal 39;
  house prices 213; knowledge
  workers 126; as million city 35; Ming
  dynasty 42; position of 17; Yuan
  dynasty 41
Hankou 45, 117
Han people 25; migrant villages 253
Haoyang 36
Harbin 45, 117, 158–9
Harvard University's Historical GIS
  (2011) 33
health: environmental hazards 237;
  pollution related disease 220, 221, 227
health care, rural areas 95
health sector, civil society groups 271
heavy industry: easing of emphasis
  on 134; and the Grand Canal 19
Hengshan, Bei 12; Nan 12
heritage tourism 165, 166
hierarchical ordering, of public/private
  places 57
High and New Technology Industry
  Development Zones 84
higher-education sector 142–3, 150, 277
high schools 143, 144
high-tech firms, subsidies/financial support
  to 126
high-technology development zones 84,
  115
historic buildings, and redevelopment
  165
Höhhot 42–3, 79, 248–50
home loans 211–12

homeowners' associations (HOA) 269,
270
home ownership: affordability 207–10,
214; and identity 211; increasing levels
of 205
homosexuality 252, 253
Hong Kong: airport 88; corporate
headquarters 119; deindustrialization
of 121; economy of 120, 121; FDI
from 118, 120; investment in Pearl
River Delta 86; one country, two
systems policy 87; situation of 19
Hong Kong Stock Exchange 139
Hongwu Atlas 62
household registration system (*hukou*) 95–
7; blue-stamp *hukou* 97; classification
as agricultural/non-agricultural 6;
converting from rural to urban 96; to
curb migration 81; as exclusionary
tool 101–2; and housing 102–6, 197;
and public services 102, 109; reform
of 141; temporary registration 106,
141; and urban classification 75; and
urban–rural divide 93
household utilities 197
housing: budget for construction 178;
cellular structure 153; choices 207;
commercial 205, 206, 207,
209; commodity housing 102;
conditions 207; insurance/finance/
loans 205; investment 198;
management 206–7; and migrants 102–
6; prices 90, 210, 212, 213;
private 209–10; public 153, 197, 198,
206, 209; reforms 204–7; rental 210;
resettlement 167, 203; shortages 198;
as social welfare 96, 197–8, 206; space
per person 207; Taiyuan 16; tenure
and income 208, 210; unsafe 210;
village 210; *see also* housing market
Housing Administration Bureau 203
housing bubbles 202, 211–14, 267
housing commodification, and sociospatial
differentiation 155
housing development, market-based 205
housing market: affordability of
homes 207–10; global economic
downturn 128; secondary 207;
segmentation of 210; Shanghai 155;
see also housing
housing provident funds 197, 205
housing provision, as social welfare 96
housing savings system 205
Huairou 231

Huashan 12
Huashui 237
Hui people 26, 253
Hu Jintao 102
hukou see household registration system
human capital, and globalization 126
Human Development Index (HDI) 149
human resources, harnessing 139–45
Huolinhe 80
hutong 165
hutong tours 165
Hu Yaobang 113
Huzhou 42
hydroelectric systems 80
hypermarkets 243

identity, and home ownership 211
image marketing 231
imperial city, Beijing *see* Beijing
incentives: to attract FDI 265; for
workers 140
income disparity 95, 98, 244; *see also*
inequality
incrementalism, and reform 112, 136
Indus River, origins of 13
industrial development: coastal treaty
ports 45; and port facilities 23
industrial districts, new 158
industrial facilities, relocation of *see*
factories, relocation of; industry,
relocation of
industrialization 4; move to balanced
industrialization 134; and new
mobility 98; and urbanization
1949–1957 80–1
industrial land, pricing of 202
industrial monoculture 121
industrial parks 88, 89, 127, 155, 159, 164
industrial restructuring 168;
financing 138–9
industrial wastewater 227
industry: heavy 19; relocation Hong Kong
to China 120; relocation of 158, 165,
168, 233
inequality: changing 95; economic 89,
147; gender wage gap 142; and
growth 145; income 95, 98, 244;
increasing 149; and infrastructure
development 187–8; regional 90;
social 93; total national 95; and urban
poverty 145–6
informal economy 137–8, 226, 245
informal market, land 200
informal public space 244–5

informal work 137–8, 226
infrastructure: transportation 168; and
    urbanization 180
infrastructure connection fees 186
infrastructure development: Beijing 266;
    public sector in 180
infrastructure-led development 178
infrastructure provision, household 197
infrastructure services, progress in 180–2
infrastructure spending, roads 162
initial public offerings (IPOs) 139
inner-city redevelopment 165–7
in situ urbanization 275, 277
institutions for elderly care 251
insurance, housing 205
Interim Provision on Administrating
    Concession Right of Chengdu 192
International Business Incubators 115–16
international non-governmental
    organizations (INGOs) 238
internet-based campaigns, pollution
    data 236
interprovincial migration 99–100
investment: competition for 86; in
    infrastructure 184–7; in R&D 126;
    redistribution of 89; sources 85; *see
    also* foreign investment; foreign direct
    investment
iron 39
iron rice bowl 140, 141
Islam 26

Jiangnan 19
Jiangsu 188
Jiankang 36
Jianye 36
Jiaozhuo 272
Jiaxing 42
Jiayuguan, city walls 53
Jinan 15
Jin dynasty 36, 59
job mobility 140
jobs–housing imbalance 163
joint-venture industries 84
joint ventures, infrastructure
    development 191

Kaifeng 35, 36, 40, 59
Kailuan 80
kinship/social ties, and FDI 118
kite flying 249
knowledge-based economies, moving
    toward 122–8
knowledge-based firms 126

knowledge-intensive industries 278
knowledge sector 277
knowledge workers 126
KOSDAQ 139
Kublai Khan 41
Kumbum monastry 254
Kunlun Mountains 13
Kunming 158, 224, 231–2
Kunming World Horticultural
    Exposition 231
Kunshan 119–20

labor: appeal of to MNCs 118; and foreign
    investment 112
labor contracting 140, 141
labor-intensive manufacturing 117, 118, 120
labor market reform 140–2, 150
labor productivity levels 140, 141
lacquerware 39
Lake Dian 224, 231, 232
Lamaism 26
Land Administration Bureau 196
land allocation 202
land, as means of production 196
land conveyance 199–200, 201
land development, and revenue
    generation 267
landfills 225
land grabs, illegal 261
land leasehold, transition into 198–203
land lease system 198–9, 201, 278
land market 63, 153, 154, 168, 200, 267
land marketization 200–3
land occupation, illegal 200
land ownership 153, 195–7, 198
land prices 88, 200, 214
land reform 84, 198–200, 214
land rent, differential 206
landscape morphology, and *fengshui* 51
land transfer revenues 186, 198, 201, 264
land use: changing 156; in cities 171;
    industrial/urban fringes 158
land values, increasing 202
Lanzhou 15, 65, 79
large cities, proliferation of 85
large urban agglomerations, in the east 85
late imperial period, cities 32, 41–4
Law on Urban and Rural Planning
    2007 235–6
laws, environmental management 235–6
legal instruments, environmental
    management 235–6
lending, to environmentally friendly
    ventures 234

Lenovo 127
Lhasa 13
Lianyungang 113
Liao dynasty 58–9
Liaoning 147
license plates control, automobile 223
light rail systems 165
Liuzhou 79
living standards 149
loans: benchmark loan rate 213; and
    environmental data 234; financing 138;
    home 211
local government: decentralization 260;
    and land lease system 201; and public
    services 184; revenue generation 184,
    214; role in land acquisition 196;
    see also administration; government
local state corporatism 265
loess dust, air pollution 223
Loess Plateau 14, 15
Luda 80
Lujiazui Central Business District of
    Pudong 124
Luoyang 15, 34, 36
luxury goods sector 160

Macao 19, 85
Madian 253–4
maglev trains 188, 190, 272
Manchu imperial hierarchy 60
Manchurian urbanization 44, 45
Manchu, the 43
manufacturing powerhouses 147
manufacturing sector: decline in 135;
    export-led 4; and FDI 117, 118;
    globalization of 117; progression
    of 278; share of GNP 98
Mao, Chairman: antiurban bias 2–3,
    29, 71, 80; cities as producers 134;
    development strategies 95;
    industrialization 154
Maoist era, urban development 78–82
mapping project, Ming dynasty 62
marketization 3; land/housing
    development 214
market reform: and consumer
    revolution 160; and housing 204–7;
    and industry 98, 111; post 1979 134;
    transformation of cities 154–6
markets: global consumer 279;
    informal 245; traditional Chinese
    cities 62–4; traditional/open-air 244;
    walled 62–3
market transformation, regional 147

marriage, and social mobility 101
McKinsey Global Institute 83, 86, 90,
    145
medieval period: cities 32; medieval urban
    revolution 38–41
mega-cities 275
mega-event planning 231
miandi 222
middle classes: growth of 276; in
    suburbs 168, 170
mid-sized cities: proliferation of 85
migrant circulation 99
migrants 4; barriers and access 101–2;
    bias against 106; characteristics
    of 100–1; housing/settlement
    patterns 102–6; in inner suburbs 171;
    and poverty 146; rural–urban 83,
    96, 98, 101, 167, 251; settlement
    patterns 102–6; socioeconomic
    mobility 106; socioeconomic standing
    of 277; and suburbanization 167;
    temporary/permanent 102; temporary
    registration 106, 141; young rural
    women 121
migrant schools 107–8
migration 97–101; to cities 4; and
    Communist Party 3; forced 97;
    and future urbanization 275; and
    household registration system 96;
    interprovincial 99–100; magnitude
    of 99; patterns of 98–100; permanent/
    temporary 98–9; rural–urban 83,
    96, 98; scale of 93; Shenzhen 73; to
    the south 24; and urban growth 77;
    women/families 101
migration policies 277
Ming dynasty 24, 25, 30, 32, 36, 39, 41–3,
    53–4, 61, 62; Minimum Living Standard
    Scheme 146
Ministry of Communications 177
Ministry of Construction 8, 181, 183, 185,
    191, 269
Ministry of Environmental Protection 223,
    227, 228, 234
Ministry of Finance 272
Ministry of Housing and Urban–Rural
    Development 223
Ministry of Land and Resources 200
Min River Delta, as Open Coastal
    Economic Area 84
mobility: and migrants 105;
    population 93, 97–8, 108
model cities programs 228
modernity, 20c. China 66

modernization, and family support for elders 250
Moganshan Road art district 255
monasteries 61
Mongols 26
monitoring role, communist state 269; *see also* control
monopolies, state 39, 40, 42, 43
monumental structures 57–61
mortgages 205, 210, 211–12, 213
mosques 61
motorized travel 162, 163–4, 276
mountains, significance of 12
multinational corporations (MNCs): attracting 115; China's appeal to 118; manufacturing 117; R&D 125–8
multiplex shopping/entertainment 160
municipal administration 260, 263–4; *see also* municipal government
municipal budgets, for infrastructure 178
Municipal Bureau of Environmental Protection, Beijing 235
Municipal Bureau of Forestry and Parks, Beijing 235
Municipal Construction Commissions 177
municipal government: and development industry 267; as housing provider 153, 197, 198; and infrastructure 177, 184; land ownership 195–6; and land use rights 200, 200–1; role of 264; *see also* municipal administration
Muslim communities 41, 253
Muslim nationalities 26

nail houses 203
Nanjing 17, 35, 36, 42, 80, 86, 165
Nantong 113
NASDAQ 139
National Bureau of Statistics 7, 71, 212
National College Entrance Examination 143
national environment laws 236
national high-tech parks 115
National Model City for Environmental Protection (NMC) 228
National Population and Family Planning Commission 99
National Sustainable Experimental Cities 231
national urban planning law 1989 85, 235
nature reserves 234
Naxi people 26
neighborhood committees (*juweihui*) 268, 269, 270

neighborhoods: good/bad 251; traditional Chinese cities 62–4; widening range of 156
networks, cities/urban regions 276
new Argonauts 127
New China Square 248–50
New York Stock Exchange 139
NIMBYism 272
Ningbo 17, 45, 113
Ningxia Hui Autonomous region 26
Niujie 253–4
nomenklatura system 260
non-governmental organizations (NGOs) 238
non-profit organizations 238
North (Bei) River 19
North China Plain 14
northeastern region: and economic restructuring 147; investment policies 89–90
northern China, agriculture 22
northern grasslands 14
Northern/Southern era, cities 36
northwestern China, agriculture 22
northwestern desert 14
notebook computer industry 120

off-budgetary funds 187
officially designated towns 5, 261
oil production, and urban expansion 80
old industrial base, Northeastern China 89–90
Olympics, Beijing 84, 223, 232–3
One Percent Population Survey 101, 207
open-air markets 244
open cities 123; coastal 84, 112, 113, 178
Open Coastal Economic Areas 84
open-door policy 112
Open Economic Zones 113–14
open economy, toward an 111–14
open government information initiatives 271
openness, of Chinese economy 122
Opium Wars 22, 45, 116
Ordinance on the Management of Urban Housing Demolition and Relocation 166
outward-oriented development strategy 114
over-building, and land transfer revenues 264
overcrowding, housing 198
overseas Chinese community: attracting back 115, 116; and FDI 118

Pan-Pearl River Delta Economic Zone 88
Pan Yue 219, 220, 237
paper production 39
parks, new activities in 245–8
participatory budgeting 271–2
Pearl River Delta 19–20; cities 13, 35;
    as city region 86; as factory of the
    world 121–2; foreign banks 124;
    manufacturing 147; as Open Coastal
    Economic Area 82; as Open Economic
    Zone 113; reducing competition in 88;
    regional economy 87–8; rise of 89
Peking *see* Beijing
Peking University 127
pensions 141
People's Bank of China 205, 213, 234
peoples of China 25–6
People's Republic of China: antiurban
    bias 29; cities as producers 134;
    cities in 36; economy in 117; export
    economy 23; population in 24, 25, 71,
    81; as rural movement 21; shopping
    in 242; urban surveys 66; *see also*
    Mao, Chairman; Communist Party
peri-urbanism 122
personal savings 211
petroleum, and urban expansion 80
pharmaceutical industry 125
physical environment of China 11–21
pilgrimage, to mountains 12
Pingcheng 36
Pingdingshan 80
pipe networks 192–3, 224
place-of-origin guilds 63
planning: centralized 134; effectiveness
    of 267; environmental 227–8;
    green 231; local economic 264–6;
    mega-event planning 231; and road
    usage 164; and transportation 181–2
policies: redevelopment 166–7; urban
    transport 161
polluting industry, and residential
    areas 237
pollution: accidents/mass
    mobilization 272; air 181, 220–3, 236,
    278; coastal waters 279; data/internet
    campaigns 236; particulate 221;
    related disease 220, 221, 227;
    rivers 219; Taiyuan 16; vehicle 220–1;
    water/waste materials 219, 278;
    waterways 224
polycentric strategy, urban
    development 277
population: aging 142, 250; current 2;

decentralization/redevelopment 167;
    growth/distribution 1, 23–5;
    measuring 6; mobility 93, 97–8, 108;
    Qing dynasty 44; urban 24, 71, 75–6,
    81, 82, 86, 278
porcelain 40
port cities 41
port facilities, and industrial
    development 23
poverty: migrant families 108, 277;
    older laid-off workers 142; and
    unemployment 146; urban 145–6, 149,
    150, 280
poverty lines 145–6
power structures, within cities 56
pragmatism, and reform 112
prefecture-level cities 261, 262, 263
premium water networks 192
private housing 209–10
privatization of SOEs 136, 138
privatization, of water supply 224
Procedures for Land Classification in
    Cities and Towns 200
pro-development agenda 267
production: globalization of 117; land as
    means of 196
property boom 211–14
property bubble 202, 267
property development: and FDI 118; and
    local government revenues 214;
    see also real estate development
property management companies 269;
    see also real estate management
Property Management Regulation 270
property rights 203
property sector, employment in 211;
    see also real estate sector
protests: environmental 237, 272;
    rights 203
provinces, in administrative hierarchy 260
provincial capitals 79
provincial-level cities 260, 261, 263
public engagement 271
public finance, and infrastructure 182–4
public-funded companies, and
    development process 267
public hearings 271
public housing: and state work units 153,
    197, 198, 206; trade in 209
public involvement, in environmental
    assessment 271
public order, privatization of
    maintaining 270
public parks 245–6

public participation 273; and
environmental issues 236, 271; spaces
for 271–3; urban governance 278
public–private partnerships, infrastructure
development 191
public sector, housing provision 205
public services: and household registration
system 102; inland provinces 188;
and local government 184; rural
areas 95
public spaces: and entertainment 244–50;
formal 245–8
public squares: new activities in 245–8;
recreation in 248–50
public utility surcharge 184
Pudong New Area project 114
Putuoshan 12

Qin dynasty 25, 31, 33, 35–6, 51
Qingdao 113
Qing dynasty: Beijing 59–60; cities 32,
36, 37, 44; economy 42; garrison
cities 44; imperial palace 57–8;
monopolies 39, 43; neighborhood
administration 268; population in
24–5; Suzhou 18; urbanization 43–4
Qinghai-Tibet Plateau 13, 14, 15
Qingming Shanghe Tu 62
Qinhuangdao 113
quota grants 183

rail transport: and economic
development 46; light rail systems/
subways 164–5; maglev trains 188,
190, 272; and urban expansion 79
railway stations 65
rationing of food/goods 242
real estate development 84–5; *see also*
property development
real estate management 206–7; *see also*
property management
real estate market 155
real estate sector: foreign investment 118,
198; state work units in 267; *see also*
property sector
recession, global 128, 146, 213
recreational activities, public parks 246
recycling 224, 226
redevelopment: and displacement 166;
inner-city 165–7; shift in policy 166–7
reform: economic 111–14; higher-
education 150; labor market 140–2,
150; land 84, 198–200, 214; and
SEZs 113

reform era 29; early years 1979–1992
83–4; environmental advocacy 238;
shopping in 243–4; trade/export
processing 124; urban system
during 82–5
reforms, housing 204–7
refugee camps, Shanghai 66
regional inequality 90
regional/inter-regional competition 86–9
regions: economic performance/
welfare 146–9; urban 276
regular higher-education institutions
(RHEIs) 143, 144
regulation, and district governments 171
religion 25, 26
religious buildings 61
relocation, of industry 156, 165, 168, 233
rental market 105, 210
reorganization, state 259–64
Republican era, cities 30, 36
research and development (R&D):
investment 89, 116, 126, 278;
Shanghai 279
research institutions 115
resettlement housing 167, 203
residency policies, foreign workers 127
resident groups 269
residential areas, and polluting
industry 237
residential care, for elders 251
residential decentralization 158
residential dispersion 168
residential management 269
residential mobility 155; and
neighborhood committees 269
residential quality 155
resource allocations, urban/rural areas 262
responsibility system 243
retailing 243
retail landscape 243
retail sector 160
returnee-founded companies 128
revenue: expanding local base 267; from
land sales 277; land transfer 198, 201,
264
revenue generation: local
governments 184
revenue sharing 183
revolving funds 138
ribbon dances 247
rice cultivation 17, 19–20, 21–2, 24
right-of-way transit solutions 164
rights protection 270
rights protests 203

Rio environment summit 231
rivers 13–19; polluted 219
road construction 162
rural administration 261
rural administrative staff 260–1
rural areas: conversion to urban areas 76–7; lack of investment 94; participatory budgeting 272
rural collective communes 98
rural development 261
rural industrialization 4
rural land, ownership of 196
rural society, village committees 268
rural townships, in county-level cities 261–2
rural unrest 261
rural–urban divide 4, 96, 145
rural urbanization 76
rural–urban migrants 83, 96, 98, 101, 167, 251
rural–urban migration 83; and spatial expansion 171
rural–urban transition areas (*chengxiang jiehebu*) 168
rural youth, socioeconomic mobility 96
Rustication of Youth 82

salt 22, 39
satellite-town program 168
savings, personal 211
scavenging 226
schools 143; high 143, 144; migrant 107–8
science parks 115, 127
seafaring age 41
secondary education 143, 144
self-raised funds 186–7
self-sufficiency, regions 81, 134, 146, 147
senior secondary schools 143
service professions 276
service sector 98, 118, 123, 135
settlement grants 183
settlement patterns, migrants 102–6
70/90 policy, housing 213
sewage treatment 227
Shandong/Liaoning, as Open Coastal Economic Area 84
Shangdu 36
Shanghai: air quality 221; artistic/cultural activities 255, 266; automobiles 163–4; automobile sector 265; CBD 158; and city regions 86; coastal open city 113; corporate headquarters 119, 147; dominance of 18, 42, 81; as dual

city 64; economic planning 265; environmental activism 279; European settlements 64–5; financial district 125; foreign banks 124; foreign investment 113; Hongqiao and Caohejing Economic and Technology Development Zones 115; house prices 213; housing market 155; infrastructure development 188; land use in 156; maglev trains 188, 190, 272; as major regional hub 279; modes of travel 162; opening up of 114; pipe networks 193; Plaza 66 160; population distribution 157; protests 272–3; public polling 271; R&D 126; refugee camps/slums 66; research and development 89, 279; returnee areas 127; Shanghai model (local administration) 263–4; share of GDP 88; size of 77–8; solid waste 226; trade economy 22–3; traffic management systems 182; transport 164, 188; as Treaty port 45, 65; universities 116; water privatization 192; World Expo 232; *Xintiandi* 165–6
Shanghai Jiaotong University 116
Shanghai Municipal Housing and Land Administration 264
Shanghai Pudong Water Supply Corporation 192
Shanghai Stock Exchange 125
Shanghai Venture Capital Corporation 139
Shangjing 36
Shang period, cities 35
Shantou 83, 114
shantytowns 66, 155
Shanxi 16
Shenyang 46, 117, 147–8, 168, 228, 268–9
Shenzhen: as city region 86; foreign exchange center 124; house prices 212, 213; land use rights 199; and market reform 114; pipe networks 193; property management 269–70; SEZ 83, 87, 113, 114; size of 72–3; trade share 124
Shenzhen Stock Exchange 124
Shijiazhuang, and city regions 86
shipping fleets 41
shoppertainment 243
shopping 241–4
shops, permanent 63
Shougang Iron and Steel complex 233

Sichuan Basin 13
sidewalks, as informal public space 244
Silicon Valley model 127
silk production 38, 40
Silk Road 26, 37
Silk Road cities *see* Silk Route
Silk Routes 26, 33, 34, 37, 39, 40, 62
Singapore Exchange 139
Singapore, home ownership 205
Singapore-Suzhou Industrial Park 18
Sino-Japanese War 66
Sino-Singapore Tianjin Eco-City 230
size, of China 11
skill formation 142–5
skills, supply of 277
Skinner, G. William 29–30, 44
skyscrapers 153, 165
slums, growth of 66
small businesses 160–1
social egalitarianism, socialist cities 153
social impact, urban redevelopment 167
social inequality 93
socialism: infrastructure legacy 177–80;
    urban administrative hierarchy 259–60;
    and urban land/housing 195–8; urban
    society under 268
socialist cities 48, 152–4
socialist development, and education 142
social mobility, and marriage 101
social space, determinants of 156
social–spatial differentiation, urban 277
social welfare, housing as 96, 197–8, 206
society, closed 5
socioeconomic mobility: migrants 106;
    rural youth 96
solid-waste management 220, 224–7
Songbei project 159
Song dynasty 25; cities 32, 36, 38, 40;
    monopolies 39; population in 24;
    trade-based economy 39;
    urbanization 30
Songjiang 42
Songshan 12
south China 13
southeast China 13, 14, 22
South Manchuria Railway Company 46
Soviet Union, breaking with 111, 125
soybeans 280
spaces: gay/lesbian 252; modern 244;
    public and entertainment 244–50; for
    public participation 271–3
spatial expansion: of cities 172; and rural–
    urban migration 171
spatial reconfiguration/expansion 165–71

special administrative regions 260
Special Development Zones 115, 202
Special Economic Zones (SEZs) 83, 87,
    112–13, 114–16, 178
specialization, of neighborhoods 243
special-purpose grants 183
spirit walls 51
Spring and Autumn era, cities 36
Spring Festival on the river 62
squatter settlements 103
stakeholder engagement 278
standards of living 280
State Asset Supervision and
    Administration Commission 137
State Environmental Protection
    Administration (SEPA) 219, 234
state investment programs 86
State Land Administration 200
state monopolies 39
State of the World report 220
state-owned banks 134, 138
state-owned enterprise (SOE): foreign
    investment in 139; layoffs/relocation of
    workers 140; monopoly of 133; reform
    of 136–8, 148, 150
state ownership, land 195–7
state-run stores 242
state–society relations 268–73
state work units 197; and administrative/
    social control 268; and development
    process 267; and housing 197, 205;
    role of 268
Statistical Yearbook of China 7
steel 39
stock market, Chinese 124, 139
streetcars 65
subdistrict offices 263
subprovincial-level cities 261
subsidies: to high-tech firms 126;
    land 267
suburbanization 167–71, 172, 276
suburbs, inner and migrants 171
subway systems 164–5
Suez 192
sugarcane 39
Sui dynasty, cities 32, 34, 36, 38, 40
Suiyuan 43
sulfur dioxide emissions 221
superstructure, in Chinese cities 51–7
Suzhou 17–18, 41, 42, 86, 119–20, 147

tai chi 246
Taihu lake 17
Taishan 12

Taiwan 118, 120, 121
Taiyuan 15–16
Tang dynasty 25; Beijing 59; cities 32, 34, 36, 38–9; population in 24; Silk Routes 40; trade 40, 63; urbanization 30
Tarim Basin 13
taxation 138; and developers 267; and infrastructure 182, 183, 184; real estate transactions 213; tax sharing system 187; tea 40; urban maintenance and construction tax 184
taxis, bread loaf 222
tea 38–40, 43
technical/vocational education 145
technology associations, ethnic Chinese 127
technology development programs, Chinese 126
technology plus capital model 127
technology policies 126
technology transfer, and FDI 112, 118
temples 61
temples to the City God 61
Temporary Regulations of Leasing and Transfer Right of Urban Land Use 199
tertiary-sector development 134–5, 150
textile industry, Suzhou 17–18
Thames Water Overseas 192
Third-Front strategy 81–2, 83
three economic belts approach 83
Three Gorges Dam project 78, 86
Three Kingdoms era, cities 36
Tian'anmen Square 248
Tianjin 13, 45, 80, 86, 113, 115, 147
Tianshan Mountains 13
Tianshui (Gansu) 42
Tiantan Park, Beijing 248
Tibetan Buddhism *see* Buddhism
Tibetan Plateau, agriculture 22
Tibetans 26, 254
Tieli 225
tobacco production 41
topography of China 12–13
Torch Program 82, 115, 116
tourism: and eco-developments 231; heritage 165, 166
towns: defining 71; designated 5; proliferation of 83
township and village enterprises 98, 99, 158
townships, in administrative hierarchy 260
trade 22–3; 6th/7th centuries 38; boom 5; with foreign countries 45–6,

123–5; in Han dynasty 38; Höhhot 43; international 80; liberal trade regimes 113; Ming dynasty 44; outside ward-designated market system 63; Qing dynasty 44; Silk Routes 37–8, 40; walled markets 62–3
traditional markets 244
traffic congestion 163, 181
traffic management systems 164, 181–2
training, human resource 142–5
transnational corporations 192
transportation: 6th/7th centuries 38; in cities 161–5, 181; and foreign investment 116, 117; infrastructure 168; socialist era 154
transportation hubs 116–17
transport, problems in cities 164
transport projects, environmental impacts 164
transport sector, central control of 177
trash *see* solid-waste
trash picking 226
travel, modes of in cities 162, 163
Treaty of Nanking 64, 116
Treaty Port cities 36, 45–6, 64–6, 81, 90, 123, 155
tricycle riding 248, 249
Tsinghua University 116, 127
two-year colleges 143

underemployment 140
unemployment 128, 143, 146, 148, 277
United Nations action plan (Agenda 21) 231
United Nations Conference on Human Environment 234
United Nations Development Program 220
universities 116, 128, 143, 145, 277
university graduates, household registration system 96
unrest, rural 261
urban administration 66, 259–62
urban bias, development strategies 94
urban development: changing processes of 266–8; and inter-urban competition 88; investment-driven 89; in Maoist era 78–82; polycentric strategy of 277
Urban Drinking Water Sources Protection Plan 2008–2020 224
urban economic bases, diversification of 85
urban economic space, remaking of 156–61

urban economy, and GDP 276
urban environment, state of 219–27
urban expansion 79–80, 170
urban forms 48, 49, 50, 152–6
urban growth 72, 277
urban ideal, Chinese 50–1
urban infrastructure: financing of 182–7;
    investing in 184–7; maintenance
    of 182; private/foreign participation
    in 188, 191–3; regional disparity 187–
    8, 189, 190; state of 177–82
urbanization 1; from the bottom
    up 76; Communist Party 2–3;
    and development policies 71;
    environmental impacts 219; and
    foreign investment 116; future 275–6;
    global environmental impact 279; and
    human–environment relationship 278;
    and industrialization 80–1; and
    infrastructure 180; and loss of
    arable land 279; Manchurian 44;
    sustainable 278; unanticipated 84;
    understanding China's 5–6, 46
urbanized countryside 76
urban land: footprint of 202; price of 200;
    reform 198–200; state ownership 195–
    7, 198, 207
urban life, contemporary 241
urban maintenance and
    construction 177–8
urban maintenance and construction
    tax 184, 185, 186
urban management, two-tier 170–1
urban parks, new activities in 245–8
urban pension scheme 141
urban plan/layout 50–7
urban planning, experimentation
    with 66
urban poor 142, 146; *see also* poverty
urban population 24, 69, 76, 81, 82, 86,
    278
urban regions 276
urban residence 54
urban–rural divide 93, 94–7, 102, 108
urban–rural transitional areas, housing
    in 210
urban social–spatial differentiation 277
urban society, organizing 268–9
urban system: 1990s onwards 84–5;
    expansion of 36; historical 28–9;
    imbalance in 89; during reform era
    82–5; today 85–90
urban systems, geographical approaches to
    analyzing 72

urban transformation, global
    implications 278–80
urban transport problems 164
urban underclass 108, 109
urban villages (*chengzhongcun*) 103, 171,
    197
Urumqi 79, 254
USA, talent flows to China 127
user rights, land 198
utilities, household 197
Uygur people 26, 254

vehicle ownership *see* automobile
    ownership
vehicle pollution 220–1
village committees 268, 269
village housing 210
Vivendi Utilities/Group 191, 192
vocational education 145

walking 161
walled markets 62–3
walls, Chinese cities 53–6
wards, traditional Chinese cities 62
Warring States era 36, 58
waste, solid *see* solid–waste
wastewater services 181, 227
water: foreign/private participation 192;
    pollution 219, 224, 278; privatization
    in Shanghai 192; quality 223–4;
    services 180–1, 191–3; 224;
    shortages 223–4, 278
Water Conservation Bureau, Xining 235
water/sanitation, and GDP 180
wealth: coastal region 188;
    urban/rural 111
Wei River 33
welfare: disability benefits 141;
    housing 96, 197–8, 206; nomenklatura
    system 260; regional 146–9; and rural
    citizens 94
Wen Jiabao 102
Wenzhou 113
Wenzhou migrants, Zhejiang
    Village 104–5
Western Development Strategy 89
western market, Chang'an 62–3
western region, development in 89
West (Xi) River 13, 19
women: employment choices 142;
    migrants 101; as new labor force 121
work: and accessibility 156–65; and
    residence 153, 155, 163; *see also*
    employment

workers: and bicycles 161–2;
    educated 126; relocation of 140;
    Western-educated 126–7
workers' villages 156
workforce, decline in urban/increased
    migrant 142
work unit compounds 153
work units (*danwei*) 197; and
    housing 205
World Bank, and pollution related
    deaths 220
World Bank project, public budgeting
    reforms 272
World Bank Urban Labor Survey 148
World Trade Organization 5, 87, 118
Worldwatch Institute 220
Worldwide Fund for Nature (WWF) 238
written language 25
Wuhan 13, 17, 276
Wu Ping's nail house 203–4
Wutaishan 12
Wuxi 147

Xiamen: as dual city 64; foreign trade 45;
    protests 272; SEZ 83, 114; Taiwanese
    investment in 113; trade share 124
Xi'an 13, 33, 54, 57–8, 61
Xianyang 36
Xining 13, 79, 235, 246, 254
Xinjiang 80
Xinjiang Uygur Autonomous region 22,
    26, 254
Xinjiang Village 253, 254
Xintiandi 165–6
Xi River 13, 19
Xuzhou 80

yamen 61

yangge dancers 246, 247
Yangtze region, agriculture 21
Yangtze River: acid rain 221; origins
    of 13; trade/industry 22
Yangtze River Delta 14, 17–19; cities 13,
    17, 42; as city region 86; and FDI 119;
    financial sector in 124; growth in 124;
    manufacturing 147; as Open Coastal
    Economic Area 84; as Open Economic
    Zone 113; rise of 90
Yantai 113
Yanzhou 80
Yellow River 2, 13, 15–16, 35
Yellow River Delta Efficient Eco-
    Economic Zone 231
yin/yang 51
Youlian Consortium 192
Yuan dynasty 24, 32, 36, 39, 41, 59
Yu the Great 15

Zhanjiang 113
Zhao Ziyang 113, 114
Zhejiang 188
Zhejiang Village 253
Zhengzhou 15
Zhenjiang 42
Zhejiang University 126
Zhongdu 59
Zhongguancun 127–8, 266
Zhongjing 59
Zhongshan 228, 279
Zhou dynasty 33, 36, 51
Zhuhai 83, 86, 88, 114
Zhuzhou 234
Ziguang 127
Zizhu Science Park 116
zone fever 115
zone policy 113